KIELCE2018 Conference

Resilient and Safe Road Infrastructure

IOP Conference Series: Materials Science and Engineering
Volume 356

Kielce, Poland
8 – 9 May 2018

ISBN: 978-1-5108-6312-5
ISSN: 1757-8981

Printed from e-media with permission by:

Curran Associates, Inc.
57 Morehouse Lane
Red Hook, NY 12571

CURRAN ASSOCIATES INC. proceedings.com

Some format issues inherent in the e-media version may also appear in this print version.

Printed by Curran Associates, Inc. (2018)

For permission requests, please contact the Institute of Physics at the address below.

Institute of Physics
Dirac House, Temple Back
Bristol BS1 6BE UK

Phone: 44 1 17 929 7481
Fax: 44 1 17 920 0979

techtracking@iop.org

Additional copies of this publication are available from:

Curran Associates, Inc.
57 Morehouse Lane
Red Hook, NY 12571 USA
Phone: 845-758-0400
Fax: 845-758-2633
Email: curran@proceedings.com
Web: www.proceedings.com

Table of contents

Volume 356

Resilient and Safe Road Infrastructure

8–9 May 2018, Kielce, Poland

Accepted papers received: 30 April 2018
Published online: 15 May 2018

Preface

Papers

Evaluation of surface properties

Materials for pavement repairs

Bearing capacity of road and airfield pavements

Non-destructive testing

IOP Conf. Series: Materials Science and Engineering **356** (2018) 011001 doi:10.1088/1757-899X/356/1/011001

Preface

It is my great honor to present to you the scientific articles from KIELCE2018 conference. This conference was organized by the Road and Bridge Research Institute (IBDiM) on May 8-9, 2018 in Kielce, Poland. The main theme of KIELCE2018 was **Resilient and Safe Road Infrastructure**. In particular, all articles were dedicated to modern pavement diagnostics and repair techniques, and are grouped into five distinct topics: Evaluation of surface properties, Materials for pavement repairs, Bearing capacity of road and airfield pavements, Modern repair techniques, and finally Non-destructive testing.

As road infrastructure is vital to every country economy, I trust you find these articles both useful from the practical perspective as well as scientifically valuable. Road diagnostic and repair techniques are constantly evolving and I hope that articles from the KIELCE2018 conference will contribute to further development in these engineering areas.

KIELCE2018 conference was a well-attended event with many representatives from construction contractors, design companies, consultants, operators, researchers and students. You can find more information about KIELCE2018 on its website: http://kielce2018.ibdim.edu.pl/

Dr. Adam Zofka
Chair of Scientific Committee
Chief Scientist, Road and Bridge Research Institute (IBDiM)
Poland

Prof. Leszek Rafalski
Director, Road and Bridge Research Institute (IBDiM)
Poland

Scientific Committee

Dr. Adam Zofka, Chair, Road and Bridge Research Institute
Dr. Wojciech Bańkowski, Road and Bridge Research Institute
Dr. Bohdan Dołżycki, Gdansk University of Technology
Dr. Władysław Gardziejczyk, Bialystok University of Technology
Dr. Mirosław Graczyk, Road and Bridge Research Institute
Prof. Marek Iwański, Kielce University of Technology
Dr. Piotr Jaskuła, Gdansk University of Technology
Dr. Karol Kowalski, Warsaw University of Technology
Dr. Lech Krysiński, Road and Bridge Research Institute
Dr. Kukiełka, Lublin University of Technology
Dr. Piotr Mackiewicz, Wroclaw University of Science and Technology
Ing. Tomasz Mechowski, Road and Bridge Research Institute
Dr. Paweł Mieczkowski, West Pomeranian University of Technology
Prof. Piotr Radziszewski, Warsaw University of Technology
Dr. Mieczysław Słowik, Poznan University of Technology
Ing. Jacek Sudyka, Road and Bridge Research Institute
Prof. Dariusz Sybilski, Road and Bridge Research Institute
Prof. Antoni Szydło, Wroclaw University of Science and Technology

IOP Conf. Series: Materials Science and Engineering **356** (2018) 011001 doi:10.1088/1757-899X/356/1/011001

List of Reviewers

All papers published in this volume of *IOP Conference Series: Materials Science and Engineering* have been peer-reviewed through processes administered by the Editors. Reviews were conducted by expert referees listed below to the professional and scientific standards expected of a proceedings journal published by IOP Publishing. Conference organizers would like to thank all the reviewers for their time and effort.

Dr. Wojciech Bańkowski
Dr. Krzysztof Błażejowski
Dr. Wiesław Dąbrowski
Dr. Bohdan Dolzycki
Dr. Wladyslaw Gardziejczyk
Dr. Mirosław Graczyk
Prof. Marek Iwanski
Dr. Andrzej Janowski
Dr. Piotr Jaskula
Dr. Robert Jurczak
Dr. Karol Kowalski
Dr. Bartlomiej Krawczyk
Dr. Lech Krysiński
Dr. Jerzy Kukielka
Dr. Elżbieta Macioszek
Dr. Piotr Mackiewicz
Dr. Stanisław Majer
Dr. Krzysztof Małecki
Ing. Maciej Maliszewski
Ing. Tomasz Mechowski
Dr. Paweł Mieczkowski
Dr. Andrzej Plewa
Prof. Piotr Radziszewski
Dr. Mieczyslaw Slowik
Ing. Jacek Sudyka
Prof. Antoni Szydło
Dr. Marta Wasilewska
Dr. Mariusz Wesołowski
Dr. Agnieszka Woszuk
Dr. Piotr Zieliński
Dr. Adam Zofka

IOP Conf. Series: Materials Science and Engineering **356** (2018) 011002 doi:10.1088/1757-899X/356/1/011002

Peer review statement

All papers published in this volume of *IOP Conference Series: Materials Science and Engineering* have been peer reviewed through processes administered by the proceedings Editors. Reviews were conducted by expert referees to the professional and scientific standards expected of a proceedings journal published by IOP Publishing.

Effect of aggregate graining compositions on skid resistance of Exposed Aggregate Concrete pavement

Marta Wasilewska*, Wladysław Gardziejczyk, Pawel Gierasimiuk,

Bialystok University of Technology, Wiejska 45A, 15-351 Bialystok, Poland

*marta.wasilewska@pb.edu.pl

Abstract The paper presents the evaluation of skid resistance of EAC (Exposed Aggregate Concrete) pavements which differ in aggregate graining compositions. The tests were carried out on concrete mixes with a maximum aggregate size of 8 mm. Three types of coarse aggregates were selected depending on their resistance to polishing which was determined on the basis of the *PSV* (Polished Stone Value). Basalt (*PSV* 48), gabbro (*PSV* 50) and trachybasalt (*PSV* 52) aggregates were chosen. For each type of aggregate three graining compositions were designed, which differed in the content of coarse aggregate > 4mm. Their content for each series was as follows: A - 38%, B - 50% and C - 68%. Evaluation of the skid resistance has been performed using the FAP (Friction After Polishing) test equipment also known as the Wehner/Schulze machine. Laboratory method enables to compare the skid resistance of different types of wearing course under specified conditions simulating polishing processes. In addition, macrotexture measurements were made on the surface of each specimen using the Elatexure laser profile. Analysis of variance showed that at significance level $\alpha = 0.05$, aggregate graining compositions as well as the *PSV* have a significant influence on the obtained values of the friction coefficient μ_m of the tested EAC pavements. The highest values of the μ_m have been obtained for EAC with the lowest amount of coarse aggregates (compositions A). In these cases the resistance to polishing of the aggregate does not significantly affect the friction coefficients. This is related to the large areas of cement mortar between the exposed coarse grains. Based on the analysis of microscope images, it was observed that the coarse aggregates were not sufficiently exposed. It has been proved that PSV significantly affected the coefficient of friction in the case of compositions B and C. This is caused by large areas of exposed coarse aggregate. The best parameters were achieved for the EAC pavements with graining composition B and C and trachybasalt aggregate.

Other paragraphs are indented (BodytextIndented style).

1. Introduction

Skid resistance plays a special role in ensuring the safety of traffic on wet road pavements. It is related to the texture of road pavement, which is responsible for draining water from the plane of contact between a tire and a pavement surface. Pavement surface texture is specified in the three-level scale, each defined by the wavelength (λ) and peak – to peak amplitude (A) of its components. According to this division there are: megatexture (λ = 50 to 500 mm; A = 0.1 to 50 mm), macrotexture (λ = 0.5 to 50 mm; A = 0.1 to 20 mm), microtexture (λ < 0.5 mm; A = 1 to 500 μm) [1]. Especially macrotexture and microtexture influence skid resistance. Microtexture depends on polishing resistance of coarse aggregate and content of fine aggregate in a mixture. Macrotexture is related to the type of asphalt mixture and the content of the individual fractions in a mineral mix. In the case of PCC pavement macrotexture depends on the technique of texturing of the upper layer.

The following texturing techniques are used on concrete pavements in Poland: burlap drag, transverse broom drag, transverse tine texturing and exposed aggregate concrete (EAC). The first three techniques do not ensure satisfactory durability on heavy traffic roads. Currently exposing of aggregate with the use of chemical agents is the only texturing method that is specified in Poland for the expressways pavements [2, 3]. This method is planned to be used on over 800 km of national roads to be built by the year 2023. The rationale for the choice of this method is the conviction that it provides both very good durability and skid resistance of road pavement.

The EAC was developed in the 1980s in Austria. In this method the process of cement hydration is retarded by applying appropriate retarding admixtures on the surface of fresh concrete. Then, after few hours, the not-hydrated cement mortar is removed with mechanical brushes or washed off with water jet [4]. Graining composition was designed on the basis of two aggregate fractions: 0/1 as fine aggregate (ca. 30%) and 4/8 as coarse aggregate (ca. 70%). This will ensure open grading of the mineral mix and optimum number of coarse particles on exposed surface. Currently, the Austrian guidelines allow the use of fractions: 0/2 and 5/8, as well as an EAC with a maximum aggregate size 11 mm. But recommendations are for the use of the maximum aggregate size of 8 mm so that the stone count or contact points should be at least 50 points per 25 cm^2 of surfacing [5]. Recommendations for maximum grain size aggregates are different in different countries. This has a direct impact on the diverse requirements of the texture depth, which are in the range from 0.90 to 1.80 mm depending on the country [6, 7, 8, 9]. EAC with the maximum aggregate grains 16 mm and 22 mm were made in Sweden and the Netherlands [7, 10, 11]. In contrast, the British requirements only allow the use of maximum grains 10 mm on highways and 8 mm on other pavements. Studies show that noise emission is much lower on pavements with such graining compositions compared to pavements with grains bigger than 11 mm [7, 10, 12]. In addition the impact of selection of aggregate graining on the proper execution of the texture of EAC pavement is indicated. It should be noted that the amount of coarse aggregate allows to achieve high macrotexture. Due to that fact pavement made of EAC has good skid resistance at high speeds [13, 14]. However skid resistance depends on both macrotexture and microtexture. Due to the low amount of fine aggregates used, microtexure of EAC is almost completely dependent on the resistance to polishing of coarse aggregates [107]. Therefore aggregates with PSV (Polished Stone Values) above 50 or 53 (depending on the country), should be used for EAC in order to meet the required level of skid resistance [5, 15, 16].

In Poland it is recommended to use aggregate with the maximum size of 8 mm and compose it from minimum three fractions. However, experiences of other countries show that in order to perform correct textures of concrete pavements it is required to compose aggregate from two fractions: fine aggregate 0/1 or 0/2 and coarse aggregate 4/8 or 5/8 [5]. The lack of such records may lead to insufficient or excessive exposure of coarse aggregates and their losses. As a consequence, it has a significant impact on the skid resistance of EAC pavement during its exploitation period.

The aim of this study was to determine the impact of aggregate graining as well as the PSV of coarse aggregate on skid resistance of EAC pavement. In actual conditions, skid resistance is associated with the characteristics of the wearing course, specially its surface texture, traffic intensity, the period of pavement life, climate, ect. Therefore, it was important that the tested EAC pavements could be subjected to the processes that simulate phenomena occurring in actual traffic conditions. Such studies in laboratory conditions are possible using FAP test equipment, popularly known as the Wehner/Schulze machine. It was developed in Germany over 30 years ago. In 2014 this method was adopted as a harmonized European Standard EN 12697-49 as the Friction after Polishing test. The device simulates the phenomenon of polishing on the wearing course. Studies show that determination of the friction coefficient before construction of wearing course allow to avoid the use of materials that could contribute to the slipperiness of the road pavement [17, 18].

IOP Conf. Series: Materials Science and Engineering **356** (2018) 012001 doi:10.1088/1757-899X/356/1/012001

This paper presents the comparison of skid resistance of EAC pavements which differ in aggregate graining compositions and the resistance to polishing of coarse aggregate measured using the FAP test equipment.

2. Research program

2.1. Materials and compositions

The tests were carried out on concretes with the maximum aggregate size of 8 mm. Three types of coarse aggregates were selected: basalt, granite and trachybasalt. Their choice depended on resistance to polishing which was determined on the basis of the *PSV*. Table 1 shows physical properties of the tested aggregates.

Table 1. Physical properties of coarse aggregate.

Properties	Test method	Basalt	Gabbro	Trachybasalt
Resistance to polishing, *PSV* [-]	EN 1097-8	48	50	52
Resistance to fragmentation, *LA* [%]	EN 1097-2	9	23	12
Resistance to freezing and thawing, F_{NaCl} [%]	EN 1367-6	1	1	1
Density, ρ_a [Mg/m^3]	EN 1097-6	3.09	2.90	2.70
Water Absorption, WA_{24} [%]	EN 1097-6	0.3	0.5	0.5

For each type of aggregate three graining compositions were designed, which differed in the content of coarse aggregate > 4 mm. The content of coarse aggregate for particular series was as follows: A – 38%, B – 50% and C – 68%. Three fractions of aggregates: 0/2, 2/5, 4/8 were used in composition A and B, whereas for composition C only two fractions 0/2 and 4/8 were used. Aggregate particle-size distribution of particular aggregate compositions are presented in Table 2.

Table 2. Aggregate particle size distribution.

Sieves [mm]	Passing fraction [%]								
	A			B			C		
	Basalt	Gabbro	Trachybasalt	Basalt	Gabbro	Trachybasalt	Basalt	Gabbro	Trachybasalt
11.2	100	100	100	100	100	100	100	100	100
8	98	97	99	95	94	98	93	92	97
5.6	82	81	86	66	67	68	44	47	46
4	63	62	62	48	49	48	32	33	32
2	38	36	39	28	34	33	28	29	28
0.125	3	4	3	2	3	3	2	3	3
0.063	1.0	1.0	1.0	0.0	1.0	1.0	1.0	1.0	1.0

Cement CEM I 42.5R was used in amount from 430 to 440 kg/m^3, the water-cement ratio w/c was in the range 0.36 - 0.37. Air-entraining and water reducing admixtures were applied. The consistence of concrete mixes has been classified to V2 according to the Vebe method. Three slabs of size 35 × 35 × 5 cm were formed for each concrete composition. In order to expose the aggregate, surfaces of slabs were sprayed with retarding admixture. The aggregate was exposed after approx. 18 hours by removing the unhydrated layer of cement mortar with a steel brush. Then slabs were stored at 20 $^{\circ}$C and RH 50%. Specimens with a diameter of 225 mm for testing in the FAP device were drilled from the slabs after 28 days. Exemplary specimens of EAC with particular aggregate graining compositions are shown in figure 1.

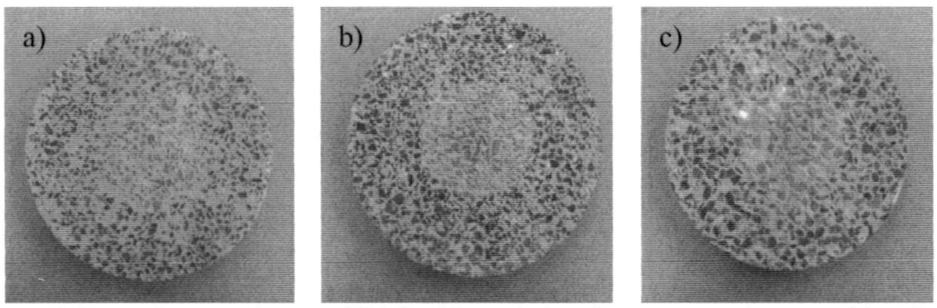

Figure 1. Exemplary specimens of EAC with trachybasalt aggregate a) composition A; b) composition B; c) composition C.

2.2. Test procedure

The FAP equipment (figure 2a) consists of two heads: for polishing and for measurement of friction coefficient μ_m. The polishing action is performed by means of three rubber cones mounted on rotary disc and rolling on the specimen surface (figure 2b). A water-quartz powder mix is projected during the polishing process. The second measuring head is composed of three small rubber sliders disposed at 120° on the rotary disc (figure 2c). The disk rotates at tangential velocities up to 100 kph. Water flows over the surface being tested. The rotating disk is then dropped onto the wet surface and the coefficient of friction μ_m is measured. Measuring of μ_m was conducted after 2000, 4000, 6000, 8000, 10 000, 20 000, 40 000, 60 000, 80 000, 100 000, 160 000, 180 0000 passes of polishing head. In this study μ_m at slip speed 60 kph was taken. Before starting the test the function of the measuring device was tested using the glass control plate (figure 2d). If the results of coefficient of friction μ_{ref} is not in the range from 0.095 to 0.125, the rubber sliders should be exchanged.

In addition, macrotexture measurements were made on the surface of each specimen using the Elatexure laser profile This is the rotating laser sensor which scans the road pavement surface with high resolution: vertical - 0.01 mm and horizontal - 0.2 mm. When the measurement starts the laser sensor rotates on a circumference of the measurement circle 400 mm. The software calculates the Mean Profile Depth (*MPD*) parameter that characterizes the profile macrotexture.

Figure 2. a) The FAP test equipment b) polishing rotary head, c) friction measuring rotary head, d) glass control plate.

3. Results and analysis

The mean values of friction coefficient μ_m obtained on the tested EAC pavements during polishing process with standard deviation are presented in figure 3.

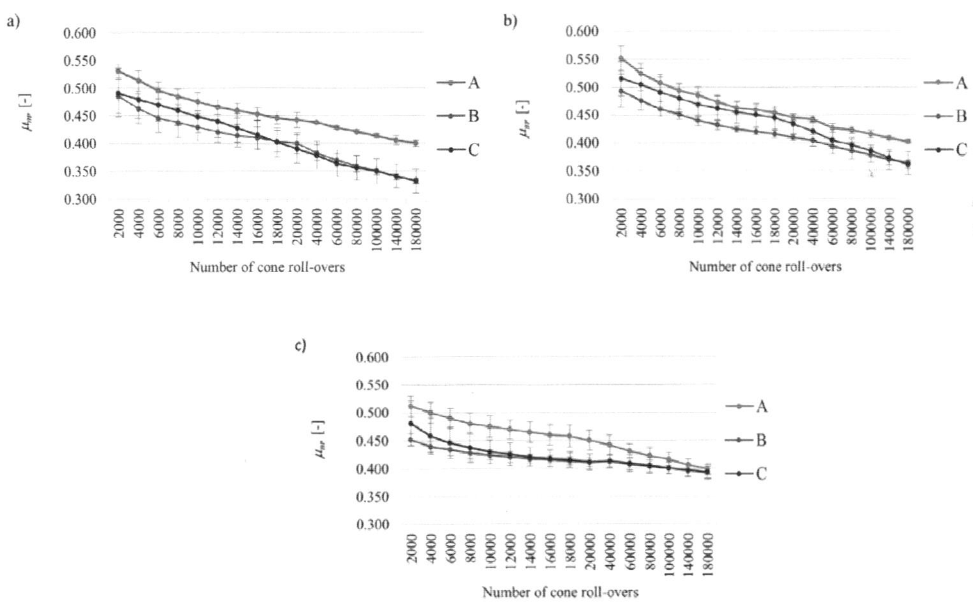

Figure 3. Changes of friction coefficients obtained on the tested EAC with individual graining composites with a) basalt; b) gabbro; c) trachybasalt aggregate.

Significant differences between changes of friction coefficients were noted for tested EAC pavements in polishing process. Specimens with graining composition C were affected by polishing more quickly than those with composition A. This is caused by large area of exposed coarse aggregate >4 mm. However, the most rapid surface changes occurred on EAC with basalt aggregate. In this case, the following decreases in friction coefficients were recorded: for composition A – 25%, B – 31% and C – 33%. In contrast, decreases of friction coefficients on EAC with trachybasalt aggregate ranged from 13 to 22% depending on the aggregate composition. Different susceptibility to polishing factors is related to the structure of rocks and properties of particular minerals. Both basalt and trachybasalt are volcanic rocks, but they differ in structure. Trachybasalt has an aphanitic structure. Therefore, small crystals contribute to better polishing resistance than aggregates from basalt rocks with a cryptocrystalline structure. Whereas gabbro is a igneous rock with crystalline structure. This is probably the reason of its better polishing resistance than basalt. But gabbro consists of minerals of similar hardness, which makes it more susceptible to polishing than trachybasalt.

The analysis of variance (ANOVA) was carried out in order to determine on the accepted significance level $\alpha = 0.05$ the significant effect of the aggregate graining composition and the PSV of coarse aggregate on the obtained coefficient of friction μ_m. The comprehensive evaluation of skid resistance should also include macrotexture measurements. The obtained results of the *MPD* parameter indicate the influence of the aggregate graining composition on macrotexture of surface EAC. Therefore, the ANOVA was used to determine on the accepted significance level $\alpha = 0.05$ the significant effect of aggregate graining compositions and the PSV of coarse aggregate on the obtained parameter *MPD.* The calculations were carried out using the Statistica 13.1 program. Tables 3 and 4 summarizes the main results of these analyses. It was found that at the level of significance 0.05, both graining the composition of tested EAC and the PSV of aggregate have the significant effect on the values of the μ_m. In the case of *MPD* results at the significance level of 0.05, the significant effect of graining composition was proved.

Table 3. ANOVA summary table (effect of aggregate granining compositon and the *PSV* on the values of μ_m).

Source of variance	Sum od squares SS	Degrees of freedom	Mean squares MS	F ratio	p - value
Composition	0.0075	2	0.0038	40.18	0.00000
PSV	0.0078	2	0.0039	41.54	0.00000
Composition* PSV	0.0044	4	0.0011	11.84	0.00007
Error	0.0017	18	0.0001		
Total	0.0214	26			

Table 4. ANOVA summary table (effect of aggregate granining compositon and the *PSV* on the values of the *MPD*.

Source of variance	Sum od squares SS	Degrees of freedom	Mean squares MS	F ratio	p - value
Composition	0.21361	2	0.10680	22.83	0.00001
PSV	0.0108	2	0.0054	1.15	0.33870
Composition* PSV	0.0077	4	0.0019	0.41	0.79789
Error	0.0842	18	0.0047		
Total	0.3163	26			

The graphic interpretation of the results is shown in the figures 4 and 5. The highest values of the μ_m have been obtained for EAC with the lowest amount of coarse aggregates > 4mm (compositions A). In these cases the resistance to polishing of the aggregate does not significantly affect the friction coefficients. This is related to the large areas of cement mortar between the exposed coarse grains because of the low content of coarse aggregate in these compositions. It can be seen on the microscope images of the surfaces of specimens (Fig. 5a, 6a, 7a). Although they were characterized by the highest value of the coefficients of friction, this solution should not be used due to problems with obtaining the correct and homogeneity of the texture surface of the upper concrete layer.

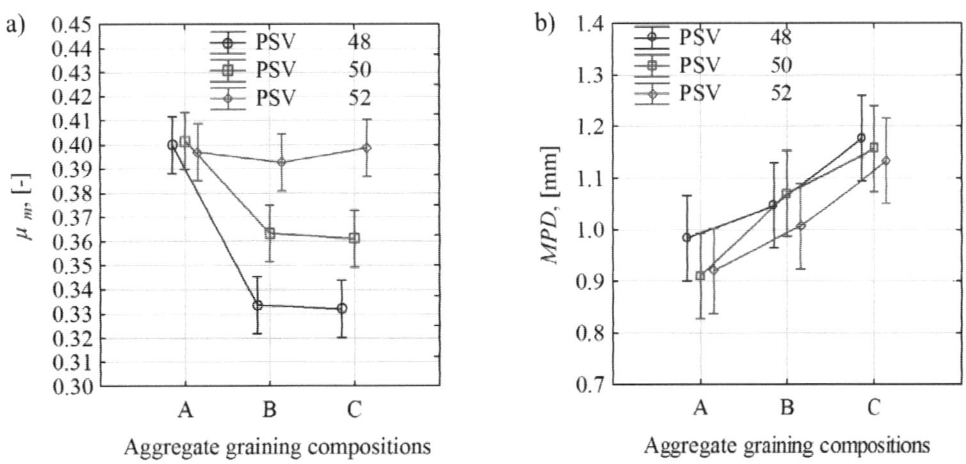

Figure 4. Mean values of the μ_m **(a)** and the *MPD* parameter **(b)** obtained on the tested EAC pavements with 95% confidence interval.

It has been proved that PSV significantly affected the coefficient of friction in the case of compositions B and C. This is caused by large areas of exposed coarse aggregate. The spaces between the exposed grains on the surface of the EAC with the graining compositions B and C are definitely smaller than in the case of the compositions A (Fig. 5 – 7). Additionally, the significant changes due to polishing can be noticed on these images. It should be noted that due to the shape of the aggregate grains, the polishing process occurs only on a these parts of aggregate grains which have the direct contact with the rubber cones. A typical gloss is visible on these areas. However, the surface of coarse aggregates grains in specimens of EAC with composition C is more exposed than in specimens of EAC with composition B. The difference in content of coarse aggregate also affects the macrotexture of the tested EACs. The Mean values of MPD were obtained in the following ranges: compositions A 0.91 – 0.98; compositions B 1.00 – 1.07 and compositions C 1.13 – 1.17. Based on the evaluation of macrotexture it could be expected that compositions B and C ensure an the optimal number of exposed aggregates grains so that the passing tire is mostly in contact with the aggregate tips, thereby allowing sufficient space between the tire and the mortar for drainage.

It should be noted that there were no significant differences between the results of the μm obtained for compositions B and C when the same type of aggregate was used. Differences in the coefficients of friction result only from the PSV of aggregates. The effect of aggregate resistance to polishing caused that the friction coefficients of EAC with trachybasalt were/are higher by about 0,08 unit than the coefficients of EAC with basalt.

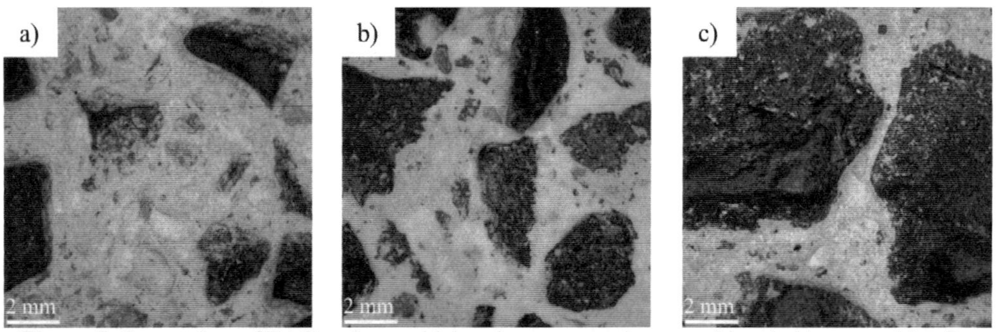

Figure 5. Comparison of EAC surfaces with basalt aggregate a) composition A; b) composition B; c) composition C;

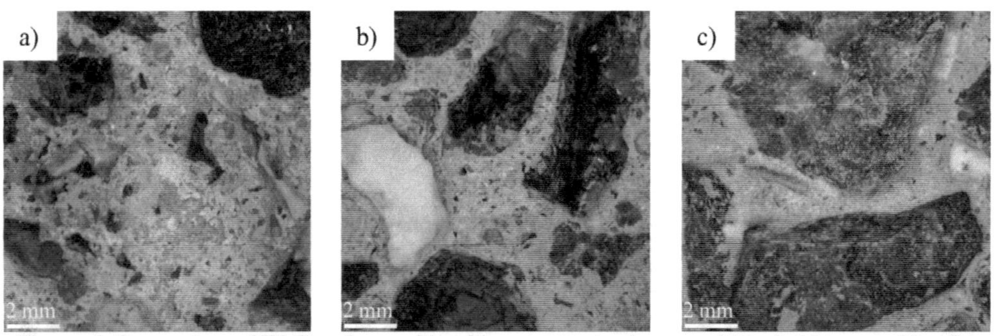

Figure 6. Comparison of EAC surfaces with gabbro aggregate a) composition A; b) composition B; c) composition C;

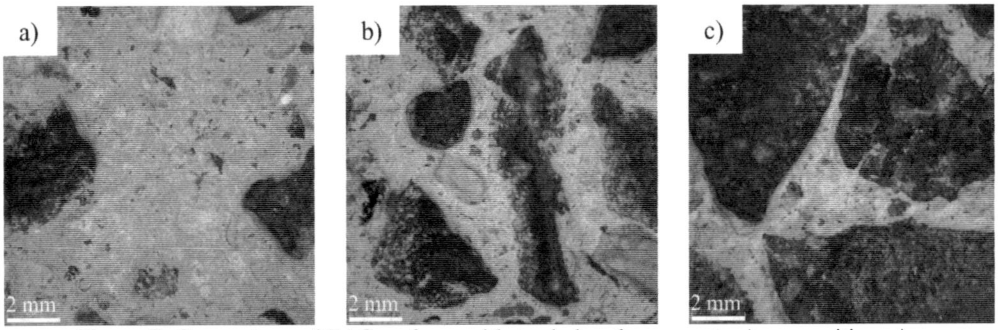

Figure 7. Comparison of EAC surfaces with trachybasalt aggregate a) compositions A; b) compositions B; c) compositions C.

Conclusions

The use of the FAP test equipment made it possible to compare EAC pavements differing in graining composition as well as the coarse aggregate resistance to polishing. The tests are valuable at the stage of designing the upper layers of the pavement, because they enable assessment of skid resistance at the design stage in laboratory conditions. However, the results obtained using the FAP test equipment should be complemented by macrotexture measurements on tested specimens. This will allow for comprehensive verification of materials for the wearing course, which can contribute to the slippery of road pavement.

It has been proven that the skid resistance is related to the aggregate graining composition as well as the resistance to polishing of the coarse aggregate used for EAC. The best parameters were achieved for the EAC pavements with about 50% content of aggregate > 4 mm (composition B and C) and coarse aggregate with the highest PSV. It should be noted that EAC is a very difficult technology for constructing upper layers of road pavements. To obtain the proper texture, it is very important to comply with the regimes during its construction, especially at the stage of its texturing and curing.

The research was supported by the projects N° S/WBiIS/1/2015 and it was financially supported by Ministry of Science and Higher Education, Poland.

References

[1] Hall J W *et al* 2009 Guide for Pavement Friction Final Report for NCHRP Project 01-43 *Transportation Research Board of the National Academes*

[2] Generalna Dyrekcja Dróg Krajowych i Autostrad 2013 Ogólne Specyfikacje Techniczne - Nawierzchnie Betonowe

[3] Wasilewska M *et al.* 2017 Evaluation of skid resistance of exposed aggregate concrete pavement in the initial exploitation period *Roads and Bridges - Drogi i Mosty* **16** pp 295-308

[4] Sommer H 1994 Developments of the exposed aggregate technique in Austria. *Session 8 – Noise reducing surfaces. Seventh International Symposium on Concrete Roads* pp 133-136

[5] RVS 08.17.02 Technical contract conditions "Concrete pavements – Pavement construction" 2011 *Forschungsgesellschaft Straße - Schiene - Verkehr (FSV)*

[6] Chandler J W E *et al* 2003 Quieter concrete roads: construction, texture, skid resistance and noise *TRL Report TRL576* **40**

[7] Teuns K C J G *et al* 2003 Full Scale Pavement Tests of Exposed Concrete Aggregates: Acoustical Aspects and Friction Characteristics *9th International Symposium on Concrete Roads*

[8] Cackler E T *et al* 2006 Evaluation of U.S. and European concrete pavement noise reduction methods. Part 1, Task 2, of the ISU-FHWA Concrete Pavement Surface Characteristics Project *Federal Highway Administration National Concrete Pavement Technology Center*

[9] Hoerner T E *et al* 2003 Current practice of portland cement concrete pavement texturing *Transportation Research Record* **1860** pp 178–186
[10] Sandberg U 2007 Noise characteristics of an exposed aggregate cement concrete surface *Proceedings 14th International Congress of Sound and Vibration*
[11] Tompkins D *et al* 2010 2008 Survey of European Composite Pavements, *Transportation Research Board*
[12] Gardziejczyk W and Gierasimiuk P 2016 Influence of texturing method on tyre/road noise of cement concrete pavement *International Journal of Pavement Engineering* (published online)
[13] Hall J W *et al* 2009 Texturing of Concrete Pavements NCHRP Report 634 *American Association of State Highway, Transportation Officials, Federal Highway Administration*
[14] Wasilewska M *et al* 2016 Evaluation of Skid Resistance Using CTM, DFT and SRT-3 Devices *Transportation Research Procedi*a **14** pp 3050-3059
[15] [15] Daskova J and Kudrna J 2012 The experience with Wehner/Schulze procedure in the Czech Republic *SIIV - 5th International Congress - Sustainability of Road Infrastructures, Procedia - Social and Behavioral Sciences* **53** pp 1035 – 1044
[16] Wasilewska M 2017 Evaluation of skid resistance of wearing course made of Stone Mastic Asphalt in laboratory conditions *IOP Conference Series-Materials Science and Engineering* **245**
[17] Do M T *et al* 2007 Pavement polishing – Develoment of dedicated laboratory test and its correlation with road results *Wear* **263** pp 36-42
[18] Kane M *et al* 2013 Long-term skid resistance of asphalt surfacings: Correlation between Wehner-Schulze friction values and the mineralogical composition of the aggregates *Wear* **303** pp 235–243

IOP Conf. Series: Materials Science and Engineering **356** (2018) 012002 doi:10.1088/1757-899X/356/1/012002

Review of modern methods for continuous friction measurement on airfield pavements

Paweł Iwanowski*, Krzysztof Blacha, Mariusz Wesołowski

Air Force Institute of Technology, Ks. Boleslawa 6, 01-494 Warsaw, Poland

*pawel.iwanowski@itwl.pl

Abstract. The safety of traffic, including both road and air traffic on a ground manoeuvre area, depends on many factors. These mainly include the anti-slip properties of a road or airfield pavement on which the traffic takes place. The basic pavement parameter that determines its characteristics in terms of anti-slip properties is the skid resistance, which constitutes the ratio of the wheel downforce and the friction on the contact surface.

There are currently many devices for continuous measurement of the skid resistance (Continuous Friction-Measuring Equipment - CFME) around the world. Most of them, in principle, do not vary much from one another. Most of the devices measure the measuring wheel's downforce on the pavement and the friction on the wheel-pavement contact surface. The skid resistance is the result of this measurement. The devices vary in many aspects, such as the type and size of the used measuring tyre, pavement-wheel slip or tyre pressure. This does not mean that the results obtained from various devices mbe directly compared. On the other hand, each device allows determining the pavement's anti-slip conditions in terms of the requirements specified for the given type of devices, thereby enabling pavement classification in these terms. The classification allows for comparing the results obtained from various measuring devices.

The paper presents an overview of equipment used in Poland and around the world to measure the skid resistance on airfield pavements. The authors draw attention to the requirements for pavements in terms of their roughness, with division into road and airfield pavements.

1. Introduction

Nowadays, much attention is paid to safety. This is mainly dictated by the high social costs of occurring incidents. In the case of transport, one of the factors that substantially affect safety are the pavement's anti-slip properties. Any safety systems, from ABS, various traction controls, to advanced anti-slip systems, are sensible only when sufficient tyre-pavement contact is ensured. As for the vehicle, the deciding factors are the selection of proper tyres or suspension configuration. As for the pavement, this includes obtaining the best possible anti-slip properties by ensuring the correct skid resistance measured by standard methods.

There are currently many devices for continuous measurement of the skid resistance (Continuous Friction-Measuring Equipment - CFME) around the world. Only a few of them were admitted measuring the skid resistance on Polish airports and only two on roads. The SRT-3 measuring kit has been used for over 20 years on national roads to assess the anti-slip properties. The TWO device was also admitted measuring campaigns a few years back. Both devices have their supporters and opponents, but there are no known procedures that would allow admitting other devices to measuring the skid resistance on national roads. As for now, there is also no institution that would certify such devices.

Some countries provide research programs such as The National Cooperative Highway Research Program (NCHRP) conducts research in problem areas that affect highways. In the result of NCHRP activity there were created "Guide for friction pavement" [1] uses in the USA. There are no known similar guide in the Poland.

The international document titled Doc. 9137 AN/898 Airport Services Manual - Part 2 - Pavement Surface Conditions [2], issued by the International Civil Aviation Organization allows using new measuring devices by providing the criteria and method of procedure in the device's certification process. The paper is to elaborate on the method of procedure with new CFME devices in the certification process for the road environment.

2. Overview of devices intended for continuous measurement of the skid resistance

There are many CFME-type devices for continuous measurement of the skid resistance around the world. Some of them are used in Poland. Despite the fact that each device measures the same parameter, their configuration affects the obtained result. This section includes presentation of typical devices used for measuring the skid resistance on Polish airfields and roads.

2.1. Airport devices

The Polish defence standard NO-17-A501:2015 *Airfield pavements - Friction testing* [3] includes a list of CFME-type devices admitted to using in Polish military facilities. The aforementioned standard specifies the requirements for such a device and indicates the possibility of using the devices specified in AC 150/5320-12C FAA [4], Doc. 9137 ICAO [2] and in Annex 14 ICAO [5]. The cited entry demonstrates that also new devices can be used, provided that they meet the specified requirements.

Table 1. presents the list of devices specified by the standard [3] including the measurement conditions and requirements applicable to airfield pavements.

Table 1. Requirements acc. to NO-17-A501:2015 Airfield pavements – Friction testing [3].

Measurement device	Test tyre		Measurement speed	Water film thickness	Friction coefficient		
	Type	Pressure			Design values for new pavements	Values for planning repair activities	Minimum (threshold) values
		kPa	km/h	mm			
Trailer	B	700	65	1.0	0.70	0.50	0.40
Surface Friction Tester (ASFT)	B	700	95	1.0	0.60	0.40	0.32
Trailer	A	70	65	1.0	0.72	0.52	0.42
Mu-meter	A	70	95	1.0	0.66	0.38	0.26
Trailer	B	210	65	1.0	0.82	0.60	0.50
Skiddometer	B	210	95	1.0	0.74	0.47	0.34
Trailer	B	210	65	1.0	0.69	0.52	0.45
RUNAR	B	210	95	1.0	0.63	0.42	0.32
Vehicle	B	700	65	1.0	0.70	0.50	0.40
Airport Surface Friction Tester (ASFT)	B	700	95	1.0	0.60	0.40	0.32
Vehicle	B	210	65	1.0	0.82	0.60	0.50
Surface Friction Tester	B	210	95	1.0	0.74	0.47	0.34

Vehicle	B	210	65	1.0	0.82	0.60	0.50
Runway Friction Tester	B	210	95	1.0	0.74	0.54	0.41
Vehicle	B	210	65	1.0	0.76	0.57	0.48
TATRA Friction Tester	B	210	95	1.0	0.67	0.52	0.42

On Polish airfields, the most popular skid resistance measuring devices are the ASFT (Airport Surface Friction Tester) devices mounted on a vehicle (Figure 1) or on the ASFT T-10 trailer hauled by a vehicle (Figure 1). Both devices allow for continuous, linear measuring of the friction between the airfield pavement and a model airplane wheel at the speed of 65 or 95 km/h. They are equipped with a water tank and a pavement humidifying mechanism that allows obtaining the water film of no less than 1 mm in thickness required by the standards.

Figure 1. The ASFT T-10 trailer, ASFT vehicle in the background, ASFT measuring system on left [10].

Smaller functional airfield elements, the geometry of which disallow measurement at the speed of 65 or 95 km/h, can be tested by using the BVS-1 device or the T2Go manual portable roughness tester (Figure 2). Currently, the relevance of the BVS-1 testing apparatus decreased in favour of the T2Go. The T2Go allows for continuous measuring of the skid resistance at low speeds and proper correlation between the results obtained by using the ASFT T-10 trailer allows for referring to the requirements specified in the aforementioned documents.

IOP Conf. Series: Materials Science and Engineering **356** (2018) 012002 doi:10.1088/1757-899X/356/1/012002

Figure 2. The measurement of the skid resistance on the airfield apron's pavement using the T2Go device.

Spot measuring of the skid resistance is possible by using a British skid resistance tester. The results provide an idea of the skid resistance of the tested pavement. However, it refers to a specific spot in contrast to the earlier presented devices which allow continuous measuring. The skid resistance tester is currently used mainly in laboratory testing. In the field, it is to compare to the results obtained by other methods. The criteria for the SRT indicator obtained from testing with the British skid resistance tester are known only for road pavements.

Less frequently, but our airfields use devices as the Griptester MK2 (Figure 3) and Skiddometer BV11 (Figure 3) as well. Both devices operate similarly to the ASFT T-10 by allowing for continuous measuring of the skid resistance, whereas the Griptester MK2 is characterised by its small weight and sizes. Each of the aforementioned devices is hauled during measuring by a wheel vehicle and requires constant feeding of water from the vehicle's tank.

Figure 3. Griptester MK2 (left) [11], Skiddometer BV11 (right) [12].

For measuring the skid resistance, the Griptester MK2 uses a 10-inch slick tyre which is in compliance with the ASTM 1844 specification. The device's running wheels also have a 10-inch diameter, however, they are treaded tyres. The measuring tyre's pressure during testing amounts to 140 kPa.

Two types of tyres can be used to test the skid resistance using the Skiddometer BV11. The first tyre type is the ASTM low-pressure tyre and the second one is the T520 high-pressure slick tyre. Both comply with the FAA and ICAO recommendation. The measuring wheel's working pressure amounts to 210 kPa and the wheel's slip in relation to the running wheels amounts to 17%.

2.2. Road devices

In the road environment, there are also many devices that are used to assess the pavement's anti-slip properties. They allow measuring the skid resistance at speeds from 30 to 120 km/h. Depending on the method of measurement; it is possible to distinguish the following device groups:

- "side force" - the measuring wheel is rotated with the device's motion direction by an angle of 7.5° - 20° and the side friction is measured. This group includes, among others, the SCRiM, SKM, Mu-Meter, Stradograf;
- "fixed slip" - the measuring wheel is set parallel to the device's motion direction and the friction is measured at a fixed slip. This group includes, among others, the DWW Trailer, Skidometer BV-8, GripTester, TWO;
- "variable slip" - the measuring wheel is set parallel to the device's motion direction and the friction is measured at a variable slip. This group is represented by the ViaFriction;
- "locked slip" - the measuring wheel is set parallel to the device's motion direction and the friction is measured with the wheel completely locked (100% slip). These devices include, among others, Adhera, Stuttgarter, Reibugsmesser, Pavement Friction Tester, SRT-3.

The SRT-3 and TWO devices were admitted to using on Polish national roads in accordance with Ordinance no. 34 of the General Director for National Roads and Motorways of 30 April 2015 [6]. The SRT-3 device (Figure 4) was constructed by the Road and Bridge Research Institute, whereas the TWO devices (Figure 5) is a foreign design.

Figure 4. SRT-3 measuring kit [13].

Figure 5. TWO measuring kit [14].

The SRT-3 measuring kit, classified as the "locked wheel", allows measuring the skid resistance with a fully locked measuring wheel. SRT-3 can't be included to CFME devices but it is worth to be mentioned due to widespread use on Polish roads. The skid resistance can be determined in two ways, i.e. either by measuring the wheel's measuring point or by direct measurement of the friction on the tyre-pavement contact point. Thanks to the reviewed device, it is also possible to measure the maximum skid resistance at the set measurement speed as well as turning resistances. The testing can be conducted both on straight road sections and on curves.

The TWO kit, classified as the "fixed slip", is intended to measure the skid resistance both on roads and on airfield facilities. The measuring device is mounted directly on a car. It is equipped with two wheels combined with each other by a chain so that they move relatively to one another with a slip of 17.8%. One of the wheels constitutes the reference point and the other is the measuring wheel. The measurement can be conducted both on a wet or dry pavement at the speed of 2 to 100 km/h. The quantity of water fed is adjustable and allows for creating a water film with a thickness of 0.1 to 1 mm. Both slick and treaded tyres can be used for measuring.

3. Procedure with new CFME devices

The international document titled Doc. 9137 AN/898 ICAO [2] specifies the criteria for new CFME-type devices, whereas appendix no. 3 [7] to the document includes the testing procedures for those devices. It is thus possible to introduce other devices for use than the ones currently specified in the standard documents, provided that specific requirements are met.

3.1. Requirements for new devices according to Doc. 9137 AN/898

In 1974, the basic technical and operating criteria for the CFME-type (designed by ICAO) devices were presented and recommended for runway skid resistance measurements. In 1991, the requirements were amended and included in the Airport Service Manual Doc. 9137 AN/898 ICAO [2]. The criteria are intended to standardise the design parameters for new friction measuring devices. They are also intended to ensure flexibility and allow for adding new devices in the future without excluding technical progress in this area.

The basic requirements for new CFME-type devices specified in the document are as follows:

- a measurement mode that should allow for continuous measuring in motion along the tested element;

- the ability to calibrate the device to ensure control over the correctness of the results. Even slight deviations during calibration have substantial impact on the test result;
- the measuring wheel's breaking point - the measuring wheel should be in a continuous slip in relation to the pavement (in the range of 10% to 20%), whereas in devices simulating the side force, the set angle should be between 5° and 10°. Currently, the most popular configuration used on Polish airfields is a 15% wheel slip without a set side force;
- limitation of excessive vibration - the device should be designed in a way that minimises or even eliminates the vibration of the absorbed or non-absorbed mass in the full range of the measuring speed, especially in relation to the measuring wheel. Vibrations affect the quality of the measuring apparatus' readings and the control of the wheel's downforce;
- device stability when driving, providing the ability to quickly leave the runway. The device cannot cause an air traffic hazard on the ground manoeuvre area and must be able to quickly move to the safe zone on the traffic controller's request;
- the recorded friction coefficient must be in the range between 0 and 1.0;
- the device should ensure continuous recording of the runway's friction coefficient chart and allow the operator to enter any observations as well as the recording's date and time. Continuous recording is a simplification, because the CFME-type devices record their results at specific intervals; the distance travelled by the device is so small in relation to the measuring section's length that it can be deemed as a continuous measurement;
- the device should be able to repeat the average friction coefficient results within ± 0,06 μ, at a confidence level of 95.5% (or two standard deviations);
- the friction coefficient should constitute the ratio of the longitudinal friction and the vertical measuring wheel load, in the case of a measurement simulating the side force, the skid resistance should correspond to the ratio of the side force and measuring wheel load. This requirement results from friction coefficient definition;
- the range of the device's possible measuring speeds should amount from 40 to 130 km/h. Despite the fact that the measuring range is broad, the standards speeds at which the skid resistance is measured are 65 and 95 km/h, while the other measuring speeds are intended for determining the correlations between devices as precisely as possible;
- the device should at least specify the average friction coefficient for: the first 100m of the runway, each 150m of the measuring section and for each one-third of the runway's length. The division of the runway to three parts is especially important for the pilot. When approaching to land, the pilot is provided with information about the skid resistance on particular runway fragments. Thanks to this, the pilot knows what to expect after touch down and how to manoeuvre the aircraft on subsequent runway parts;
- the measuring tyre should correspond to the types specified in the document; the tyre can be replaced with a different one, provided that reliable and certain results are ensured. The standardisation of measuring tyres is intended to eliminate the impact of the type of tyre used on the measurement conditions;
- the devices must be able to conduct the measurements regardless of the atmospheric conditions. The skid resistance measurements on airfields are most often conducted if there is suspicion that the pavement's anti-slip properties have deteriorated. Usually, such situation occurs as results of the impact of disadvantageous atmospheric conditions. For this reason, the device must be ready for operation in any atmospheric conditions;
- the device must be sufficiently easy in maintenance to ensure safety during measurement and transport;
- the device must be equipped with a pavement rinsing system to ensure a water film with a thickness of at least 1 mm.

3.2. New device testing procedure

Appendix 3 [7] specifies four basic assumptions that should be met during the testing of a new CFME-type device:

- it is necessary to determine whether the equipment and the device's water distribution systems calibration procedures applied by the manufacturer are satisfactory. The document does not specify concrete requirements, which is why an appropriate assessment of the experts participating in the testing is extremely important;
- the measurements must be conducted on at least four different types of pavements ensuring a broad range of skid resistance results. The document does not specify concrete parameters to be met by a pavement, such as, for example, its micro- and macro-structure, however it is worth selecting at least four types of pavements varying from one another with the obtained skid resistances;
- the measurements should be conducted at least two measuring speeds, i.e. 65 and 95 km/h. These are the standards speeds of the skid resistance measurement reproduced in many documents, whereas to ensure a broader range of results, it is also necessary to conduct the measurements at other speeds. This will increase the quantity of data and broaden the range of results;
- for each combination of measurement conditions, it is necessary to conduct the measurements in repeatable conditions and, if possible, conduct a measurement in the same conditions with another, certified CFME-type device to check the compliance of the results.

Meeting the aforementioned conditions will ensure significant degree of repeatability of the results and allow for controlling the obtained results.

Despite the fact that the document does not include such an entry, it is worth ensuring cleanliness of the pavement on which the measurement is conducted. The authors' experiences demonstrate that the spread of the results obtained on a contaminated pavement is broad. Even certified devices can be unable to ensure repeatability of ± 0.06 μ, at a confidence degree of 95.5%. To maintain proper pavement cleanliness during the measurements, it is advantageous to clean the pavement prior to each passage of the measuring device. At airfields, the quickest solution is to use an airfield cleaner.

The next steps of the CFME devices testing procedure, according to Appendix 3 [7] are as follows:

- checking the device's components, tyre and data recording system in terms of correct configuration and operating conditions.
- conducting the device's calibration according to the manufacturer's instructions and recording the results.
- checking the water distribution system in terms of compliance of the distributed water quantity with the assumptions and whether the water is properly fed under the measuring wheel at various speeds. The document does not specify the method of performing such a check. In practice, the length and width of the water footprint created during the measurement are measured and the water film thickness created in front of the wheel is determined based on the device's water consumption.
- repeating the first three steps for a second certified CFME-type device used to determine the correlation of results, if such device is used.
- if the testing features more than a single device, it is necessary to determine the order of the devices during passages and maintain the order throughout the testing. This is important, because the determination of the correlation features a comparison of the results obtained from particular passages of the measuring devices. Despite the fact that the document does not specify so, it is necessary to take care that the time between the passages of comparable devices is as short as possible.

- performance of two or three test passages with each of the CFME-type devices on the assumed pavement to stabilise the friction coefficient for a wet pavement and achieve an acceptable repeatability of ±0.03. If an acceptable repeatability is not achieved after two passages at similar speeds, it is necessary to check the measuring and data recording systems to identify the incorrectnesses. If necessary, correct any incorrectnesses and repeat the device's calibration before continuing testing.
- at each analysed speed, it is necessary to perform at least 6 measurement passages for each device and for each of at least 4 pavement types. Naturally, the probability of the device's correct assessment increases along with the increase of the number of measurements.
- each of the selected pavements must be subjected to measurements at a minimum of two measuring speeds - 65 and 95 km/h. Other measuring speeds allow for more precise determination of the dependency of the friction coefficient on speed.
- analysing the recorded data to determine the accuracy, repeatability and compliance of the results for each pavement and measuring speed.
- for the purpose of comparison with the friction coefficient obtained on a wet pavement, it is necessary to conduct two dry tests for each pavement and at similar measuring speeds as used in the case of wet pavement measurements.
- performance of repeated calibration of the devices according to the manufacturer's guidelines at the end of testing and results recording. The values obtained during the calibration should be similar to those from before the testing. If this isn't the case, it is necessary to identify the cause and repeat the testing, if necessary.
- development of charts for the dependency of friction coefficient on speed for each pavement. In the case of using more than one device, it is necessary to apply the charts for each device to a single chart.
- designation of the linear dependency and correlation coefficient for each case. The values should be presented on each of the charts.
- development of the list of main test parameters, including: test date, number of test passages, measurement times, type of pavement, measurement speed, measurement direction, water film depth, average friction coefficient,
- creation of a copy of historical test data, general notes, weather conditions and other observations to present them to the proper certifying institution. In general, the documents are sent to the ICAO, which checks the correctness of the conducted procedure and approves the device. In Poland, the unit that can admit a device for use on domestic airfields is the Civil Aviation Authority.

The document recommends taking photographs or making films during the testing, although it is not mandatory. Experiences show that it is an important procedure which facilitates reproducing the device's testing process.

4. Summary

The spectrum of continuous friction-measuring equipment available around the world is very broad. There are many such devices and new ones are created from time to time. One of the factors that favour this process is ensuring proper anti-slip conditions on road and airfield pavements by controlling them and quickly reacting to observed incorrectnesses. This is to ensure the safety of road and airport users.

Based on international documents, the Polish defence standard [3] specifies a list of devices that allow for designating the skid resistance on functional elements of airfields. The list includes eight of the most popular devices, some of which are used by operating services of the Polish aviation facilities. The standard also allows using two alternative devices (BVS-1 and ASFT T2Go) when the element's geometrical dimensions prevent from using the standard testing apparatuses.

IOP Conf. Series: Materials Science and Engineering **356** (2018) 012002 doi:10.1088/1757-899X/356/1/012002

On national roads, it is standard to use SRT-3 device, which is used by the General Directorate for National Roads and Motorways on a daily basis. Moreover, it is possible to use the TWO device in measuring campaigns in accordance with the ordinance [6].

The aviation environment is not closed to devices specified currently in the standard documents and enables introducing new CFME-type devices for testing. The document titled Doc. 9137 AN/898 [2] specifies the procedure for such cases in fair detail. It includes requirements for new devices and their testing procedure cited in this article. Appendix 3 [7] to the aforementioned document specifies further steps of procedure along with the testing conditions, basic assumptions, requirements for results interpretation and correct test documentation.

The paper is intended to present the road environment with the testing procedure for a new CFME-type device in the process of its certification at airfield facilities and the possibilities of introducing new measuring apparatuses into the list of devices admitted by national and international aviation institutions.

The authors point out the possibilities of adapting the presented method for use by the road environment to admit alternative devices intended for testing the anti-slip properties of road pavements.

Bibliography (numbered in the order of appearance in the text)

[1] J.W. Hall, K.L. Smith, L. Titus-Glover, J.C. Wambold, T.J. Yager, Z. Rado, *Guide for pavement friction,* Contractor's final report for NCHRP Project, 2009

[2] Doc. 9137 AN/898 *Airport Services Manual - Part 2 - Pavement Surface Conditions,*

[3] NO-17-A501:2015 *Airport pavements - Friction testing,*

[4] AC 150/5320-12C FAA *Measurement, Construction, and Maintenance of Skid Resistant Airport Pavement Surfaces*

[5] Annex 14 to the Convention on International Civil Aviation, *Airfields - Volume I - Engineering and operation of airfields*

[6] Ordinance no. 34 of the General Director of National Roads and Motorways of 30 April 2015

[7] Appendix 3 *NASA Certification Test Procedure for New Continuous Friction-Measuring Equipment Used at Airport Facilities*

[8] W. Gardziejczyk, M. Wasilewska, P. Gierasimiuk, M. Motylewicz, Przegląd urządzeń do oceny właściwości przeciwpoślizgowych nawierzchni drogowych [Overview of devices intended for assessing the anti-slip properties of road pavements], *VIII Techniczne dni drogowe,* Białystok 2015

[9] W. Gardziejczyk, M. Wasilewska, P. Gierasimiuk, M. Motylewicz, Ocena właściwości przeciwpoślizgowych nawierzchni drogowych przy wykorzystaniu urządzeń two, ctm i dft, *journal of civil engineering, environment and architecture,* 2016

[10] http://www.asft.se

[11] http://www.griptester.us

[12] http://www.moventor.com

[13] http://www.docplayer.pl

[14] http://www.two-friction.com

IOP Conf. Series: Materials Science and Engineering **356** (2018) 012003 doi:10.1088/1757-899X/356/1/012003

Evaluation of airfield pavement evenness

Paweł Pietruszewski[1], Adam Poświata[1], Mariusz Wesołowski[1]

[1] Air Force Institute of Technology

E-mail: pawel.pietruszewski@itwl.pl

Abstract. The evenness of airfield pavements is one of the basic operating parameters, which characterize them. The evenness determines not only comfort of traffic along an airfield pavement, but also influences the size of dynamic effect on the pavement, hence, the safety of air operations. In addition, the evenness condition changing as a result of dynamic loads, adverse weather conditions or inappropriate airfield pavement construction technology, lead to deviations from the desired condition in the form of longitudinal and transverse unevenness. As a result, systematic and correct performance of tests is a very significant and required factor impacting the improvement of traffic safety on airfield pavements. If the data obtained through the measurements are not sufficiently reliable, they may consequently lead to making incorrect decisions, which can ultimately impact the safety of air operations.

Keyword: airfield pavements, airfield pavement evenness, airfield pavement evenness state evaluation

1. Introduction

One of the main operating features that characterises each airfield or road pavements is its evenness, which as a result of dynamic loads, adverse weather conditions or inappropriate construction technology leads to deviations from the desired condition, in the form of longitudinal or transverse unevenness. Pavement evenness, or rather lack of, is the only operating parameter, which might be directly felt and noticed by users through inadequate driving comfort, bad rainwater drainage, the formation of isolated water bodies or increased dynamic loads. Formed pavement unevenness also adversely impacts the rolling resistance of vehicle wheels and accelerated surface deterioration, hence, the degradation of safety conditions.

The evenness measurements in the road industry, conducted by the General Directorate for National Roads and Motorways (GDDKiA) are carried out with multi-sensor laser profilometers. The obtained test results are converted into, so called, *International Roughness Index IRI* expressed in mm/m or m/km, that characterizes the suspension work in conventionally applied computational model [1]. However, in conformity with the International Civil Aviation Organization (ICAO) and Federal Aviation Administration (FAA) requirements, the measurements may be performed using profilometer or leveling rod and a wedge. Measuring devices differ in their dimensions and acceptable single unevenness on the airfield pavement. In accordance with existing provisions of ICAO, the single acceptable unevenness shall be 3 mm over the length of 3 metres, whereas according to FAA it is 6,4 mm over the length of 5 metres. The example of unevenness measurement device used on road pavements in Poland, as well as the unevenness measurement device used on airfield pavements in The United States of America in accordance to FAA, is presented in Figure 1 below.

Figure 1. Surface evenness measuring devices a) RSP [1],
b) California Profilograph [2].

In the road industry, in a situation of deviations from the desired surface condition there is a possibility of placing (installing) signs warning about the formed road unevenness and limiting the speed over an entire road section. Moreover, removing the formed damage and restoring an appropriate surface evenness condition is possible without entirely putting a given road section out of operation, and only limiting it, through temporary traffic and using, e.g. alternating traffic. Unfortunately, such actions are impossible in the case of air traffic, and therefore, the requirements for airfield pavements must be much higher than the criteria used for road pavements. In the light of the above, it is not recommended to apply road standards, requirements or criteria to evaluate the technical condition of airfield pavements.

Trying to meet the presented issue half-way, the author, along with a broader research team, drawing from more than 60 years of experience gained during the execution of investments and repairs at airfield facilities, developed Defence Standard NO-17-A502:20175 *Airfield pavements. Evenness testing*, which is applied by military, as well as civil airport services. The standard in question determines the procedure for evaluating airfield pavement evenness condition on individual airfield functional elements (AFE), the evaluation criteria and the manner of presenting the obtained field test results. In addition, the a/m standard draws attention to the need to determine the evenness condition of newly constructed and renovated pavements prior to handing them over to operation, and to systematic periodic inspections, over the course of their operation process. It will enable tracking the changes undergoing during operation, determining the areas in which degradation is faster, and also forecast their rate. Fig. 2 shows examples of airfield pavement damage regarding their unevenness and having significant impact on the safety of air operations executed by aircraft.

Figure 2. Examples of formed airfield pavement unevenness.

2. General information

The measurement of airfield pavement evenness as per NO-17-A502:2015 *Airfield pavements. Evenness testing* shall be conducted with the use of a 4 m or 3 m long planograph, which enables unevenness recording. In case the geometrical dimensions of an evaluated AFE make it impossible to use a planograph for the measurements, measuring with a 4 m or 3 m long staff is permitted.

2.1. Measurement with a staff

A staff should be made from metal or wood. However, it is recommended to use staff made from aluminium alloy sections. The bottom plane (measuring) aimed towards the pavement should be smooth and even. A 3 m long staff shall be made with a tolerance of ± 3 mm, and a 4 m with a tolerance of ± 4 mm. The rigidity of the staff should be prepared in a way so that the deflection in the centre, when supporting on the ends, did not exceed 0.5 mm [3,4,5]. A 3 m long measurement staff is shown in figure 3. A 4 m long measurement staff is identical to a 3 m one.

Figure 3. A 3 m long measurement staff.

2.2. Measurement with a planograph

A planograph used to measure airfield pavement evenness condition records the displacement of a mobile measuring wheel relative to a fourth-metre or a three-metre baseline, moving along the pavement on 14 or 10 rolling wheels. The planograph design should ensure adequate rigidity, which is defined by its deflection in the central part when supported only on the extreme wheels. The deflection of a planograph structure should not exceed 0.3 mm. The axes of extreme wheels in a 4 m long planograph should be placed at a distance (4000± 70) mm, in a 3 m long planograph – (3 000 ± 50) mm, with the spacing of the other wheel with a tolerance of ± 5 mm. The diameter of rolling wheels should be (200 ± 10) mm, with the diameter of a measuring wheel being (159 ± 5) mm [3]. A general diagram of a 4 m long planograph is shown in figure 4. A diagram of a 3 m long planograph is identical as of a 4 m long one, with the remarks as discussed above.

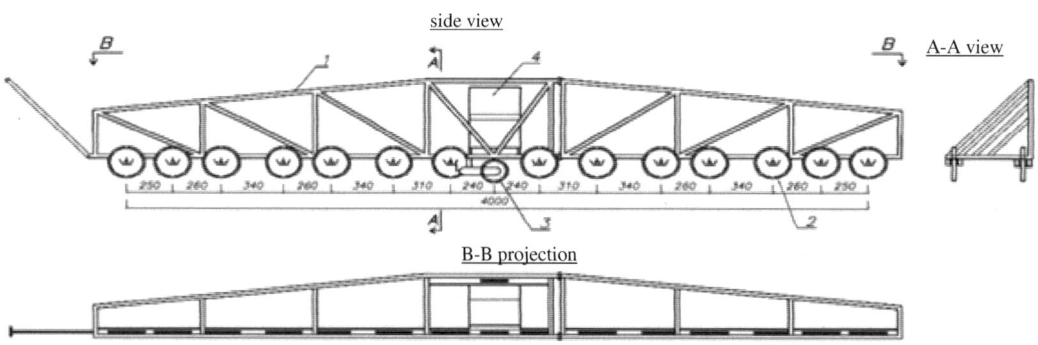

Explanations
1 – trolley frame, 2 – rolling wheel, 3 – measuring wheel, 4 – measuring table
(a recording set, transmitting the results to a computer)

Figure 4. A diagram of a 4 m long planograph [3].

A planograph used for measuring airfield pavement evenness is a modernized device made in Poland. The modernization involved adding an assembly recording pavement unevenness, an unevenness sensor, a road increment sensor, a serial interface and an analysis computer. Recently, in order to increase measuring precision, another modernization has been performed, involving changing the rolling wheels, adding a new linear displacement converter transformer, and a new interface and software developed at the Air Force Institute of Technology (AFIT). Figure 5 shows a planograph after the modernization.

Figure 5. A post-modernization P-3Z planography.

This set allows to measure the unevenness present on a pavement, in the form of road length increment, with an accuracy of 0.3 mm and a frequency every 10 cm.

3. Test methodology and evaluation

Airfield pavement evenness, pursuant to the NO-17-A502:2015 standard, should be expressed with the defectiveness degree *W*. This term is understood as a percentage share of the number of 5 m long route sections, with at least one event of exceeding the permissible value between the theoretical connecting line, formed by contact points of planograph rolling wheels, and the upper pavement surface.

The 5 m section length adopted for analysis and the developed criteria are not accidental. In [6], the author determined the impact of measurement route length and the variability of the module length on assessed parameters, such as: mean unevenness size, standard deviation or defectiveness. In order to determine their impact, an analysis was conducted, which adopted measuring route lengths of: 100 m, 200 m, 500 m, 1000 m, 2000 m and module lengths: 0.1 m, 0.2 m, 0.5 m, 1.0 m to 10.0 m, with an increment of 1.0 m. An example analysis result is shown in figure 6.

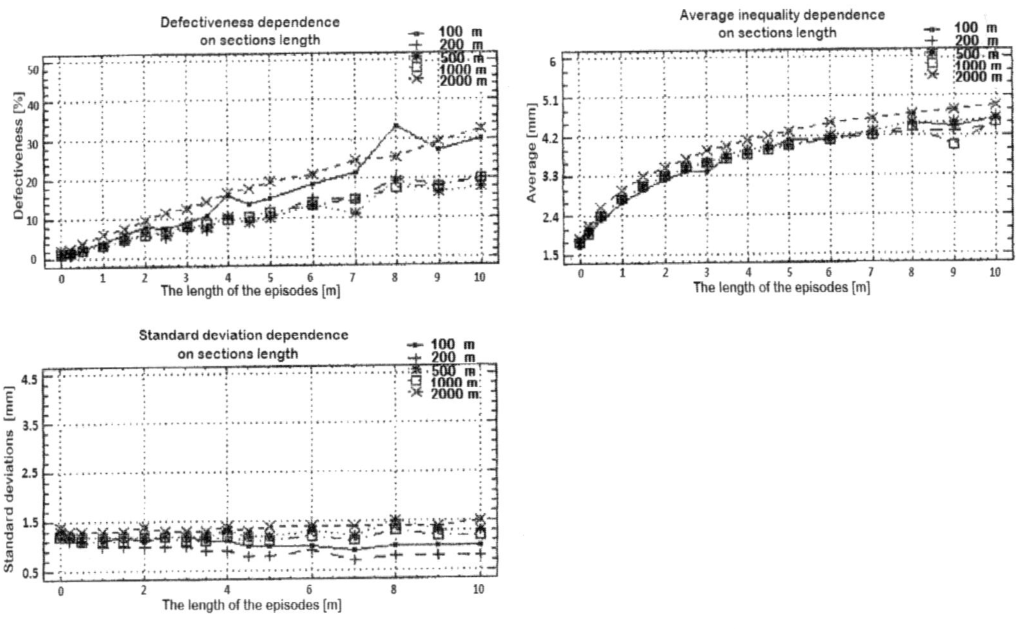

Figure 6. Measurement route length and module relationships [6].

Based on the conducted analyses, a measurement route division into 5 m modules was adopted. The adopted length of 5 m corresponds to a pavement slab most frequently used at airports. As a result, a set of a defined number of readings is taken for the evaluation, which makes the measurement itself and the later analysis more accurate.

A studied AFE is divided into 100 m long (1 hectometre) test areas. Pavement areas, which can be assigned a certain rating, are obtained as a consequence of that division. A road structure is evaluated in terms of evenness, by analysing the mean defectiveness value *W*, which is determined from a relationship:

$$W = \frac{\sum_{i=1}^{n} w_i \cdot F_i}{\sum_{i=1}^{n} F_i} \tag{1}$$

where:
W – defectiveness of the evaluated area or zone [%],
w_i – defectiveness of the "ith" evaluated test area [%],
F_i – length of the evaluated route section (test area) adopted for assessment [m].

The airfield pavement evenness condition is evaluated through a defectiveness criteria, which is determined by using a five-grade scale presented in table 1.

Table. 1. Airfield pavement evenness condition evaluation criteria.

Evenness condition according to defectiveness evaluation	Planograph length 4 m	Planograph length 3 m	Colours of an evaluated area
Very good	W≤5%		
Good	5%<W≤10%		
Sufficient	10%<W≤20%		
Unsatisfactory	20%<W≤50%		
Insufficient	W>50%		

A test result is deemed positive when the number of measurement route 5 m sections exceeding permissible unevenness is lower than 20% for newly constructed or renovated pavements, and lower than 50% for operated pavements.

According to [3,7,8] the permissible unevenness for newly constructed pavements and the maximum unevenness values for operated pavements are shown in table 2.

Table. 2. Airfield pavement maximum and permissible unevenness.

Device type	Maximum unevenness [mm]	Permissible unevenness [mm]
Planograph or a 4 m staff	12	5
Planograph or a 3 m staff	9	3

Moreover, in order to confirm the suitability of a 3 and 4 m planograph measuring system to evaluate an airfield pavement unevenness condition, repeatability and reproducibility coefficients with the use of the Repeatability & Reproducibility (R&R) method were determined. [9] contains an analysis of repeatability and reproducibility, together with the total spread taking into account the impact of individuals operators on the measurement. A block diagram of the R&R method is shown in fig. 7.

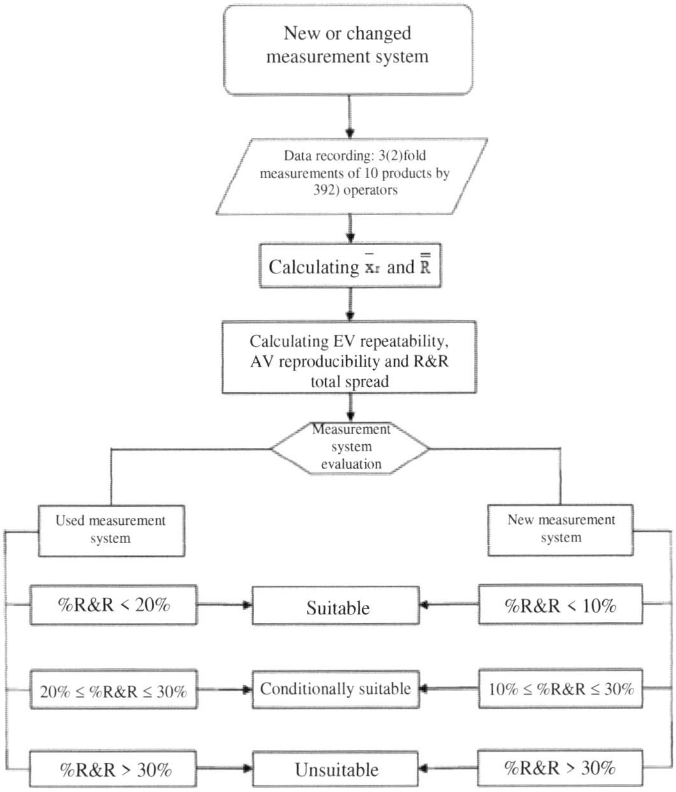

Figure 7. R&R method block diagram [10,11].

The results of a conducted repeatability and reproducibility analysis are shown in table 3.

Table. 3. Common table of calculated values after a conducted analysis.

Quantity	3 m planograph [%]	4 m planograph [%]
Equipment Variation %EV	5.8	3.2
Appraiser Variation %AV	1.6	1.5
%R&R	7.4	4.7
Measurement system evaluation	**System fit**	**System fit**

Based on a conducted analysis of a measurement system suitability, it was concluded that the calculated %R&R indicators fall within a range of up to 10%. It confirmed the fact that both a 3 m, as well as a 4 m long planograph, are suitable for evaluating airfield pavement evenness conditions. However, when analysis the components o individual determined quantities, they indicate a higher suitability of a 4 m long planograph, which in consequence leads to a more accurate reflection of the actual condition of an airfield pavement.

4. Test result presentation

According to the defence standard NO-17-A502:2015, an airfield pavement evenness analysis is performed on all operated AFEs, however the scope and presentation manner of the results depends on their function during air operations. As a consequence, evaluating evenness condition and the test scope depends on the fact, whether the measurements are performed on runways, taxiways or aprons. However, regardless of the AFE being measured, a result of an airfield pavement evenness condition test is a mean value of calculated defectiveness and mean unevenness values, standard deviations and numerical values of unevenness for the adopted sections.

Measurements on a runway are performed on all strips in the longitudinal direction and every 50 m in the transverse direction. A 100 m long module, with calculated defectiveness *W* is adopted to evaluate the evenness condition of runways. The obtained values of defectiveness in the longitudinal and transverse directions are plotted on accurately determined test areas, which are marked with colours adopted for individual evenness conditions as per table 1, with the resulting *Evenness condition evaluation metrics*. Test areas constitute separated long runway pavement surface areas, 100 m (1 hectometre) long and usually 5 m wide (strip or slab width).

Measurements on taxiways and aprons are performed in all strips in the longitudinal directions and every 50 m in the transverse direction (transverse measurements on taxiways up to 25 m wide are not required, unless the presence of water pools is identified, or it is required by the Ordering Party). Taxiways and aprons are evaluated separately in the longitudinal direction, whereas in the transverse direction, a common assessment is presented. The final rating of a road structure, apart from the mean defectiveness value, contains average unevenness, standard deviation, percentage share of individual exceedings and unevenness profiles in the road increment function [12].

An example analysis of test results, together with the manner of their presentation is shown in figures 8-11 and table no. 4.

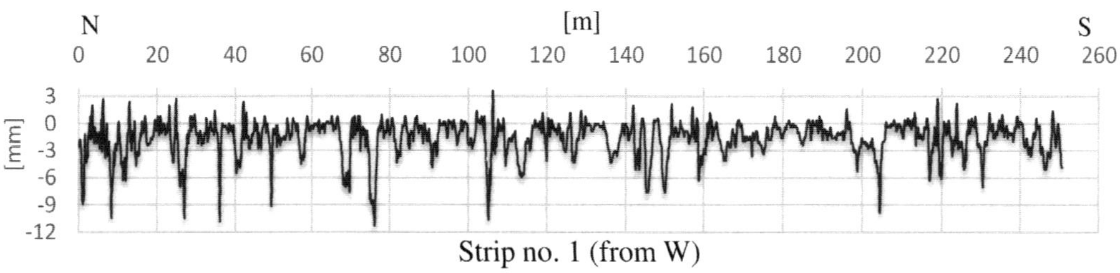

Figure 8. Longitudinal unevenness profile on a strip example.

Table 4. Mean values of longitudinal unevenness, standard deviation, defectiveness and the number of unevenness within individual unevenness ranges and individual strips.

STRIP No.	Averaged values "X" [mm]	Standard Deviation "S" [mm]	Defectiveness "W" [%]	Number of slabs with max. unevenness: up to -5.0 mm [%]	from -5.1 to -9.0 mm [%]	from -9.1 to -12.0 mm [%]	above -12.0 mm [%]
Longitudinal unevenness							
1	-2.4	1.5	6.8	93.2	6.6	0.0	0.2
2	-1.9	1.2	2.8	97.2	2.6	0.2	0.0
3	-2.1	1.2	2.8	97.2	2.6	0.2	0.0
4	-2.2	1.3	3.2	96.8	2.6	0.4	0.2
5	-2.3	1.3	3.4	96.6	2.8	0.6	0.0
6	-2.7	1.3	5.8	94.2	5.4	0.4	0.0
7	-2.4	1.5	5.4	94.6	4.4	0.6	0.4
8	-2.1	1.4	4.0	96.0	3.6	0.4	0.0
9	-2.1	1.3	4.8	95.2	4.8	0.0	0.0
10	-2.4	1.3	5.6	94.4	5.6	0.0	0.0
Mean	-2.3	1.3	4.5	95.5	4.1	0.3	0.1
Transverse alignment							
Mean	-3.3	1.5	9.4	90.4	9.2	0.2	0.0

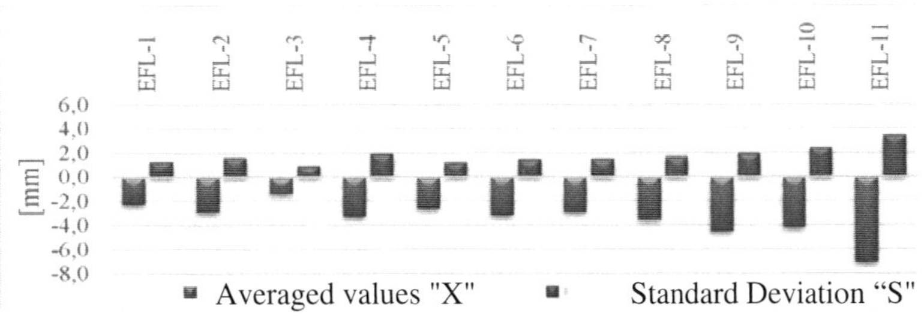

Figure 9. Average quantities of longitudinal unevenness and standard deviations on individual AFEs.

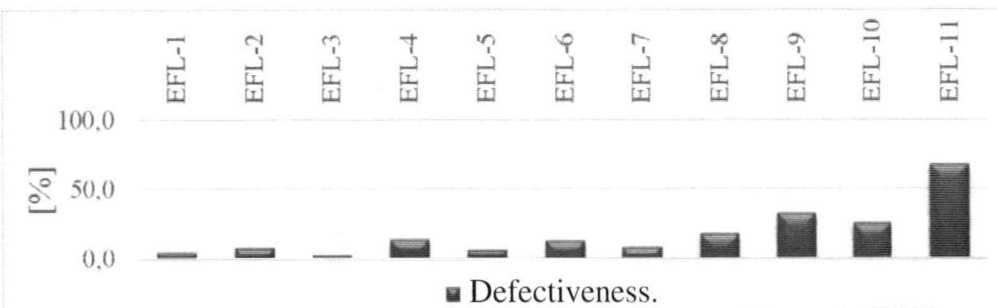

Figure 10. Percentage number of slabs exceeding normative requirements (defectiveness) on individual AFEs.

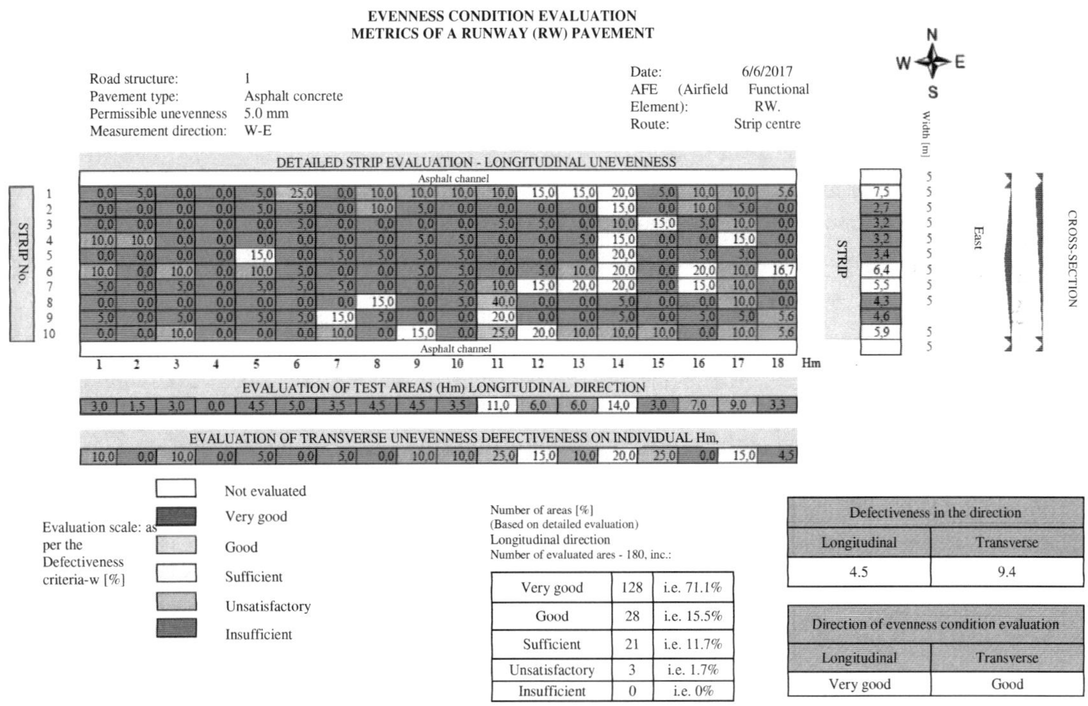

Figure 11. Runway airfield pavement evenness condition evaluation.

5. Final conclusions

Evenness is one of the most important operating parameters, which each airfield pavement is characterised by. It determines not only the comfort of an aircraft movement over an airfield pavement, but also impacts the size of dynamic loads on it. Achieving required evenness is also a prerequisite for a quick and efficient drainage of rainwater or defrosting agents from airfield pavements. Even small unevenness of a pavement may be a base for the formation of water pools, which deteriorate safety conditions through the possibility of *aquaplaning*, which is a lost of adhesion between a wheel and a surface.

The modern, extremely dynamic development of aviation and the constantly growing intensity of air operations, which has increased several times over the last 12 years, forces the conducted tests to be as accurate as possible, and the obtained results to be reliable and reflecting the existing pavement status as closely as possible. The applied measuring devices must guarantee the correctness of obtained results, since the safety of air operations belong on them.

Having regard to the foregoing, there is a constant need of monitoring the condition of airfield pavement evenness, which will enable the segregation of zones (areas) with the fastest degradation, and forecast its course over time. Furthermore, information about the airfield pavement condition collected from measurements performed in accordance with defence standard NO-17-A502:2015 *Airfield pavements. Evenness testing* are crucial for the facility Manager, who may take action to enable determining the causes of its forming due to exact diagnostics, thereby reducing the degradation tempo by adjusting the proper repair technology.

Further publications regarding the discussed operational parameter of airfield pavements will present the results with the analysis of evenness condition analysis, coming from the conducted field tests utilizing the new testing instruments.

References

[1] Dynatest 5051 Mark III/IV Road Surface Profiler. Test systems. Owner's manual. Edition 2013

[2] An IPRF Research Report Innovative Pavement Research Foundation Airport Concrete Pavement Technology Program. Airfield Concrete Pavement Smoothness – A Handbook, April, 2007

[3] NO-17-A502:2015 *Nawierzchnie lotniskowe - Badanie równości [Airfield pavements - Evenness testing]*

[4] Annex14 ICAO to the convention on international civil aviation, Volume I, Aerodrome design and operations, Edition 6, lipiec 2013

[5] Federal Aviation Administration AC 150/5300-13A - Airport Design

[6] Poświata A., *Analiza nierówności w ocenie eksploatacyjnej betonowych nawierzchni lotniskowych [Unevenness analysis in the operational evaluation of concrete airfield pavements]*, Doctoral thesis, WAT, Warszawa 1997

[7] Annex to ED Decision 2017/021/R (CS-ADR-DSN Issue 4)

[8] COMMISSION REGULATION (EU) No 139/2014 of 12 February 2014 laying down requirements and administrative procedures related to aerodromes pursuant to Regulation (EC) No 216/2008 of the European Parliament and of the Counci

[9] Pietruszewski P., Poświata A., Wesołowski M., *Wykorzystanie metody R&R do oceny przydatności systemu pomiarowego planografu 3 i 4m podczas badania równości nawierzchni lotniskowych [Using the R&R method for evaluating the suitability of a 3m and 4m planograph measurement system for testing airfield pavement evenness]*, Autobusy, no. 12, 2017

[10] Sałaciński T., *Statystyczne sterowanie procesami produkcji [Statistical control of manufacturing processes]*, Oficyna Wydawnicza Politechniki Warszawskiej

[11] C Chrysler, Ford, General Motors, Measurement System Analysis", (2d ed.), Southfield, MI, Automotive Industry Action Group, 1998 oraz Statsoft® , 2002

[12] Poświata A. Pietruszewski P. *Wpływ stanu równości nawierzchni lotniskowych na bezpieczeństwo wykonywania operacji lotniczych [The impact of airfield pavement evenness condition on the safety of air operations]*. Autobusy, no. 12, 2016

IOP Conf. Series: Materials Science and Engineering **356** (2018) 012004 doi:10.1088/1757-899X/356/1/012004

Influencing factors on the visibility of surface distresses

Wojciech SMĘT[1], Marek SKAKUJ[2]

[1] Heller Consult sp. z o.o., ul. Reformacka 10, 21-500 Biała Podlaska, Poland, wsmet@heller-consult.pl

[2] HELLER Ingenieurgesellschaft mbH, Otto-Hesse-Straße 19 / T9, 64293 Darmstadt, Germany, marek.skakuj@heller-ig.com

marek.skakuj@heller-ig.com

Abstract. A systematic and future-oriented maintenance management requires quality-assured data regarding the condition of the roadway. An important characteristic for assessing the condition of the road surface are cracks [1]. Experience has proven that the road surface temperature and residual humidity can influence the visibility of cracks. To investigate and quantify these influencing factors in more detail, the Federal Ministry of Transport and Digital Infrastructure has implemented a comprehensive research project [2], during which several measurement campaigns were carried out at different seasons and at different temperatures and humidity degrees of the road surface. The visibility of the cracks was assessed using the German approach ZTV ZEB-StB [3]. After evaluating the cracks, the data was synchronized and the network allocation of the geo-referenced raw data was realized. The relevant condition performance indicators were calculated subsequently. For each evaluation section of 100 meters length, the mean road surface temperature was calculated and, if available, the type and year of construction of the surface course was determined. The evaluated data was visualized to establish correlations between the data and the influencing factors. For the statistical analysis of the results, a total of six comparative scenarios were defined in which the influencing factors of road surface temperature and residual humidity were assessed. This paper documents the results of the research project. In order to obtain a uniform and comparable picture of the condition of the road surface over the years and thus to be able to observe the aging process of the road surface, the measurement campaigns are to be carried out under defined conditions. The research project included proposals for defining and further specifying the measurement conditions.

1. Research needs and scope

The aim of the pavement condition survey and assessment (ZEB) on federal highways is to provide quality-assured data for maintenance management. The acquisition and evaluation rules for this data are described in the ZTV ZEB-StB [3]. Cracks, including accumulation of cracks, individual cracks as well as longitudinal and transverse cracks, on the road surface are an important characteristic for assessing the condition of both the structure and the surface of the pavement [1], [4]. In the ZTV ZEB-StB [3], there are currently no restrictions regarding the permissible temperature ranges for the condition survey. However, it is prescribed that measurements of evenness within the ZEB should be carried out on a dry and clean road surface. Yet, previous projects showed that residual humidity on the road surface can improve the visibility of fine cracks [5], [6]. This and the influence of the road surface temperature on the visibility of cracks were studied and quantified within this research project.

IOP Conf. Series: Materials Science and Engineering **356** (2018) 012004 doi:10.1088/1757-899X/356/1/012004

2. State of research

More than 25 years ago, a federal initiative launched the pavement condition survey and its assessment (ZEB). The condition data are combined into a usage condition index and a substance condition index [3]. Within the ZEB, the damage on the road surface is recorded in the so-called subproject 3 by measuring vehicles using imaging methods and evaluated manually on-screen (state of the art). Rules for the survey and evaluation of surface damage are defined in the regulations ZTV ZEB StB [3]. Weather conditions are not addressed, but the measurements of evenness indirectly indicate that the measurements must be carried out on clean and dry road surfaces.

The Working Group 4.3.1 [7] of the Road and Transport Research Association is currently working on the technical test specifications for acquiring surface images using high-speed imaging systems. According to this, measurements should only be performed on surfaces that appear to be dry and clean. Specific temperature ranges are not required. However, the US non-profit NORTHWEST PAVEMENT MANAGEMENT ASSOCIATION recommends that the road condition survey and assessment should always be carried out at the same time of the year [8].

In the years 2012 to 2014, HELLER Ingenieurgesellschaft mbH conducted a research project on the influence of temperature on the visibility of cracks, in which surface distresses were evaluated in detail. The final report of the project [5], [6] states that the visibility of cracks can be dependent on temperature and humidity.

LEHMANN + PARTNER [9] carried out measuring campaigns with high-resolution surface cameras on dry and drying roads (after wetting the lanes artificially). While fine cracks could not be detected in the dry state, they became visible when wetted. However, the road surface dried at different speeds due to weather conditions, which affected the results.

As VOS and BOUMAN [10] report, the illumination of the road surface and the observation angle are crucial for determining the type and extent of damage correctly. The quality of the data collected depends directly on the quality of the underlying surface images. Weather conditions can significantly influence the image quality and output data when the lighting is unfavorable [11]. Xu [12] used artificial lighting in the condition survey to eliminate all shadows in the image and thus improve the uniformity of the data regardless of the weather conditions. Artificial lighting allowed consistent crack detection results under different natural lighting conditions (at night, daytime with shadows and daytime without shadows).

3. Planning and execution of the experiment

3.1. Measurement campaigns

To investigate whether the road surface temperature and residual humidity influence the visibility of cracks, a total of nine measurement campaigns were carried out on selected asphalt and concrete roads under different conditions.

- Campaign 1 – Fall 2015 [H15]. Survey of four test sections (3 asphalt, 1 concrete), dry and clean roadway, road surface temperatures 10-20 °C.
- Campaign 2 – Winter 2016 [W16]. Survey of four test sections (3 asphalt, 1 concrete), dry and clean roadway, road surface temperatures < 0 °C.
- Campaign 3 – Spring 2016 [F16]. Survey of four test sections (3 asphalt, 1 concrete), dry and clean roadway, road surface temperatures 10-20 °C.
- Campaign 4 – Summer 2016 [S16]. Survey of four test sections (3 asphalt, 1 concrete), dry and clean roadway, road surface temperatures 30-40 °C.
- Campaign 5 – Fall 2016 [H16]. Survey of four test sections (3 asphalt, 1 concrete), dry and clean roadway, road surface temperatures 10-20 °C.
- Campaign 6 – Winter 2017 [W17]. Survey of four test sections (3 asphalt, 1 concrete), dry and clean roadway, road surface temperatures < 0 °C.
- Campaign 7 – Wet roadway [N16]. Survey of four continuously wet test sections (3 asphalt, 1 concrete), road surface temperatures 10-20 °C.

- Campaign 8 – Diurnal curve I (Spring 2016) [TGL1]. Multiple (at least four times each) survey of two test sections (1 asphalt, 1 concrete) throughout the day, dry and clean roadway, road surface temperatures approx. 10-30 °C.
- Campaign 9 – Diurnal curve II (Summer 2016) [TGL2]. Multiple (at least four times each) survey of two test sections (1 asphalt, 1 concrete) throughout the day, dry and clean roadway, road surface temperatures approx. 20-40 °C.

The campaigns 1 to 6 were designed to collect measurement data on the visibility of cracks at different road surface temperatures. With the diurnal curves (campaigns 8 and 9) it was investigated to what extent the natural variations of the road surface temperature during the day influence the detectability of cracks. In addition, the degree of residual humidity, if any, was also evaluated during the day.

3.2. Selection of the test sections

The test sections were selected by using existing condition data from previous years and by a virtual inspection [13].

All test sections had to show cracks but not to an extent that would have made the comparison of the distresses impossible. Furthermore, comparable traffic densities without obstructions should prevail so that traffic could also be examined as a possible influencing factor. Other influencing factors had to be minimized. Therefore, the construction data was analyzed in an engineering manner when selecting the test sections. Moreover, no maintenance treatments should be carried out during the research project.

Three asphalt sections and one in concrete construction, i.e. a total of four test sections of 25 km each, were selected that met these criteria.

3.3. Measuring vehicle

The acquisition was carried out with a high-speed (80 km/h) measuring system equipped with an area scan camera system. Both front and surface images were taken to capture the surface distresses. The surface camera system at the back of the measuring vehicle can capture a total width of up to 4.50 m and uses stroboscopes to illuminate the road surface. With a second, the so-called secondary front camera, additional road images were taken. As the camera is attached to the beam to detect transverse evenness, it captures the road from a different, significantly flatter angle than the (primary) front camera. In addition, the measuring vehicle was equipped with an infrared industrial thermometer to measure the road surface temperature.

It had the time-limited operating approval of the Federal Highway Research Institute, a license required to guarantee high measuring quality und conformity with the standards. Laser-based systems for the automatic identification of surface distresses could not be used for acquiring the data in this research project, as there are currently no vehicles in Germany that fulfil the conditions for the license required.

3.4. Evaluation of the measurement data

The evaluation of the surface images was carried out only by experienced inspectors with at least three years of experience in detecting surface distresses who had been additionally trained at the beginning of the project. Each measurement campaign was assessed twice by two independent inspectors. Both results were compared, possible differences discussed and, if necessary, the data was changed accordingly. To see how often the inspectors agreed upon the visibility of cracks, contingency tables were assembled. Those tables show the frequencies of different results assigned by two inspectors. The picture-based assessment is a highly repeatable way of quantifying the visibility of cracks. The agreement rate between experienced inspectors is on average 95% [5].

Since no suitable method is known to date to reliably quantify the degree of humidity on a road surface, the following classification was determined for this research project: dry surface, drying surface (residual humidity is only visible in the area of cracks) and wet surface (the road surface is continuously wet). In order to determine which degree of humidity prevailed on a certain section, the test sections were divided into segments of 100 meters each and visually examined by inspectors.

To compare the results of the assessment, it is necessary to compare the data of two or more measurement campaigns which requires a very accurate position of each picture. The GPS position and the odometer length collected during the data acquisition, however, are not accurate enough [5], [14]. Therefore, the measurement data was finely synchronized based on the chainage, the front and secondary camera images and suitable unique features (e.g. peaks in the longitudinal profile).

After that, the raw data was converted into condition values and compiled as a result table. Afterwards, the condition parameters for cracking were evaluated and the overall condition indexes were calculated. The evaluation sections with maintenance treatments between the first and last measurement were marked and excluded from further analyses. For each evaluation section, the mean road surface temperature was calculated, the degree of residual humidity defined and, if available, the type and year of construction of the surface course was determined.

All data was compiled in a project database, which provided the basis for further visualizations and analyses.

3.5. Visualization of measurement data

The evaluated data was visualized in multiple ways to allow for engineering analyses and for establishing correlations between the data and the road surface temperature or humidity, provided that identical conditions existed when comparing two locations.

Firstly, the linear profiles and the results from the crack assessment were compared with each other. The linear profile with the raw data can be used for the object-specific analysis of changes in cracks on the raw data level. The cracks were coded with respect to grid fields of 1 m x 1/3 of the lane width for asphalt roads or per concrete slab. Grid fields with observed distresses are marked accordingly on the profile.

Secondly, the aggregated crack performance indicators were compared to each other. The linear profile with the aggregated condition data (Figure 1) allows the changes of the relevant condition performance indicators to be displayed. However, not only the condition performance indicators can be read off, but also the total length of a crack within an evaluation section, which results from the sum of the crack lengths in the individual grid fields or concrete slabs.

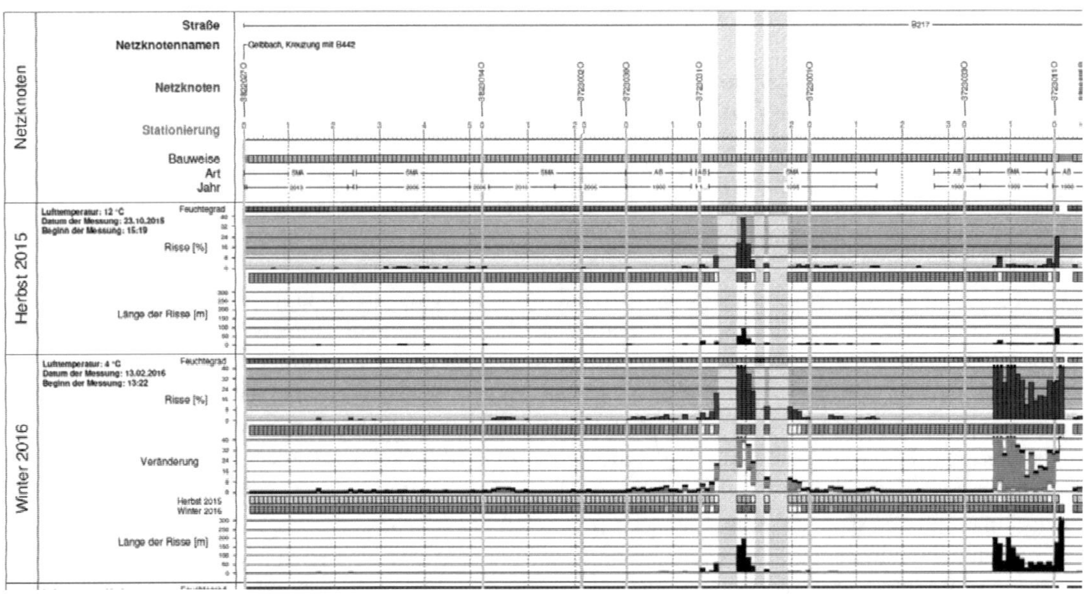

Figure 1. Example of a linear profile with aggregated crack performance indicators.

Finally, the cracks were analyzed visually and their development was illustrated [2]. Within this research project, several measurements were carried out in a relatively short period of time. Due to this frequency, the development of specific cracks could be observed and documented based on road and surface images. However, this image-based method cannot be compared with in situ crack monitoring which allows to look at the damage from different angles and to investigate the environment for possible causes. An online visualization information system was therefore used. The database was searched for relevant examples of cracks and the matches were provided with further information such as date and time of the measurement, air temperature and temperature of the road surface.

4. Results and evaluation

4.1. Statistical and scientific evaluation

To be able to differentiate between the influences of the road surface temperature and the degree of residual humidity, a total of six comparative analysis scenarios were developed iteratively for the statistical and scientific analyses of the results.

4.1.1. Scenario 1: Comparison of the results of all seven seasonal measurements. In the first comparative scenario, seven measurement campaigns (H15, W16, F16, S16, H16, N16 and W17) and all degrees of road surface humidity were taken into account. This scenario thus represents the broadest spectrum of measuring conditions in terms of temperature and humidity.

For the statistical analysis, the condition values ZGRISS (area affected by cracks on asphalt) and ZGLQRP (slab area affected by longitudinal and transverse cracks) were calculated for the sum of all common comparable evaluation sections of a construction method (585 asphalt and 137 concrete sections). Figure 2 shows the affected areas in percent and the average road surface temperatures $T_{FOf,asphalt}$ and $T_{FOf,concrete}$ for the respective measurement campaigns.

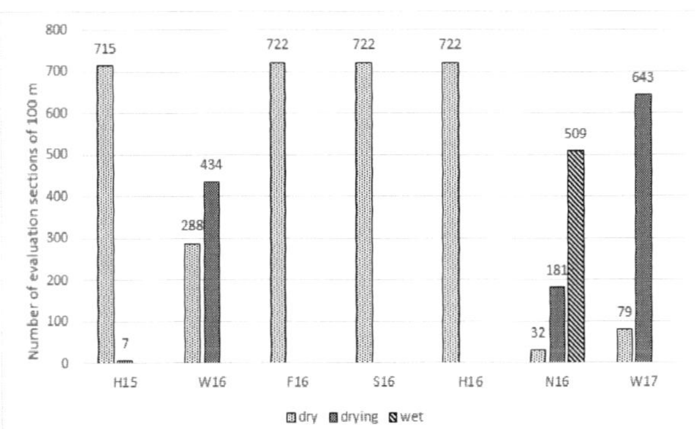

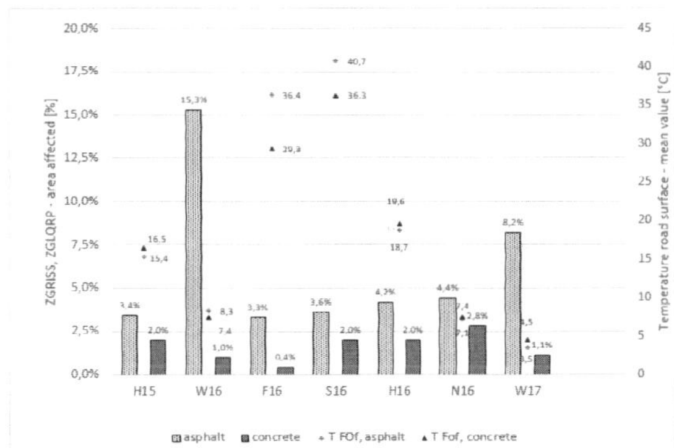

Figure 2. ZGRISS and ZGLQRP, plus mean values of the road surface temperatures per measurement campaign in scenario 1.

Figure 3. Degree of humidity of the road surface in scenario 1.

Figure 2 clearly shows that in the campaign W16, significantly more cracks were detected on the asphalt sections than in H15. The campaign W17 also revealed considerably more cracks than all other seasonal measurements that did not take place in winter. Such considerable differences cannot be observed for the concrete section, except for the campaign F16. This suggests that the lower road surface temperatures in winter and the evaluation sections with drying road surfaces are responsible for the higher areas affected. However, Figure 3 shows that more cracks were detected in the measurement campaign W16, which in contrast to W17 had higher road surface temperatures and a lower proportion of drying evaluation sections. A differentiation of the influences of road surface temperature and humidity can therefore not be deduced.

The transition diagrams, which refer to the condition indexes of the individual evaluation sections, confirm the previous results. The campaign W16, which was carried out mostly on a drying road surface, shows the largest proportion of evaluation sections in poor condition (condition index greater than 4.5).

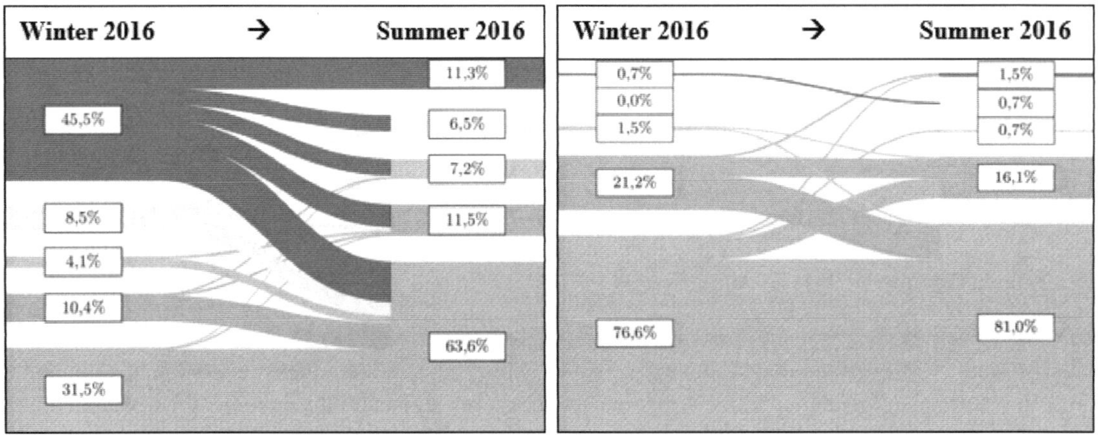

Figure 4. Transition of ZWRISS (left) and ZWLQRP (right) in scenario 1.

As shown in the transition diagram W16 to S16 (Figure 4) for the asphalt sections, the evaluation sections change predominantly from a poor to a better condition. In other measurement campaigns on the asphalt test tracks, there were only minor changes in the condition indexes. Both from H15 to H16 and from F16 to H16, however, a significant decrease in evaluation sections with condition indexes better than 1.5 could be observed. The transition diagrams of the concrete section reveal a deterioration in condition from W16 to S16. This development was not expected.

4.1.2. Scenario 2: Comparison of the results of measurements on dry roadways. To determine the influence of road surface temperature on the visibility of cracks in asphalt and concrete sections, an analysis of all dry evaluation sections of the measurement campaigns H15, F16, S16 and H16 was carried out. The campaigns N16, W16 and W17 were not taken into account, as there were no or too few dry evaluation sections.

Results:

- Comparing the campaign S16 with the two measurements H15 and H16 conducted in fall, it becomes apparent that the ZGRISS of S16 (3.5 %) lies between the ZGRISS of H15 and H16 (3.2 % and 4.0 %). The range ZWLQRP of the campaigns H15, S16 and H16 is 0.6 %, with the campaign S16 having the lowest ZWLQRP (2.2 %) of the three measurement campaigns. This suggests that medium (>15.0 °C) and high (>35 °C) road surface temperatures have no direct influence on the detectability of cracks. The evaluation of the crack lengths of all common comparable evaluation sections of the test sections in asphalt supports this conclusion.
- The same is valid for longitudinal and transverse cracks on the concrete section. The mean total length of the cracks remains constant at 1.2 m during all four measurement campaigns, while the proportion of slabs affected varies between 0.8 % (F16) and 2.9 % (H15). This again indicates that there is no direct influence of medium and high road surface temperatures on the visibility of cracks.
- Both the mean indexes ZWRISS and ZWLQRP are almost constant during all measurement campaigns and do not indicate a direct influence of medium (>15.0 °C) and high (>35 °C) road surface temperatures on the detectability of cracks.

4.1.3. Scenario 3: Comparison of the results on drying roadways. In this scenario, the effects of a drying road surface on the results of the condition survey were investigated. For this purpose, only evaluation sections of the campaigns W16 and W17 were evaluated which were marked as drying. No comparable data were available for the concrete section, as all evaluation sections were classified as dry.

Results:

- Due to the higher percentage of ZGRISS in the measurement campaigns W16 and W17, compared to scenario 1, it can be assumed that the degree of humidity has a considerable influence on the visibility of cracks on asphalt pavements.
- Comparing the condition indexes ZGRISS and ZGLQRP with the sums of the crack lengths shows that the indexes in the campaign W16 are approximately twice as high as in W17. A possible explanation could be variations in the degree of humidity "drying", which can have a positive or negative effect on the detectability of cracks.
- The transition diagrams show that almost 23% of the evaluation sections that were above the threshold level (condition value greater than 4.5) in campaign W16 transitioned into a better condition area. The relatively big change of the condition indexes leads to the assumption that the road surface was humid to varying degrees, although the examined evaluation sections were uniformly classified as drying.

4.1.4. Scenario 4: Comparison of the results on dry roadways at different road surface temperatures. In this comparative scenario, the effects of different road surface temperatures on the visibility of cracks at identical degrees of humidity of the road (here: dry) were examined in more detail. The campaign S16, which had the highest road surface temperatures, and the campaign W16 with the lowest road surface temperatures were used for this purpose.

Results:

- There were significant differences in the ZGRISS for the sum of the common comparable evaluation sections of the asphalt track between the campaigns S16 (0.9 %) and W16 (9.3 %).
- There is an opposite trend in the concrete section. The range of the ZGLQRP for the sum of all common comparable evaluation sections between the campaigns S16 and W16 is 0.5 % and is thus relatively small compared to the range for the asphalt sections (8.4 %). The low value in campaign S16 suggests that at high road surface temperatures the crack lengths decrease or are less visible than at low road surface temperatures.
- The examination of longitudinal and transverse cracks on the concrete section also shows that slightly more slabs were affected in summer at high road surface temperatures than in winter. The mean total length of all cracks, on the other hand, remains largely the same.
- In contrast to scenario 2, there was a considerable difference in this scenario between low (< 10 °C) and high (> 35 °C) road surface temperatures. It can be assumed that the difference is due to differences in assessing the degree of humidity of the road surface.

4.1.5. Scenario 5 and 6: Comparison of the results on a dry roadway for the examination of the diurnal curve in spring and summer. For scenarios 5 and 6, two of the four test tracks, the motorway A7 (concrete construction) and the federal highway B217 (asphalt construction), were surveyed five times during the day. The TGL1 measurement campaign took place in spring 2016 and the TGL2 in summer 2016. Both campaigns should take place on dry roads. The TGL1 was carried out at road surface temperatures of 17.1 °C to 31.7 °C. The TGL2 was conducted at slightly higher road surface temperatures (21.9 °C to 33.9 °C).

Results:

- The ZGRISS of the asphalt section showed ranges of 0.8 % (TGL1) and 0.6 % (TGL2). The condition index ZGLQRP for the test track in concrete construction resulted in ranges of 0.5% for the TGL1 and 0.7% for the TGL2.

- No correlation between the varying road surface temperatures and the visibility of cracks could be observed during the day, neither in the asphalt nor in the concrete section.
- Since all evaluation sections had a dry surface, it could be excluded in this scenario that residual humidity had an influence on the results.
- Both the mean values ZWLQRP and ZWRISS remain at an almost constant level throughout the day and show no influence due to the road surface temperature.

4.2. Conclusions

The evaluations reveal that the temperature and humidity degree of the road surface actually affect the visibility of cracks. In the campaigns W16 and W17, where the average temperatures were the lowest and the number of drying sections was highest, the cracks were generally more visible than in any other campaign.

Residual humidity as an influencing factor was investigated in more detail by analyzing only drying sections from the two winter measurements. The results clearly show that on average more cracks were visible on these drying sections than in scenarios where also sections with other humidity degrees were considered. However, it is not possible to establish a fundamental correlation between the degree of residual humidity and the improved visibility of the cracks from this data. In the measurement campaign W17, significantly fewer cracks were detected than in the measurement campaign W16, although only the drying sections were considered in each case. This suggests that in the sections defined as "drying", the road surface was humid to varying degrees and thus the visibility of the cracks was promoted differently. When analyzing the acquired data and especially when evaluating the raw data, it has become clear that the degree of humidity of a road surface is difficult to define since it is based on purely subjective estimations.

In order to analyze the extent to which different road surface temperatures affect the detectability of cracks, the dry evaluation sections of the measurement campaigns H15, F16, S16 and H16 were first compared with each other. It became obvious that mean (> 15.0 °C) and high (> 35 °C) road surface temperatures have no direct influence on the visibility of cracks. However, when comparing the measurement campaigns W16 and S16 with a temperature variance of 31.7 K, differences were found. Cracks in asphalt roads were more visible at lower road surface temperatures. This influence could not be proven on the test section in concrete construction. The improved visibility of the cracks on asphalt roads is probably due to differences in the determination of the residual humidity degree. Although the evaluation sections were classified as dry during the measurement campaign W16, a certain percentage of residual humidity in the crack areas could not be excluded, which improved the detection of cracks.

As a result, the temperature does influence the visibility of cracks indirectly, as residual humidity in crack areas is favored by low temperatures. The visibility of cracks on concrete roads decreased slightly during the measurement campaign W16 compared to the measurement campaign S16. The campaigns W16 and W17 were carried out in winter conditions and with apparently dry road surfaces at the beginning of the measurement. Nevertheless, evaluation sections with residual humidity were detected which will always be the case during the winter months.

5. Summary and outlook

The objectives of the research project were to further specify the acquisition method, to determine and quantify the influence of temperature and humidity on the visibility of cracks and to formulate technical requirements for the acquisition as a supplement to the regulations.

5.1. Influence of residual humidity

The evaluations have confirmed that the degree of residual humidity on the road surface has an influence on the visibility of cracks. However, this could not be quantified within the research project. Compared to dry road surfaces, cracks become more visible when the road surface is drying off and residual humidity remains in the cracks. Since residual humidity on the road surface cannot be controlled and is difficult to quantify, however, surveys on drying roadways cannot be recommended. Measurements in rain as well as in case of closed and large water films on the road surface are also to be avoided, as this greatly reduces the visibility of cracks.

5.2. Influence of road surface temperature

The extent to which the temperature affects the detectability of cracks could not be determined with certainty. Although cracks were increasingly detected during winter measurements, it was not possible to differentiate precisely whether the visibility of the cracks could be attributed exclusively to low road surface temperatures. Due to the low air temperatures in winter, there is a high probability that sections identified as dry still contain residual humidity that cannot be seen visually (mainly in the crack area). Consequently, the measurement results can lead to different conclusions. At higher air and road surface temperatures, areas with residual humidity quickly dry off and thus comparable conditions are more frequent. Therefore, measurements to investigate the surface distresses should not be carried out in winter. However, it was not possible to determine an exact lower temperature limit for measurements based on the measurement campaigns analyzed in this research project.

A significant influence of mean and high road surface temperatures on the visibility of cracks could not be proven for the test sections in asphalt construction. On the test section in concrete construction, cracks became more visible at higher road surface temperatures and less visible at lower temperatures. A possible explanation for the unexpected behavior of the test track in concrete construction may be the different temperature at the top and bottom of the concrete slabs at different seasons. At high air temperatures and direct sunlight, the concrete slabs heat up faster than the underside of the concrete slab. This temperature gradient has the consequence that at the top of the road a positive strain and at the bottom a negative strain (compression) takes place. As a result, the concrete slab arches upwards. Cracks can thus be better visible, for example in summer. On the contrary, the concrete slab may behave in cold temperatures. If the temperature of the road surface is lower than the temperature at the bottom of the concrete slab due to low air temperatures, the top will have a negative (compression) and the bottom will have a positive strain, causing the concrete slab to bulge down. Since this research project investigated only one concrete test section, no general conclusions can be drawn from this.

6. Acknowledgment

This report is based on parts of the research project carried out at the request of the Federal Ministry of Transport and Digital Infrastructure, represented by the Federal Highway Research Institute, under research project No. 04.0275/2013/FGB [2]. The author is solely responsible for the content.

The research has been conducted in a cooperation between HELLER Ingenieurgesellschaft mbH, responsible for planning the experiment and visualizing the data, and the University of Wuppertal, responsible for the statistical and scientific evaluation of the data. The data acquisition was conducted by IFI Consult GmbH, the pavement distresses were inspected by Heller Consult.

References

[1] R. Haas und R. Hudson, Pavement Management Systems, New York: McGraw-Hill, 1978.

[2] M. Skakuj, H. Balck, H. J. Beckedahl, T. Schrödter, S. Koppers, E. Ramadan, P. Brzezińska, W. Smęt und M. Worpenberg, „Temperatureinfluss auf die Sichtbarkeit von Substanzmerkmalen (Oberfläche) FE 04.0275," Bundesanstalt für Straßenwesen, Bergisch Gladbach, 2017.

[3] Forschungsgesellschaft für Straßen- und Verkehrswesen, „Zusätzliche Technische Vertragsbedingungen und Richtlinien zur Zustandserfassung und -bewertung von Straßen," Forschungsgesellschaft für Straßen- und Verkehrswesen, Köln, 2011.

[4] S. Bonnière und P. Villotte, „Diagnostic du réseau routier départemental de la Mayenne, Auscultation, évaluation de l'état et programmation de l'entretien," *RGRA*, Bd. 920, 2014.

[5] M. Skakuj, „Observation of the Development of Pavement Demages Over the Winter," HELLER Ingenieurgesellschaft mbH, Darmstadt, 2015.

[6] M. Stöckner, E. de Jong, R. Demmer, M. Skakuj, K. Opara and H. Balck, "Defy Winter: A New

Approach to Road Winter Demage Prevention," Transportation Research Board 94th Annual Meeting, Washington, 2015.

[7] Forschungsgesellschaft für Straßen- und Verkehrswesen, „Schaffung und Weiterführung des Regelwerks für die Erfassung und Auswertung von Oberflächenbildern," 2013.

[8] Northwest Pavement Management Association, „Pavement Surface Condition Field Rating Manual for Asphalt Pavements," Washington, DC, 1999.

[9] Lehmann + Partner GmbH, „Bericht - Projekt P15-154 / Maßnahme 1 - B 81 : Durchführung Funkti-onsinspektion für den Fahrbahnoberbau gemäß ZTV Funktion - StB 01 und MtZEB - StB 01," Erfurt, 2015.

[10] E. Vos und F. Bouman, „Oberflächeneigenschaften in den Niederlanden - Methoden und Ziele," *Straße + Autobahn,* Bd. Heft 1, pp. 23-27, 2015.

[11] N. Vitillo, N. Gucunski, C. Rascoe und S. Zaghloul, „Evaluation of the Automated Distress Survey Equipment," New Jersey, 2009.

[12] B. Xu, „Artificial Lighting for the Automated Pavement," Austin, 2005.

[13] Bundesanstalt für Straßenwesen , „IT-ZEB Server," [Online]. Available: http://itzeb.bast.de/index.html. [Zugriff am 06 02 2018].

[14] J. Bergmann-Syren, W. Smęt, C. Komma und M. Skakuj, „The Long-Term Observation Section on Motorway BAB A5 in Hesse as a Data Source for Research in the Field of Road Maintenance," World Conference on Pavement and Asset Management (WCPAM), Baveno, 2017.

IOP Conf. Series: Materials Science and Engineering **356** (2018) 012005 doi:10.1088/1757-899X/356/1/012005

Proactive pavement asset management with climate change aspects

Adam Zofka

Instytutowa 1, 03-302 Warszawa, Poland

azofka@ibdim.edu.pl

Abstract. Pavement Asset Management System is a systematic and objective tool to manage pavement network based on the rational, engineering and economic principles. Once implemented and mature Pavement Asset Management System serves the entire range of users starting with the maintenance engineers and ending with the decision-makers. Such a system is necessary to coordinate agency management strategy including proactive maintenance. Basic inputs in the majority of existing Pavement Asset Management System approaches comprise the actual pavement inventory with associated construction history and condition, traffic information as well as various economical parameters. Some Pavement Management System approaches include also weather aspects which is of particular importance considering ongoing climate changes. This paper presents challenges in implementing the Pavement Asset Management System for those National Road Administrations that manage their pavement assets using more traditional strategies, e.g. worse-first approach. Special considerations are given to weather-related inputs and associated analysis to demonstrate the effects of climate change in a short- and long-term range. Based on the presented examples this paper concludes that National Road Administrations should account for the weather-related factors in their Pavement Management Systems as this has a significant impact on the system outcomes from the safety and economical perspective.

1. Introduction

Transport infrastructure, with pavements and bridges in particular, represents an enormous national asset that must be properly managed using modern economical and engineering principles. Traditional constrains in the efficient management of these assets have been associated with increasing traffic, annual weather variations, and material and construction variability. While these aspects are challenging, majority of the NRAs have accommodated and successfully developed management systems to support their decision-making processes. However, the last two decades brought an increasing concern about the effects of climate change and extreme events (such as flooding, land sliding, extreme raining) on the infrastructure assets [1] [2] [3]. While both aspects are valid and should be addressed, their implementation in the traditional management process can have significant consequences. Therefore implementation of both climate change and extreme events require a strategic planning and preparation of the appropriate adaptation measures. In other words, NRAs should aware that climate change and extreme events will inevitably affect their assets and operations [4] [5]. Each NRA should decide what actions to undertake. Such a judgement should be based on the logical and comprehensive cost-benefit analysis that would accommodate stochastic nature of climate-related events together with the probability of asset failure and its impact magnitude [4]. Such an approach allows the NRS to assess the risk as follows [1]:

$$\text{Risk} = \text{P(Climate Event Occurrence)} \times \text{P(Asset Failure)} \times \text{Consequence} \qquad \text{(Eq. 1)}$$

Risk analysis identifies and classifies potential vulnerabilities of the NRA transport assets to climate change. This issue has been presented in the number of recent reports and publications including [2] [6] [7] [8]. In addition to available reports, there is also a significant number of research projects that have been initiated to address climate-change and its implications on the transport assets. Some examples include projects sponsored by the European Commission under FP7 (WEATHER, STREST, RAIN, InfraRisk, EWENT) and projects sponsored by the Conference of European Directors of Roads, CEDR (ROADAPT, RIMAROCC, MIRAVEC, DETECTOR).

2. Proactive maintenance

Proactive maintenance is a management strategy to provide and maintain serviceable roads. It is a multi-year planned strategy to select most effective treatments to preserve pavement assets, to retard their future deterioration and to maintain or to improve their functional condition without increasing their structural capacity [9]. Proactive maintenance typically includes corrective and preventive maintenance as well as minor rehabilitation. The concept behind proactive maintenance is presented in Fig. 1. As compared to reactive approach, proactive approach promotes carefully selected maintenance activities applied at optimal timing that allow to keep pavements in very good condition. Such an approach is always more beneficial from the user, owner and environmental perspective that has been proven in numerous studies and put in practice [10]-[23]. In order to demonstrate the benefits of proactive maintenance, these studies utilize various economical tools to perform life-cycle planning (LCP) and they often incorporate not only technical aspects but also include social aspects (e.g. accidents, user delay, vehicle operation costs) as well as environmental aspects (e.g. air pollution, noise emission, energy consumption).

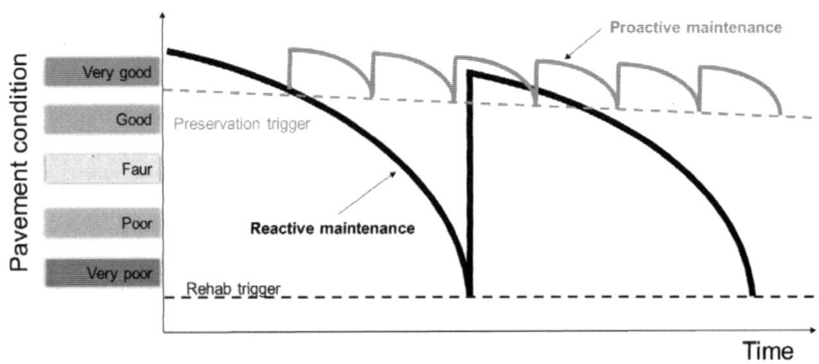

Figure 1. Concept of proactive and reactive maintenance approach (both condition-based).

It should be emphasized that optimized proactive strategy supports sustainable development and mitigates future maintenance burden. Among other advantages, it reduces carbon footprint, contributes to economy growth and improves road safety [24]. Proactive strategy is of paramount importance for the newer pavements before they reach significant deterioration. Proactive strategy should be defined and incorporated in the PMS tools as soon as the new roads are built to determine reasonable and objective funding allocation between newer and older network using life-cycle planning (LCP). Properly defined strategies in the PMS should allow to perform multi-aspect what/if scenarios to determined optimal timing for right proactive activities and to assess deferred maintenance and associated stop-gap needs.

3. Weather and climate changes vs. transport assets

Weather and climate refer to two distinct terms [25]. Weather is defined as conditions in the atmosphere and the space around us at a particular instance. Observed phenomena include temperature, rain, sun, snow, fog, clouds and wind. On the other hand, climate is the average weather collected over various time scales. Climate is simulated and predicted using global circulation models (GCM) [1]. These models operate at relatively low spatial resolution and very often do not provide sufficient information required for transport asset analysis. Therefore, regional climate models (RCMs) have been developed that cover only specific regions and operate at higher resolution. It should be noted that all climate models estimate average trends and thus contain natural variability. Their predictions are accompanied with the probability that contributes to the risk analysis presented in Equation 1.

Climate change can be also defined [25] as a change in the climate state that can be assessed using robust statistical tests. The change can be tested on average values or on the variability of properties and it should persist over significant time span (e.g. one decade). Climate changes may have a significant impact on transport infrastructure [1] [4] [6]. The impact magnitude depends on the specific asset and its vulnerability to a specific climatic phenomenon (temperature, rainfall, etc.) as well as other factors. Most common pairs of the climate phenomenon and infrastructure asset considered in the literature are as follows [1] [4] [6]:

- temperature – pavements,
- temperature – bridges,
- rainfall – pavements, drainage and integrity of the entire road structure,
- sea-level rise – drainage, integrity of coastal roads,
- hurricane intensity – bridges.

Next paragraph present several studies on the effect of the climate change on pavement performance. It is followed by more general considerations of the climate-related aspects in the PMS.

4. Climate-related aspects in PMS

Pavements, both asphalt and concrete, are susceptible to weather conditions. However, various conditions do not equally affect pavement performance. Therefore it is necessary to use realistic models that capture the dependence of material/structure behaviour in terms of climatic factors, such as temperature and rainfall. One such a study [2] demonstrates the impact of climate change on pavement performance using NOAA and MEPDG (Mechanistic-Empirical Pavement Design Guide). Analysis showed that climate changes contributed to the premature pavement rutting. Economic analysis showed that agency costs may increase but it can be offset in savings for the road users. Other study [26] was based on the LTPP (Long Term Pavement Performance) data and again the NOAA model. Presented examples showed that accounting for flooding and snowing in the pavement roughness model gives more accurate predictions which in turn can allow for more reliable assessment and timing of the Maintenance and Rehabilitation (M&R) activities. Two other studies [27] [5] investigated the rainfall and resulting moisture in pavement layers. Both studies concluded that increased moisture content will shorten pavement service life and that rainfall is one of the most important factors affecting the design and operation of the road network. Finally, recent study in Germany [28] focused on the importance of air temperature on the pavement life design. The analysis, using mechanistic pavement response model, showed that in some areas of Europe's pavement design (and associated PMS models) does not need to be changed because of current predictions in climate changes. Nonetheless presented methodology can be used to assess the need in other parts of the world.

These few studies confirmed that the effect of climate change on pavement performance varies depending on the climatic phenomenon and geographical region. However, if one wants to consider a broader view on pavements and consider their life-cycle PMS, there is a number of potential aspects that need to be considered. These aspects are presented in Fig. 2.

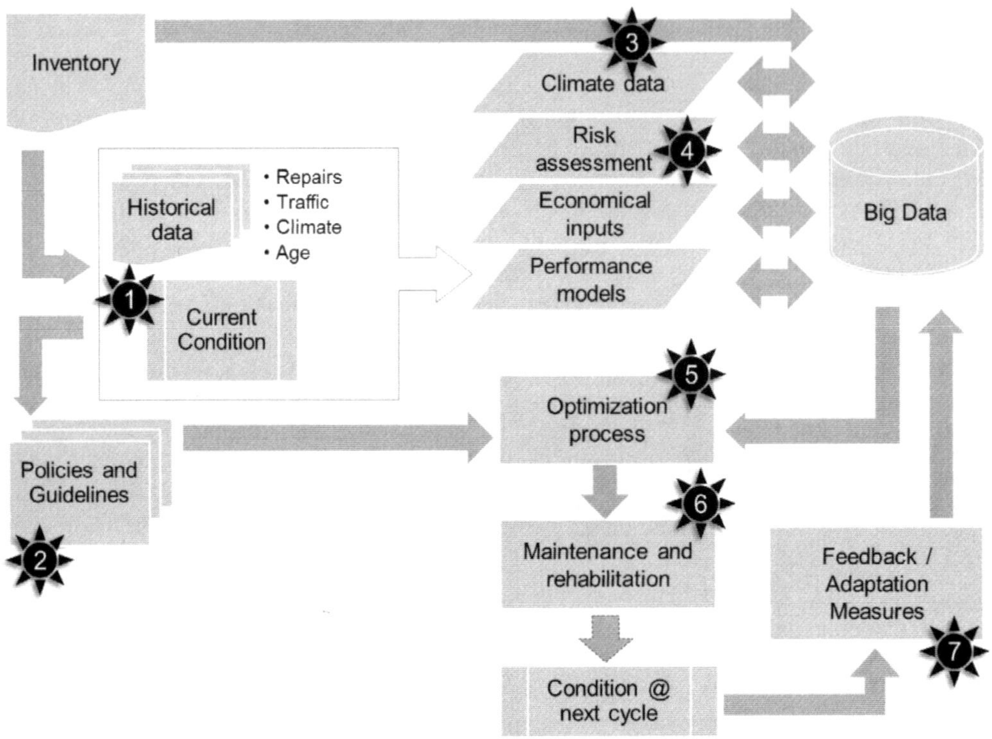

Figure 2. Potential climate aspects in the global PMS workflow.

Figure 2 presents climate-related aspects (depicted by the numbered stars) in the general PMS workflow. These aspects span across the entire PMS process and some of them are already present in the commercial PMS platforms and have been also identified in the aforementioned studies [1] [4] [29]. The following list presents in more details each of the seven aspects:

(1) Historical weather data should be associated with pavement sections to link weather and performance; climate change and extreme events may affect operations of pavement conditions assessment and may require adaptation changes in measuring techniques.

(2) Climate change and extreme events will demand new M&R techniques and alternative construction materials; decision tree and tables need to accommodate relevant climate information and criteria.

(3) Extensive weather data should be collected together with other PMS data; predictive weather models should be developed and constantly validated.

(4) Risk of extreme weather events should be incorporated into predictive weather, traffic, safety and performance models to assess the vulnerability potential and be used in the benefit/cost analysis.

(5) Optimization process should combine previous aspects (risk, performance, M&R policies) and weather-related constrains should be incorporated into priority tables to rank recommended projects.

(6) Climate change will affect not only M&R techniques (included in aspect #2) but it will also influence their timing, planning and operation to maintain construction safety and efficiency.

(7) Relevant Key Performance Indicators (KPIs) should be introduced to assess and track adaption measures; KPIs should include safety and economical parameters; feedback loop should comprise all corresponding models including weather; monitor KPIs to update NRA policies and strategies.

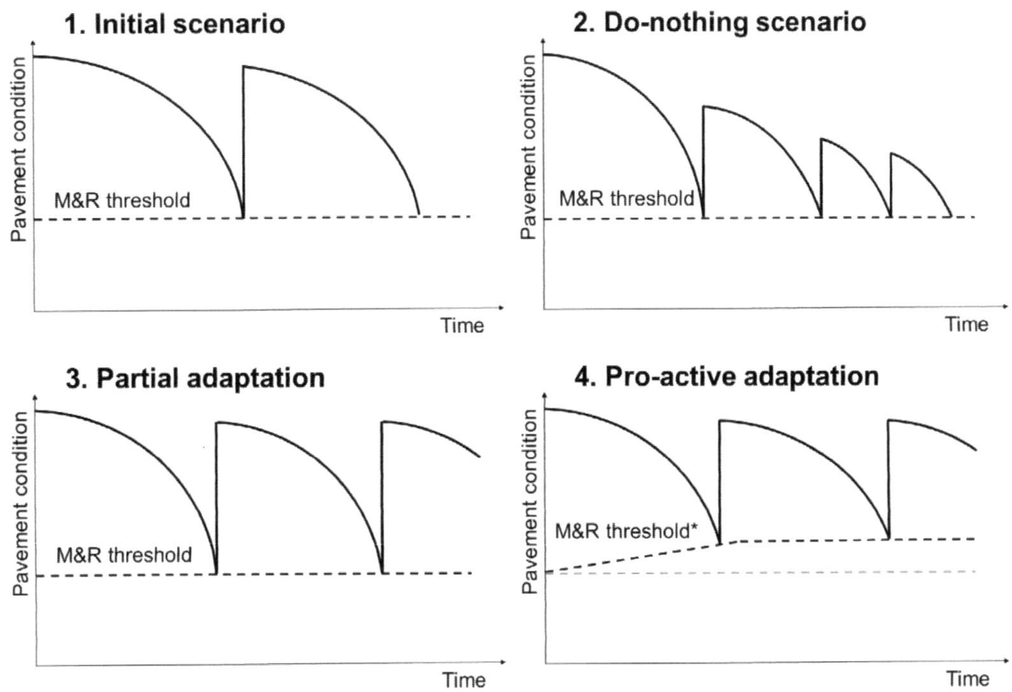

Figure 3. Mitigation of climate-related effects.

In order to complement the global effects presented in Fig. 2, Fig. 3 demonstrates the potential impact of climate change on the pavement performance. Different stages represent a potential progress of the NRA to mitigate the climate-related effects:

(1) Initial scenario (no climate change): traditional PMS approach (no pavement preservation approach); performance-based trigger for the M&R activities.

(2) Do-nothing (climate change): inadequate M&R activities; shorter initial performance gains and steeper deterioration curves; passive adaptation – no change in M&R threshold; results: poor pavement condition, high user and agency costs, safety concerns.

(3) Partial adaptation (climate change): adaptation in M&R techniques; full performance recovery and much better resilient performance; still no change in M&R threshold.

(4) Pro-active adaptation (climate change): adaptation in M&R techniques; full performance recovery and fully adapted resilient performance; adjusted condition threshold; inevitable higher agency costs as compared to scenario [1] but enhanced overall condition that leads to improved safety levels.

Figure 3 presents one potential sequence of different scenarios associated with climate adaptation in the PMS. It is certain that each NRA will try to embrace this challenge with different strategy that may produce different results across the transport network. Each strategy will depend on the NRA policy and available resources as well as the timing of adaptation measures. Several adaptation plans and strategies are presented in the next section.

5. Conclusions

(1) Cost benefits and adaptation strategies are implemented by various NRAs: Number of NRA organizations have been examining and preparing various adaptation plans to face climate-change (FHWA, Highways Agency, Transport Scotland or Trefikverket). For example, the NCHRP study in the USA prepared the benefit/cost tool for climate adaptation strategies and also listed 8-step adaption diagnostic framework which is based on the risk analysis presented

earlier in Equation (1). A UK-based study focuses on pavements and proposed several recommendations to better handle climate change including:

- cost benefit of adaptation to hotter, drier summers,
- need for wider analysis on climate change impacts,
- need for improved data on impacts of weather conditions.

(2) Development of adaptation action plans: Another report from the UK conducted by the Highways Agency discusses in more details the Agency's Adaptation Framework Model (HAAFM). It is a seven stage process that prepares adaptation action plans for all identified vulnerabilities.

(3) Need for measurable KPIs for asset management and climate change: Other strategy is presented in the report prepared by the FHWA explicitly on climate change adaptations for pavements. According to this study, climate changes will occur slowly and no immediate changes to current PMS practice is necessary. The key point is to identify when to modify the current practice and what KPIs should be used to monitor and trigger these modifications.

(4) Optimisation strategies for asset management: In another example provided by the Scottish NRA, the Agency reviewed their strategy developed in 2005 after 6 years of operations. All 28 recommendations that were prepared in the original study were found to be still valid. It should be mentioned that the Scottish strategy comprise a number of pro-active activities related to their road network addressing temperature, rainfall/flooding and wind aspects of the climate change.

In summary, given the ongoing drive for efficiency and resilience in transport infrastructure, the NRAs must acknowledge climate changes and implement appropriate adaptation plans in the foreseeable future. In particular, NRAs should account for the weather-related factors in their PMS as they have a significant impact on the PMS outcomes from the safety and economical perspective.

References

[1] Meyer M., et al., Climate Change, Extreme Weather Events, and the Highway System: Practitioner's Guide and Research Report, NCHRP Report 750, 2014.

[2] Flintsch G. W., Qiao Y., Impact of Climate Change on the Performance, Maintenance, and Life-Cycle Costs of Flexible Pavements, 11th National Conference on Transportation Asset Management July 10–12, 2016 Minneapolis Marriott City Center, Minneapolis, Minnesota.

[3] Transport Scotland, Scottish Road Network Climate Change Study, Report, 2005.

[4] Highways Agency by Atkins, Economics of Climate Change Adaptation and Risks, Final report, 2013.

[5] Highways Agency, Climate Change Adaptation Strategy and Framework, 2009.

[6] Willway T., et al., The effects of climate change on highway pavements and how to minimize them: Technical report. Project Report PPR 184, TRL Limited 2008.

[7] FHWA, Climate change adaptation for pavements, FHWA-HIF-15-015, FHWA TechBrief, August 2015.

[8] Crow R., et al., District 1 Climate Change Vulnerability Assessment and Pilot Studies: FHWA Climate Resilience Pilot Final Report, 2015.

[9] G. R. Hicks, S. B. Seeds, and D. G. Peshkin, "Selecting a preventive maintenance treatment for flexible pavements," Washington, D.C, 2000.

[10] S. Chan, B. Lane, T. Kazmierowski, and W. Lee, "Pavement preservation a solution for sustainability," *Transp. Res. Rec. J. Transp. Res. Board*, p. 7, 2011.

[11] J. J. Hajek and W. A. Phang, "Prioritization and optimization of pavement preservation treatments," Transp. Res. Rec., p. 11, 2017.

[12] L. Cesar, "Pavement preservation theory and strategies," in Northwest Pavement Management Association, 2015, p. 56.

[13] D. Peshkin, "Pavement preservation on highly trafficked roads: the new normal," in National Pavement Preservation Conference 2016, 2016, p. 26.

[14] T. Scheinberg and P. C. Anastasopoulos, "Pavement preservation prog ramming: a multi-year multi-constraint optimization methodology," in 89th TRB Annual Meeting, 2010, p. 16.
[15] NAPA, "The pavement preservation tool of choice NAPA position on thin asphalt overlays for pavement preservation, Report number SR-210," Lanham, MD, 2014.
[16] D. Pittenger, D. D. Gransberg, M. Zaman, and C. Riemer, "Stochastic life cycle cost analysis for pavement preservation treatments," in Transportation Research Board, 2012, p. 15.
[17] K. L. Smith, L. Titus-Glover, M. I. Darter, H. Von Quintus, R. Stubstad, and L. Scofield, "Cost–benefit analysis of continuous pavement preservation design strategies versus reconstruction," Transp. Res. Rec. J. Transp. Res. Board, p. 11, 2005.
[18] B. Visintine, G. R. Hicks, D. X. Cheng, G. Elkins, and J. Groeger, "Factors affecting the performance of pavement preservation treatments," in 9th International Conference on Managing Pavement Assets, 2015, p. 23.
[19] B. Aktaş, D. Gransberg, C. Riemer, and D. Pittenger, "Comparative Analysis of Macrotexture Measurement Tests for Pavement Preservation Treatments," Transp. Res. Rec. J. Transp. Res. Board, p. 7, 2011.
[20] D. F. Report, Performance-Related Specifications for Pavement Preservation Treatments, no. July 2016. National Academies Press.
[21] J. Wiegmann and B. Yelchuru, "Resource allocation logic framework to meet highway asset preservation," National Academies Press, 2012.
[22] Zimmerman K. A., Ram P. V., Pavement Management's Role in an Asset Management World, 9th International Conference on Managing Pavement Assets (ICMPA9), Washington, DC, May 18-21, 2015.
[23] Zofka A., Josen R., Paliukaite M., Vaitkus A., Mechowski T., Maliszewski M., Elements of Pavement Management System: Case Study, Baltic Journal of Road and Bridge Engineering, Vol. 9, Issue 1, pp: 1-9, 2014.
[24] World Road Association PIARC, "Preserve your country's roads to drive development, Report number 016R07EN," Paris, 2016.
[25] FHWA, Climate Change & Extreme Weather Vulnerability Assessment Framework. FHWA-HEP-13-005. FHWA, 2012.
[26] Shamsabadi S. S., Tari Y.S. H., Birken R., Wang M., Deterioration forecasting in flexible pavements due to floods and snow storms, 7th European Workshop on Structural Health Monitoring, July 8-11, 2014. La Cité, Nantes, France.
[27] Bilodeau, J-P., Dore G., Thiam P. M., Drolet F. P., Long-Term Performance of Flexible Pavement Structures in a Changing Climate. Proceedings, Annual Conference of the Transportation Association of Canada: Better - Faster – Safer. September 22-25, 2013, Winnipeg, Manitoba, Canada.
[28] Wistuba, M. P. and A. Walther. 2013. "Consideration of Climate Change in the Mechanistic Pavement Design." Road Materials and Pavement Design, Vol. 14, Sup. 1, Special Issue: EATA 2013.
[29] Shahin, M.Y.; Walther, J.A. 1990. Pavement maintenance management for roads and streets using the PAVER system. USACERL Technical Report M-90-05.

IOP Conf. Series: Materials Science and Engineering **356** (2018) 012006 doi:10.1088/1757-899X/356/1/012006

Durability of base courses with mineral-cement-emulsion mixes (MCEM)

J Kukiełka[1] and D Sybilski[2]

[1] Department of Roads and Bridges, Technical University of Lublin, Nadbystrzycka 40, 20-816 Lublin, Poland
[2] Pavement Technology Division, Road and Bridge Research Institute, 1 Instytutowa str., 03-302 Warsaw, Poland

E-mail: jerzy.kukielka@pollub.pl and d.sybilski@ibdim.edu.pl

Abstract. Base courses with mineral-cement-emulsion mixes (MCEM) have been the subject of research, surveys and development of e.g. new requirements included in the Guidelines of 2014 [15]. In this paper the results of sample test and survey of road sections, assessment of transverse cracks and load-bearing capacity with FWD after 13 years of exploitation are presented. On the MCEM samples the following tests were carried out: resilient modulus using NAT, complex stiffness modulus (E*), phase shift angle at various temperatures and loading frequencies thereby obtaining master curves, fatigue life and low-temperature resistance by identifying the tensile stress restrained (TSRST) which allowed for general assessment of constructed base courses.

1. Introduction

Re-use of materials also in roads was defined by the Americans as "recycling" [10, 11, 23, 27]. The first manual in USA entitled "Basic Asphalt Recycling" was published by Asphalt Recycling and Reclaiming Association in 2001.

In England recycling with cement or asphalt emulsion was used at first [22]. A combination of both binders was introduced to the regulations after a three-year research program was completed in 2004 [1].

In 2002, German guidelines for cold "in situ" recycling were published [21] which included the recommended thickness of structural layers with the use of the material reclaimed from the old surface. The first manual in Germany entitled "The Wirtgen Cold Recycling Manual" was published in 1998.

For instance, in the French classification of 2004 [13] class 5 of the base course was adopted, 10÷30 cm thick from deep "cold" recycling and containing the total of 3÷7% of cement and asphalt emulsion. The first manual in France entitled "Fayat Recycling Book" was published in 2007.

Also, in many European countries only asphalt emulsion is used in "cold" recycling. At the end of the 20th century the total use of cement and asphalt emulsion in recycled road bases was at the experiment stage.

The need to reinforce and repair rutted road pavements in Poland was the reason for the search of an appropriate technology which would allow for re-use of the incorporated materials [3, 24, 25].

The authors of works which preceded the development of new Polish guidelines obtained information from 91 national road sections with MCEM base on the basis of a survey carried out in

2012. In Poland the most frequent are base courses with added 3% of asphalt emulsion, 4% of cement and up to 50% of aggregate improving gradation. Numerous and varied FWD tests on asphalt pavement sections with MCEM base and quite frequent transverse cracks in certain regions of Poland were also analyzed [16].

The analysis and assessment of the MMCE base course made according to ST [8] is related to the planed design, scheduled for 2018, of redevelopment national road DK 19 on the Międzyrzec - Lubartów section upgrading the road to an expressway.

2. The survey and tests of the first road sections with bases of mineral-cement-emulsion mixtures (MCEM) in the Lublin region

Elastic deflection tests with FWD plate were carried out in 2001 on a 5-kilometer section on the national road no 19 (DK 19). The subbase of MCEM base course contains tar concrete and asphalt concrete 2÷10 cm thick on a layer of crushed stone, also with variable thickness: 21÷31 cm. The 19 cm thick MCEM base made of a mix with 0/31.5 mm gradation contained 54% of reclaimed asphalt pavement, 5% of cement and 3% of cement emulsion. The 8 cm thick binder course and the 5 cm thick wearing course were made of asphalt concrete.

The results of FWD measurements at 5°C carried out in March 2001 after autumn and winter use of the road put in operation are presented on Fig. 1.

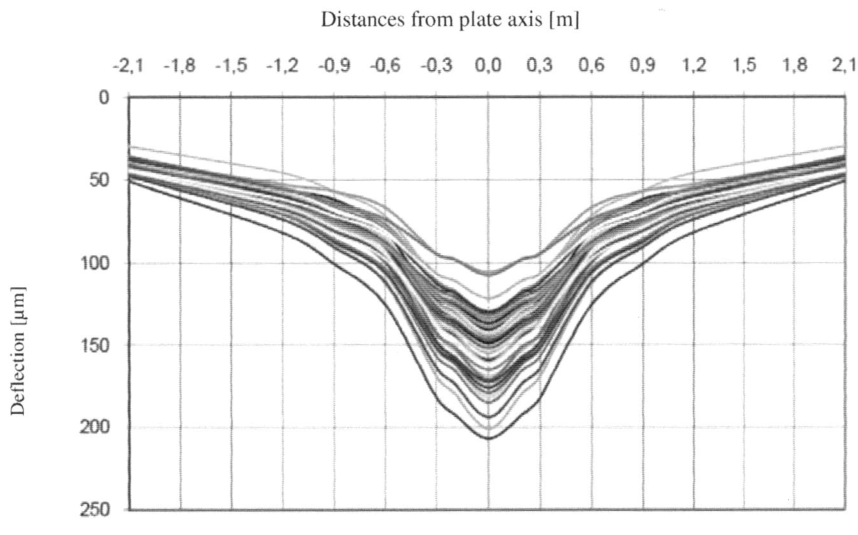

Figure 1. FWD deflections on road no 19 measured every 100 m at 5°C on the section from 348+200 km to 353+200 (Radzyń - Kock).

The deflection basin curves obtained from measurements with FWD plate demonstrate lack of shape uniformity and the maximum deflection in the load axis was approximately two times larger in relation to the minimum deflection.

In October 2001 another FWD measurement was taken in the same places after 6-month's use. The tests were carried out at fixed spots by measuring the temperature inside the layer, and the temperature was between 4°C and 34°C. The results were not significantly different and the October measurement showed slightly lower deflections than the ones measured in March. The MCEM base underwent the process of hardening at high temperatures in the summer in 2001 and its bearing capacity was not reduced due to HGV traffic loading. Moreover, the measured deflections of MCEM base course indicate its sensitivity to temperature changes [19].

The results of elastic deflection measurements with FWD allowed calculating, using ELMOD

IOP Conf. Series: Materials Science and Engineering **356** (2018) 012006 doi:10.1088/1757-899X/356/1/012006

program, stiffness moduli for the base course and pavement courses of the road surface. The mean elastic modulus of the subbase course was E_0=119 MPa and of the top layers package and the base course layer E_1=5016 MPa at the maximum research temperature of 34°C.

In July 2013 at 20°C another measurement with FWD plate was carried out. Mean deflection in the load axis was U_m=242 μm, standard deviation was σ=63 μm which leads to the conclusion that the variability coefficient was τ=26%. In the research carried out in 2001 at a similar temperature the mean deflection in the axis was U_m=195 μm, standard deviation was σ=44 μm and the variability coefficient was τ=22%. After 12 years of use the deflection increased by approximately 24%. In accordance with the requirements concerning the load-bearing capacity limit state [9], road pavement structure suffers fatigue when "the value of the equivalent elastic modulus of the pavement constitutes less than 50% of the initial value". Therefore, fatigue of the pavement structure in 2013 was much less than the limit value.

In the spring of 2017 site visit on road no 19 was conducted and on the analyzed 5-kilometer section only two transverse crack was identified. The limit state in relation to pavement cracks [9] is defined in the following way: "not less than 20% of the pavement is covered with fatigue cracks over 2 mm wide". It may therefore be concluded that also this limit is observed, even with a considerable reserve.

The main drawback of the pavement was poor durability of the wearing course made of asphalt concrete which needed to be renovated with SMA. The new wearing course was laid in 2017.

Analysis of a 5-kilometer section on the national road no 82 (DK 82) with the MMCE base course consisted of examining the elastic stiffness modules on samples taken from the mixture in the process of implementation.

The resilient moduli with NAT made in line with the PN-EN 12697-27 and BSJ DD 231:1993 standards in the case of samples with MCEM depend on the temperature of their testing. The results of selected tests are presented in Fig. 2.

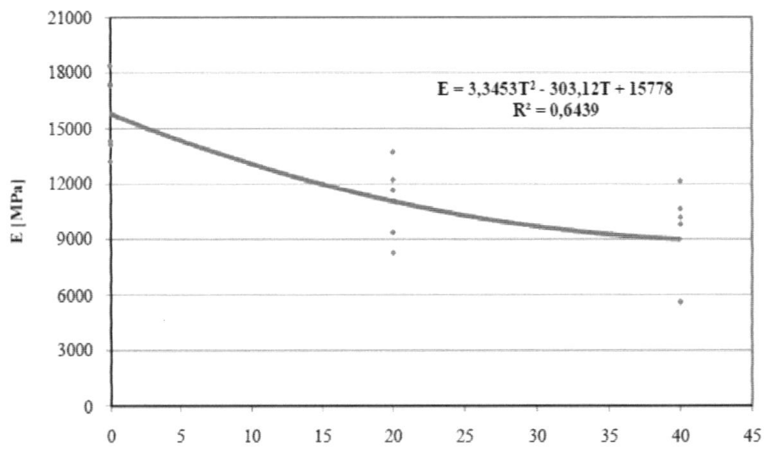

Figure 2. Dependence of the resilient moduli in NAT on the temperature of MCEM samples with added 5.5% of cement and 3% of asphalt emulsion with 0/31.5 mm gradation from a section of road no 82 after maturing for 1 year in air-dry conditions.

On the section of road no 82 (DK 82) no transverse cracks were identified in spite of the occurrence of relatively large stiffness moduli.

3. Stiffness, fatigue life and resistance to low-temperature cracks of lab samples from MCE mixes

Tests of complex stiffness modulus (E*) and of phase shift angle were carried out in line with ASTM D 3497-97 [4] in MTS press on cylindrical samples Ø 98.8x160 mm cut out of a 300x300x180 mm plate compacted by pressing. The total of 6 MCEM samples with added 5% of cement and 3% of asphalt emulsion with 0/25 mm gradation were tested.

Relatively low values of phase shift angle at different temperatures (0°C, 20° C and 40° C) and loading frequencies (0.1÷30 Hz) are presented in Fig. 3; they lead to a conclusion that the real (elastic) component (E_1) of the complex modulus has an advantage over the imaginary (viscous) component (E_2).

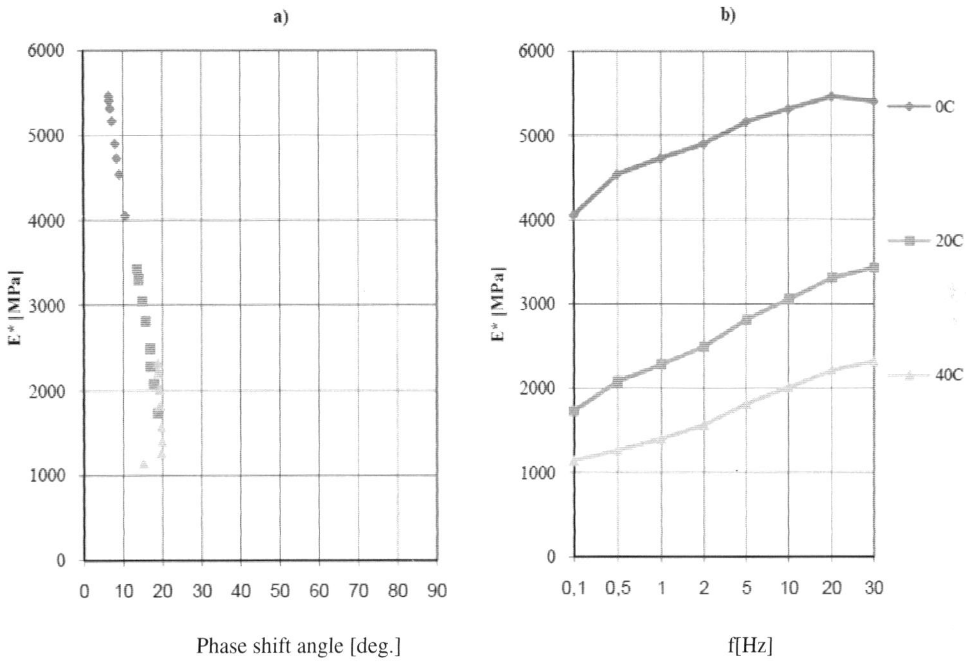

Figure 3. The results of tests of complex stiffness modulus E* of a MCE mixture sample with added 5% of cement and 3% of asphalt emulsion and with 0/25 mm gradation; a) Black's curve, b) isotherm.

On the basis of the Cole-Cole diagram presented in Fig. 4 it may be concluded that the samples' elasticity changes significantly depending on temperature. At 0°C the real component (E_1) of the complex stiffness modulus (E*) is approximately ten times higher compared to the imaginary component (E_2) and at least three times higher at 40°C.

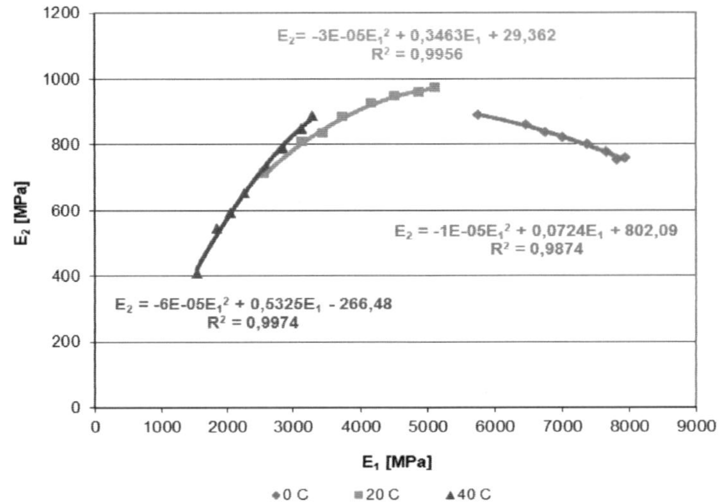

Figure 4. Cole-Cole diagram showing the relations between the viscous and the elastic components of the complex modulus of MCEM samples with added 5% of concrete and 3% of asphalt emulsion with 0/25 mm gradation.

An example of master curve at the reference temperature of T=20°C of MCE mix with 0/25 mm gradation containing 5% of cement and 3% of asphalt emulsion, presented in Fig. 5, leads to the conclusion that the volume of complex stiffness modulus (E*) depends to a large extent on the loading frequency.

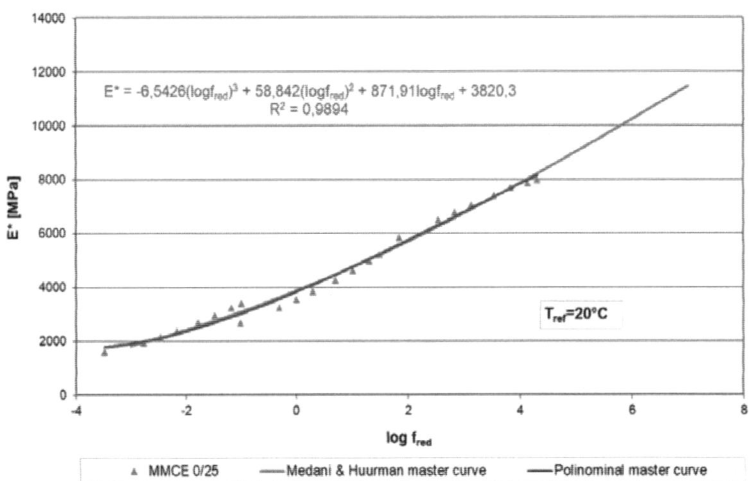

Figure 5. Master curve according to Medani & Huurman equation and the polynomial of the MCE mix with added 5% of cement and 3% of asphalt emulsion and with 0/25 gradation

The master curve equation by Medani & Huurman and the shift factor by Arrhenius were applied based on the example of research carried out by D. Sybilski [28, 29].

The impact of binding agents on stiffness of mixtures made in line with the Guidelines of 2014 was analyzed in the paper [7].

Tensile stresses which cause the predefined deformations in correlation with the number of loading cycles presented in Fig. 6 refer to the example of the research carried out in 2007. A characteristic feature of all the research is sudden reduction in the value of stresses and in the complex stiffness modulus during the initial period of loading. A reason may be destruction of the weakest cement connections with MCEM. Therefore, it is justified to apply pre-loading prior to the beginning of the relevant fatigue tests until the module achieves 50% of the initial value.

The fatigue life of mixes with asphalt emulsion and cement was analyzed in the paper [18] and according to the Polish Guidelines it was the subject of publication [20]; both works were based on lab tests only.

Fatigue life tests were carried out on cylindrical samples in axial compression at defined deformation. Results of the obtained stresses in correlation with the number of load cycles are presented in Fig. 6. The fatigue life of MCEM samples with addition 5% cement and 3% asphalt emulsion rapidly falls during the initial loading period with a fixed deflection $\varepsilon_r \geq 100 \cdot 10^{-6}$, below the permissible value of 50% of the primary resilient modulus however without destruction of the sample. With deformations below $\varepsilon_r < 100 \cdot 10^{-6}$ MCEM samples meet the slope modulus criterion i.e. no more than 50% of primary resilient modulus after 1 million load cycles.

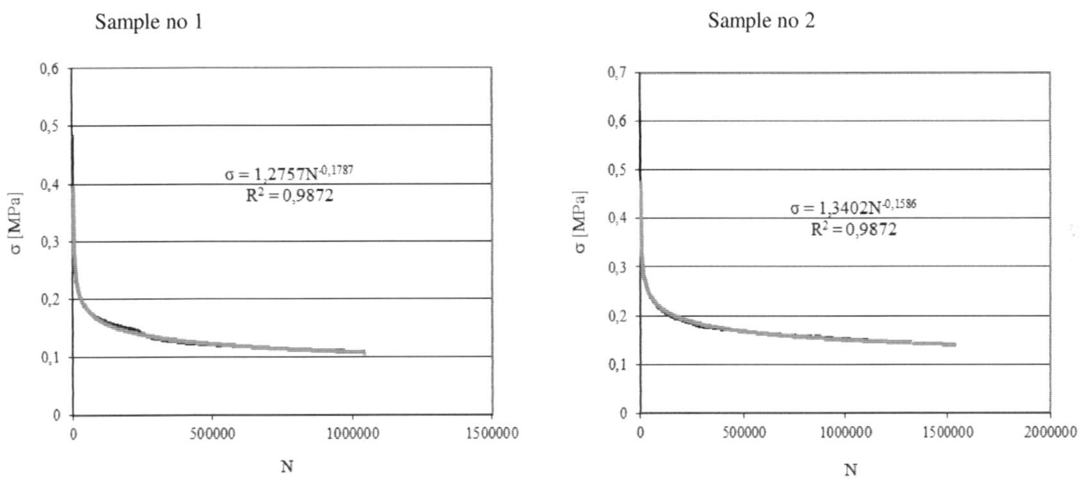

Figure 6. Correlation between the σ stress which causes the determined permanent deformation ε_6=196 μm/m (sample no 1) and ε_6=80 μm/m (sample no 2), and the number of loading cycles on MCEM samples with 0/25 mm gradation.

The low-temperature resistance tests were carried out using TSRST method with restrained deformation in line with the AASHTO TP10 - 93 standard [2]. The ends of the sample were attached to steel discs fixed in the handles of MTS press. MCEM samples with added 5% of cement and 3% of asphalt emulsion and with 0/16 mm, 0/20 mm, 0/25 mm gradation were subject to tensile stresses restraining the contraction occurring when the temperature in the climate chamber is lowered (Fig. 7 and 8). The recorded maximum stresses were achieved at approx. -30°C and their level changed depending on the gradation of MCE mixes, respectively: 0/16 mm approx. σ_{max}=1.2 MPa, 0/20 mm approx. σ_{max}=0.44 MPa, while the samples with 0/25 mm gradation got destroyed after the temperature reached -8°C and the stresses σ_{max}=0.44 MPa.

IOP Conf. Series: Materials Science and Engineering **356** (2018) 012006 doi:10.1088/1757-899X/356/1/012006

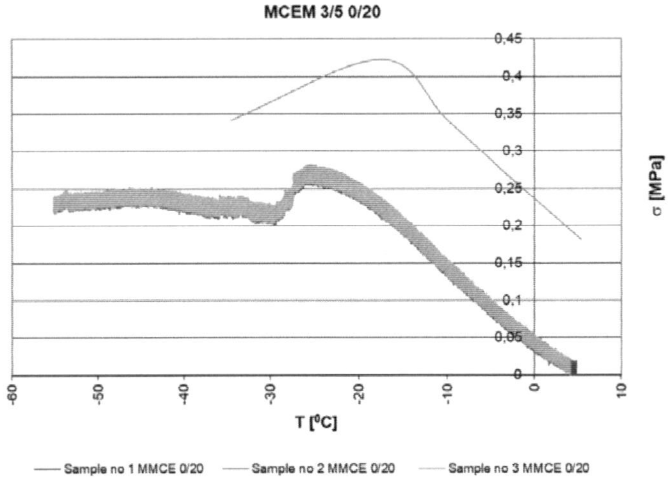

Figure 7. The recorded tensile stresses in the MCEM sample with 0/20 mm gradation caused by lowering of temperature during TSRST.

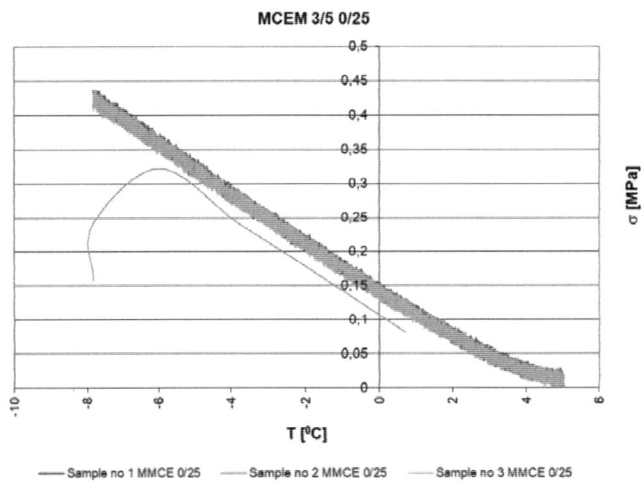

Figure 8. The recorded tensile stresses in the MCEM sample with 0/25 mm gradation caused by lowering of temperature during TSRST.

MCE mixes with finer aggregate (0/16 and 0/22) are more resistant to low-temperature cracks than the coarse aggregates. However conducted tests on MCEM samples are not sufficient enough for the unambiguous determination of the low-temperature strength of these mixtures.

Transverse cracks in MCEM base courses on road sections in South-Eastern Poland were not a major problem, if reclaimed asphalt pavement contained a relatively large amount of old asphalt and the addition of asphalt emulsion also included the participation of aggregate improving gradation in the MCEM.

The structure of concrete and other composites is known from the works of A.M. Brandt [5]. The following are identified: hardened cement paste, particles, pores and voids, cracks and fractures. Micro-cracks of less than 0.1 mm are, for instance, a feature of all the cement concretes and the cracks morphology has been presented in many research works [5, 6, 17]. Cracks length increases after exceeding the stresses which equal 0.85 cube compressive strength [12, 26]. The cracks width at the point where matrix and coarse aggregate meet in the Interfacial Transition Zone (ITZ) has a considerable impact on the propagation of cracks in concretes.

Strength of individual phases of concrete composite according to J. Hoła [14] is expressed in the formula:

$$f_{e1}>f_{e3}>f_{e2}$$

where: f_{e1} - aggregate strength
f_{e2} - ITZ strength
f_{e3} - matrix (cement paste) strength.

The transition layer is more porous and weaker than e.g. the matrix in cement concretes.

In MCEM asphalt mortar (mastic) is present on the surface of reclaimed asphalt pavement grains and of the asphalt from emulsion, and on the surface of aggregate improving gradation – only asphalt from emulsion. If the temperature of the surroundings is low, there may be no adhesion of the matrix to the aggregate, contrary to high temperatures when asphalt is soft. Lack of adhesion may be associated with linear cracks on the grain surface which are a starting point for micro-cracks and for the propagation of wing cracks in the matrix [12]. The presence of ITZ layer in MCEM may be merely fragmentary.

4. Conclusion

When designing MCEM it was common practice in Poland to add a large amount of crushed-stone aggregate improving gradation and relatively large amount of cement, as in B2 group in accordance with [21] regulations. In exceptional cases only cement (as in group C) and only asphalt emulsion (group A) were added to the mix. MCEM bases were at that time e.g. at the initial stage in France [13].

In the first decade of the 21st century MCE mixes in Poland were made in accordance with the provisions of General Technical Specifications D-04.10.01 of 2001 [8] which complied with the requirements provided in vol. 61 of the Scientific Journals of Roads and Bridges Research Institute (RBRI) in Warsaw of 1999. Excessive requirements concerning Marshall's stability (8÷20 kN at the temperature of 60°C), as in Scientific Journal vol. 61 RBRI and the General Technical Specifications, were very difficult to be complied with when designing MCEM while adhering to the guidelines concerning the dosage of emulsion and cement, even more so if the material was collected from the base layer being constructed.

(1) Service durability of a pavement with MCEM base course, constructed in 2000 on a 5-kilometre long section of road no 19, was evaluated on the basis of FWD tests at the temperature of 20°C in 2001 and in 2013. An increase of the mean deflection by 24% after 12 years leads to the conclusion that the load-bearing capacity limit state will occur after 25 years of use.

(2) If the reclaimed asphalt pavement contains a large amount of asphalt and a small amount of crushed-stone aggregate, as e.g. in Eastern Poland, it may be justified to increase the amount of cement in MCEM in relation to the amounts recommended in the new Instruction [15] and earlier regulations.

(3) Slow decrease of the stiffness modulus after the initial period of loading MCEM samples with 0/25 mm gradation and repeatability of fatigue life results justify the possibility to use in pavements of roads mixes of smaller particle size than recommended in the Instruction [15].

(4) The tests of complex stiffness modulus E*, phase shift angle and the obtained Cole-Cole diagrams, presented in this paper, allow to draw a conclusion that the elastic component has prevalence over the viscous one and its scale depends on test temperature.

(5) An analysis of 5-kilometers long sections of roads no 19 and no 82 with MCEM base courses which contain 5% of cement and 3% of asphalt emulsion shows that their condition complies with the design assumptions. The occurrence of two transverse cracks in section of DK 19 would require an analysis of the causes of their uprising.

References

[1] *A guide to the use and specification of cold recycled materials for the maintenance of road pavements*. 2004 TRL Report 611 Crowthorne, UK

[2] AASHTO Provisional Standard TP10 1993 *Standard Test Method for Thermal Stress Restrained Specimen Tensile Test Strength*

[3] Alenowicz J , Judycki J 1984 Powtórne użycie materiałów odzyskanych z nawierzchni asfaltowych. Drogownictwo **2**

[4] ASTM 3497-97. 2003 *Standard test method for dynamic modulus of asphalt mixtures*. ASTM International, West Conshohocken, PA.

[5] Brandt A M 2010 Diagnostyka betonu na podstawie badania struktury (*Concrete diagnosis based on structure testing*) **56** Konferencja Naukowa Komitetu Inżynierii Lądowej i Wodnej PAN oraz Komitetu Nauki PZITB. Kielce – Krynica. p 57–71

[6] Brandt A M , Kasperkiewicz J 2003 Metody diagnozowania betonów i betonów wysokowartościowych na podstawie badań strukturalnych (*Methods for diagnosing concretes and high value concretes based on structural tests*) IPPT PAN, Warsaw;

[7] Dołżycki B, Jaczewski M, Szydłowski C 2017 *The influence of binding agents on stiffness of mineral-cement-emulsion mixtures*. Modern Building Materials, Structures and Techniques, Procedia Engineering **172**

[8] D-04.10.01 2001 Podbudowa z mieszanki mineralno-cementowo-emulsyjnej (*Base made of mineral-cement-emulsion mixture*). Generalna Dyrekcja Dróg Publicznych, Warsaw.

[9] Dz.U poz. 124 Obwieszczenie Ministra Infrastruktury i Budownictwa z dnia 23 grudnia 2015 r. w sprawie ogłoszenia jednolitego tekstu rozporządzenia Ministra Transportu i Gospodarki Morskiej w sprawie warunków technicznych, jakim powinny odpowiadać drogi publiczne i ich usytuowanie (*Technical conditions that should be met by public roads and their location*)

[10] Epps J A 1978 Program Synthesis of Highway Practice No. **54**: *Recycling Materials for Highways*, National Cooperative Highway Research Program, Transportation Research Board, Washington, DC

[11] Epps J A , Little D N , Holmgreen R J , Terrel R L 1980 *Guidelines for Recycling Pavement Materials*, NCHRP Report No. **224**, Transportation Research Board, Washington, DC;

[12] Golewski G L 2015 Procesy pękania w betonie z dodatkiem krzemionkowych popiołów lotnych. Lublin University of Technology

[13] Guide technique Retraitement des chaussées et recyclage des matériaux bitumineux de chaussées, Service d'Études Techniques des Routeset Autoroutes, Juillet 2004

[14] Hoła J 2000, Naprężenia inicjujące i krytyczne a destrukcja naprężeniowa w betonie ściskanym (*Initial and critical stresses and stress destruction in compressed concrete*). Wroclaw University of Technology, Wroclaw

[15] Instrukcja projektowania i wbudowywania mieszanek mineralno-cementowo-emulsyjnych. (*Instruction for designing and incorporating of mineral-cement-emulsion mixtures*) . Gdańsk University of Technology. Generalna Dyrekcja Dróg Krajowych i Autostrad, Warszawa 2014 r.

[16] Judycki J , Dołżycki B , Szydłowski B , Ossowski C 2012 Opracowanie procedury do projektowania mieszanek mineralno-cementowo-emulsyjnych (MCE). (*Development of a procedure for designing mineral-cement-emulsion mixtures*) Raport z drugiego etapu prac. Gdańsk University of Technology. Zlecenie Generalnej Dyrekcji Dróg Krajowych i Autostrad

[17] Kasperkiewicz J 1983 Struktura wewnętrzna a procesy pękania w kompozytach z kruchą matrycą (*The internal structure and processes of cracking in brittle-matrix composites*). Prace IPPT PAN, no **39**, Warsaw

[18] Kavussi A 2010 *Modarre Laboratory fatigue models for recycled mixes with bitumen emulsion and cement*, Construction and Building Materials **24**

[19] Kukiełka J 2007 Trwałość podbudów z mieszanek mineralno-cementowo-emulsyjnych MMCE (*Durability of mineral-cement-emulsion mixtures bases*). Budownictwo i Architektura **1**
[20] Kuźniewski J , Skotnicki Ł , Szydło A 2015 *Fatigue durability of asphalt-cement mixtures*, Bulletin Of The Polish Academy Of Sciences Technical Sciences, Vol. **63**, No. 1,
[21] Merkblatt für Kaltrecycling in situ im Strassenoberbau 2002, Köln
[22] Milton L J , Earland M 1999 *Design guide and specification for structural maintenance of highway pavements by cold in-situ recycling*. Transport Research Laboratory Report **386**. Crowthorne, UK;
[23] Report on Cold Recycling of Asphalt Pavements 1998 American Association of State Highway and Transportation Officials AASHTO-AGC-ARTBA Joint Committee Task Force **38** Report. Washington, D.C.,
[24] Rolla S 1994 Recykling po 15 latach (*Recycling after 15 years*) Drogownictwo no **3**
[25] Rolla S 1979 Powtórne użycie materiałów nawierzchni (*Reuse of pavement materials*) Drogownictwo no **10**
[26] Santiago S D , Hilsdorf H K 1973 *Fracture mechanisms of concrete under compressive loads*, Cement and Concrete Research, vol. **3**
[27] Smith R 1979 W Energy, *Material Conservation and Economics of Recycling Asphalt Pavements*, Federal Highway Administration Conference - Use of waste material for highways, Williamsburg, Virginia
[28] Sybilski D 2004 Ocena właściwości niskotemperaturowych lepiszczy asfaltowych i mieszanek mineralno-asfaltowych (*Evaluation of low temperature properties of bituminous binders and asphalt mixtures*). Drogi i Mosty IBDiM no **2** , Warsaw
[29] Sybilski D , Kukiełka J 2006 Cechy reologiczne mieszanki mineralno-cementowo-emulsyjnej na podstawie krzywej wiodącej (*Rheological characteristics of mineral-cement-emulsion mixture based on a master curve*), I Polski Kongres Drogowy, Warsaw, p. 461-468

IOP Conf. Series: Materials Science and Engineering **356** (2018) 012007 doi:10.1088/1757-899X/356/1/012007

Natural asphalt modified binders used for high stiffness modulus asphalt concrete

Marcin Bilski[1] and Mieczysław Słowik[2]

[1, 2] Poznan University of Technology, Institute of Civil Engineering, Poznan, Poland

[1] marcin.bilski@put.poznan.pl
[2] mieczyslaw.slowik@put.poznan.pl

Abstract. This paper presents a set of test results supporting the possibility of replacing, in Polish climate conditions, hard road 20/30 penetration grade bitumen used in the binder course and/or base course made of high stiffness modulus asphalt concrete with binders comprising of 35/50 or 50/70 penetration grade bitumens and additives in the form of natural Gilsonite or Trinidad Epuré asphalts. For the purpose of comparing the properties of the discussed asphalt binders, values of the Performance Grade have been determined according to the American Superpave system criteria.

1. Introduction

In the case of pavement repairs being undertaken in special load conditions such as slow traffic or parking of heavy vehicles, it is necessary to adopt special solutions with regard to materials and technology. A situation of this kind may involve e.g. intersections with traffic lights with high levels of traffic, roundabouts, or bus stops. In order to increase the resistance of pavement under repairs to permanent deformations in case of special load conditions, it is recommended (e.g. according to [1]) to apply a thin wearing course (e.g. made of BBTM or SMA mixtures) and a binder course and/or base course made of high stiffness modulus asphalt concrete. It is advised to use for the wearing and binder courses polymer modified asphalt binders or binders with special additives.

An alternative solution for hard road bitumens and multigrade bitumens used in high stiffness modulus asphalt concrete could be natural asphalt modified binders. The tests run on asphalt mixtures comprising of natural Gilsonite asphalt from the United States of America or extracted from Pitch Lake on the island of Trinidad (Trinidad Lake Asphalt), discussed by authors in [2, 3], indicate that the application of selected additives increases the stiffness of the pavement course, making it less susceptible to permanent deformations at high temperatures. Thanks to the application of modified asphalt binders, it is possible to achieve a binder course or base course characterised by a higher resistance to rutting, which is particularly essential in case of high level of heavy vehicle traffic. As the authors of [4] have remarked, there is a current trend for designers of asphalt mixtures to pay attention to resistance to rutting, whereas fatigue and thermally induced cracking are equally destructive factors contributing to pavement degradation. Premature pavement cracking is a severe problem, not appropriately accounted for by designers.

According to the Polish technical requirements, the use of 20/30 penetration grade bitumen in asphalt pavements on national roads [5] in the binder course and base course is only allowed on the territory of Poland in the climatic zones 1 and 2, as shown in Figure 1. This condition results from small resistance of 20/30 penetration grade bitumen to low temperature cracking. Moreover, it is not allowed to leave pavement courses made of 20/30 penetration grade bitumen uncovered with another layer for the winter season; the layer in question should compose of an asphalt mixture applying a binder softer than the 20/30 one. Figure 1 shows how, in line with Polish technical requirements, hard road 20/30 penetration grade bitumen may only be used in pavement structures of national roads in the area constituting one third of the country's territory, prevailingly in North-West and West regions of Poland.

Figure 1. Poland divided into climatic zones - as per [5].

2. Purpose of the study

Due to the limited scope of Poland's territory where hard road 20/30 penetration grade bitumen can be applied in high stiffness modulus asphalt concrete, the authors of the paper decided to establish whether selected bitumens could be replaced with binders manufactured on the basis of softer ones, e.g. 35/50 and 50/70 penetration grade bitumens, modified with natural Gilsonite or Tinidad Epuré asphalts. Furthermore, they considered the possibility of applying them on a larger area of the country. 35/50 and 50/70 penetration grade bitumens are characterised by higher resistance to cracking at low temperatures than 20/30 penetration grade bitumen, and lower stiffness at high temperatures. The authors of the paper believed that by use of appropriate amounts of natural asphalts, which produce a stiffening effect [6, 7], it is possible to achieve asphalt binders with good resistance to rutting at high temperatures and unimpaired low-temperature properties at the same time.

3. Methodology of the studies

In order to verify the theory outlined above, the authors of the paper used a classification of bitumens conforming with the AASHTO M320-17 *Standard Specification for Performance-Graded Asphalt Binder*, which has been designed based on the premise that asphalt paving mixtures are resistant to rutting, as well as fatigue and low-temperature cracking, in given pavement temperature ranges. Performance Grade (PG) is a method of categorising asphalt binders used in road pavements according to their properties in working temperatures from -46°C to 82°C. The advantage of the PG classification is that bitumens are categorised on account of their functional features, which makes it possible to determine a temperature range for a given pavement and simultaneously retain its resistance to wheel loads. Bitumen type is selected basing on climate conditions and traffic speed and volume. The functional type of asphalt binder is marked as PG x – y (e.g. PG 58–28), where x means the maximum pavement temperature, and y the minimum one. Based on the archival climate data, the average 7-day maximum air temperature and the minimum temperature measured in the area of the planned asphalt pavement are estimated. While determining the PG value, the probability of occurrence of the maximum and minimum temperatures is assumed as 50% or 98%. The maximum pavement temperature is assumed on the basis of calculations, while the minimum is assumed according to the lowest air temperature, then attributed to the existing PG value ranges. Whenever the predicted traffic volume exceeds 10 or 30 million ESAL (Equivalent Single Axle Load), which is a standard single axle load of 80 kN, the maximum pavement temperature should be increased by 6°C or 12°C accordingly. An analogous action should be taken in the case of slow traffic (velocity from 20 km/h to 70 km/h) or parking (velocity up to 20 km/h) on the designed road, by increasing the temperature by 6°C in the former case, and by 12°C in the latter. For example, when traffic volume ranges from 10 to 30 million ESAL and there is slow traffic involved, the maximum pavement temperature should be increased by 12°C [8, 9].

For the purposes of determining the Performance Grade (PG) (Table 1) for asphalt binders discussed in the paper, in line with the premises of the American Superpave system (SUperior PERforming Asphalt PAVEments), the tested materials have been submitted to ageing processes simulated in laboratory conditions: in case of short-term ageing by use of RTFOT method according to standard EN 12607-1:2015-01 *Bitumen and bituminous binders - Determination of the resistance to hardening under influence of heat and air - RTFOT method*, and in case of long-term ageing by use of PAV method, according to standard EN 14769:2012 *Bitumen and bituminous binders - Accelerated long-term ageing conditioning by a Pressure Ageing Vessel (PAV).*

Table 1. Pavement temperature values corresponding to the Performance Grade [10].

average 7-day max pavement design temp., °C	< 46	< 52	< 58	< 64	< 70	< 76	< 82
min 1-day pavement design temp., °C	> -34 > -40 > -46	> -10 > -16 > -22 > -28 > -34 > -40 > -46	> -16 > -22 > -28 > -34 > -40	> -10 > -16 > -22 > -28 > -34 > -40	> -10 > -16 > -22 > -28 > -34 > -40	> -10 > -16 > -22 > -28 > -34	> -10 > -16 > -22 > -28 > -34

The next step was to conduct tests on original and aged asphalt binders by use of a bending beam rheometer (BBR), according to standard EN 14771:2012 *Bitumen and bituminous binders - Determination of the flexural creep stiffness - Bending Beam Rheometer (BBR)*, as well as the dynamic shear rheometer (DSR), according to standard EN 14770:2012 *Bitumen and bituminous binders - Determination of complex shear modulus and phase angle - Dynamic Shear Rheometer (DSR).*

The change in the bitumen mass resulting from short-term ageing simulated by the RTOFT method according to AASHTO M320 should not exceed 1.0%. In case of long-term ageing of binders in the PAV chamber following the AASHTO M320 recommendations, the temperature of 100°C was applied, as advised in case of asphalt binders used in road pavements produced and exploited in the temperate climate.

The designations of low-temperature properties in the BBR rheometer at the temperatures of -8°C, -16°C, -24°C and -32°C for asphalt binders, discussed in this paper, submitted to the short-term and long-term ageing processes, were made on the basis of four independent samples of bitumen beams. The designation results for the S(60s) stiffness modulus and m(60s) m-value arrived at by use of a BBR rheometer were applied to calculations of the T_{cr} critical temperature value of the tested asphalt binders, using the formula [7]:

$$T_{cr} = \max (T_{S=300\,\mathrm{MPa}} \,;\, T_{m=0.3}) - 10\text{ , °C} \qquad (1)$$

where:
T_{cr} – critical temperature, °C,
$T_{S=300\ MPa}$ – temperature in which the value of stiffness modulus indicated in the BBR rheometer after loading time t = 60 s reaches value equal to 300 MPa, °C,
$T_{m=0.3}$ temperature in which the m-value indicated in the BBR rheometer after loading time t = 60 s reaches value equal to 0.3, °C.

According to the American Superpave system, the critical temperature marks the resistance of pavement to low-temperature cracking depending on the properties of the asphalt binder applied.

The values of the dynamic shear modulus $|G^*|$ ($|G^*|$ is an absolute value of complex shear modulus G^*) and the phase angle δ have been determined by use of the DSR rheometer for all asphalt binders studied in this paper within the range of temperatures from -40°C to 82°C. The conducted tests used a measuring plate with an 8 mm diameter, and an aperture of the height of 1.5 mm. The test temperatures ranged between -40°C and 82°C, with measurements beginning with the highest temperature and subsequent determination being made every 60 s at a temperature lower by (1.0±0.1)°C than the preceding one. The tests were conducted at the constant angular frequency of oscillation equal to ω = 10 rad/s (≈ 1.59 Hz). The values for the oscillation amplitude (deflection angle) were changing on a logarithmic scale in the range of 10 mrad for temperature of 82°C to 0.1 mrad for the temperature of -40°C. Using determined values of a dynamic shear modulus $|G^*|$ and phase angle δ, it was possible to determine the values of equivalent temperature relating to resistance to rutting of pavement (designated in this paper as ET_R). For this purpose, the highest temperature meeting the conditions for asphalt binders not submitted to ageing – $|G^*|/\sin(\delta) \geq 1.0$ kPa and asphalt binders submitted to ageing in accordance with the RTFOT method – $|G^*|/\sin(\delta) \geq 2.2$ kPa [11]. The values of equivalent temperature relating to pavement resistance to fatigue cracking (designated in the paper as ET_F) were determined at the lowest temperature satisfying the conditions for asphalt binders submitted to short-term and then long-term ageing – $|G^*| \cdot \sin(\delta) \leq 5000$ kPa.

The PG value relating to the minimum 1-day pavement temperature was assumed as T_{cr} given in Table 2, which is equal to or greater than the value achieved from Formula 1 for the selected asphalt binder. Determining the PG value related to the average 7-day maximum pavement temperature consists in assuming ET_R value given in Table 2, which is equal to or lesser than the value presented in Figures 4 and 5 (the lower of temperatures pertains to conditions $|G^*|/\sin(\delta)$ = 1.0 kPa and $|G^*|/|\sin(\delta)$ = 2.2 kPa) for the selected asphalt binder. The next step is to verify the ET_R value by checking if the value ET_F is equal to or greater than the value presented in Figures 4 or 5 ($|G^*| \cdot \sin(\delta)$ = 5000 kPa condition) for the selected asphalt binder. Where ET_F is lesser, ET_R needs to be decreased to the level in which ET_F value is equal to or greater than the value presented in Figures 4 or 5 for a selected asphalt binder.

Table 2. Requirements for asphalt binders in accordance with the Performance Grade - as per [10].

ET_R, °C		**46**					**52**						**58**		
ET_F, °C	10	7	4	25	22	19	16	13	10	7	25	22	19	16	13
T_{cr}, °C	-34	-40	-46	-10	-16	-22	-28	-34	-40	-46	-16	-22	-28	-34	-40

ET_R, °C			**64**						**70**			
ET_F, °C	31	28	25	22	19	16	34	31	28	25	22	19
T_{cr}, °C	-10	-16	-22	-28	-34	-40	-10	-16	-22	-28	-34	-40

ET_R, °C			**76**					**82**		
ET_F, °C	37	34	31	28	25	40	37	34	31	28
T_{cr}, °C	-10	-16	-22	-28	-34	-10	-16	-22	-28	-34

4. The subject of the tests

The tests were conducted on three different penetration grade road bitumens - 20/30, 35/50 and 50/70, produced in a Polish refinery from Russian petroleum and natural asphalts - Gilsonite in powdered form and Trinidad Epuré as 0/8 mm aggregate. Their basic properties, that is penetration values

IOP Conf. Series: Materials Science and Engineering **356** (2018) 012007 doi:10.1088/1757-899X/356/1/012007

determined at 25°C and at the softening point for asphalt binders studied in this paper have been presented in Table 3. The content of Gilsonite and Trinidad Epuré natural asphalt additives in 35/50 and 50/70 road bitumens was calculated to produce upon modification an asphalt binder with similar penetration grade, that is, it was assumed that the differences between the determined penetration grade will not exceed 2 mm/10. After conducting a series of trial determinations of penetration at 25°C for asphalt binders containing natural asphalts of varying composition, the following samples were selected for further research, with the corresponding content of natural asphalt (in relation to the mass of modified asphalt binders): 3%, 5%, 7% for Gilsonite and 15%, 25%, 35% for Trinidad Epuré. The produced asphalt binders were marked, first, with the type of road bitumen, second, with the type and content of the applied additive. Gilsonite was described as GIL for short, and Trinidad Epuré as TE. For example, a 35/50 penetration grade bitumen with a 15% Trinidad Epuré additive would be – 35/50+15% TE.

Table 3. Basic properties of the asphalt binders studied in the paper.

Asphalt binder	Penetration at 25°C mm/10	Softening Point °C	Asphalt binder	Penetration at 25°C mm/10	Softening Point °C
20/30	25.6±0.5	63.6±0.2			
35/50	40.8±0.3	55.9±0.5	**50/70**	55.9±0.7	51.6±0.2
35/50+3% GIL	27.2±0.3	61.1±0.2	**50/70+3% GIL**	38.6±0.3	55.8±0.4
35/50+5% GIL	24.4±0.5	62.9±0.3	**50/70+5% GIL**	32.6±0.4	57.2±0.3
35/50+7% GIL	19.8±0.3	65.3±0.2	**50/70+7% GIL**	26.6±0.7	60.4±0.4
35/50+15% TE	28.6±0.7	57.5±0.3	**50/70+15% TE**	39.7±0.6	53.3±0.3
35/50+25% TE	24.8±0.3	59.5±0.2	**50/70+25% TE**	32.0±0.6	55.7±0.3
35/50+35% TE	19.2±0.5	62.0±0.2	**50/70+35% TE**	25.8±0.6	58.6±0.4

Gilsonite is an asphaltite extracted from deposits in North-East part of the Utah state (USA). This material is a natural mixture of (per product mass) coal - 85%, hydrogen - 10%, nitrogen - 3% and oxygen, sulphur and other components in an amount of about 2% [12]. Gilsonite resembles bituminous coal in appearance, and is usually sold after having been granulated in the form of 0/2 mm aggregate or powdered.

Trinidad Epuré is a natural asphalt submitted to the refinery process, loosened from the surface of the lake (Pitch Lake) in the vicinity of the town of La Brea in the island of Trinidad. The process consists in i.a. heating the acquired material in order to soften it, evaporate free water and chemical compounds dissolved in it, as well as to separate mineral and organic impurities, e.g. in the form of stones or wood [14].

5. Results

The requirements of Superpave specifications relating to changes in mass following the induction of ageing process by use of the RTFOT method (mass decrease ≤ 1.0%) have been fulfilled by all asphalt binders discussed in the paper except for asphalt binders containing the Trinidad Epuré additive in the amount of 35%. As far as such high-modified asphalt binders are concerned, and considering the value of expanded uncertainty, the mean mass decrease equals the boundary value of 1.0%. Figure 2 shows mean values for mass decrease in ageing processes induced by the RTFOT method. It is to be assumed that the 35% content of the Trinidad Epuré additive in road asphalt binder is the absolute maximum, considering the excessive change in mass of the binder as a result of short-term ageing. For bitumens modified with the Trinidad Epuré additive, submitting them to the ageing process by use of the RTFOT method results in a significant decrease of mass, which intensifies, the higher the content of that additive in the modified asphalt binder. This is the consequence of the Trinidad Epuré's chemical composition, which includes water trapped in crystalline structure and volatile substances that are released only at temperatures exceeding 150 °C [14].

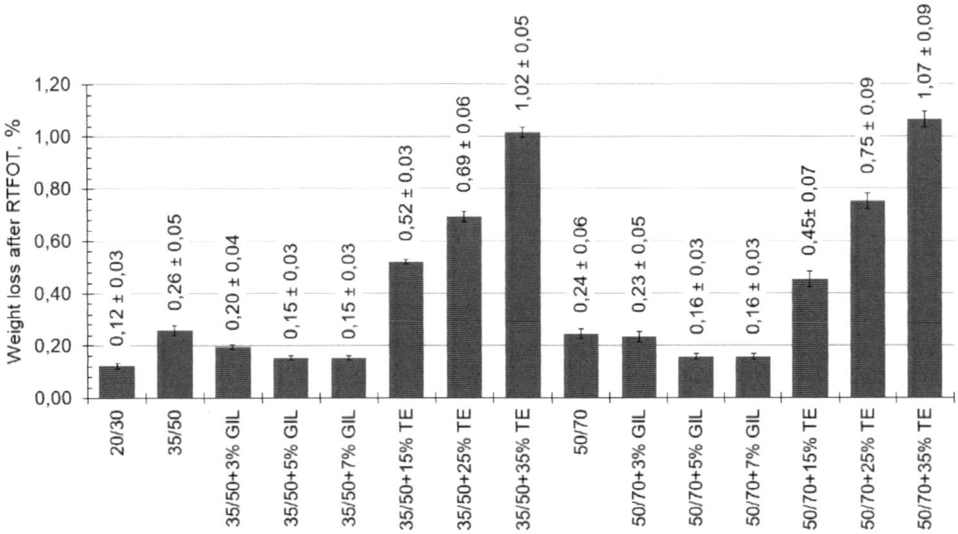

Figure 2. Weight loss of the asphalt binders as a result of short-term ageing induced with the RTFOT method.

The values of the critical temperature determined for asphalt binders submitted to both short-term and long-term ageing are presented in Figure 3. It has been observed that critical temperatures of 35/50 and 50/70 penetration grade road bitumens with and without the natural Gilsonite additive in the amount of 3% and 7%, as well as with the Trinidad Epuré additive, in the amount of 15% exclusively, are lower than that of 20/30 penetration grade bitumen. The application of Gilsonite reflects on a substantially lesser increase of critical temperature as a result of 35/50 or 50/70 penetration grade bitumen modification than in the case of Trinidad Epuré.

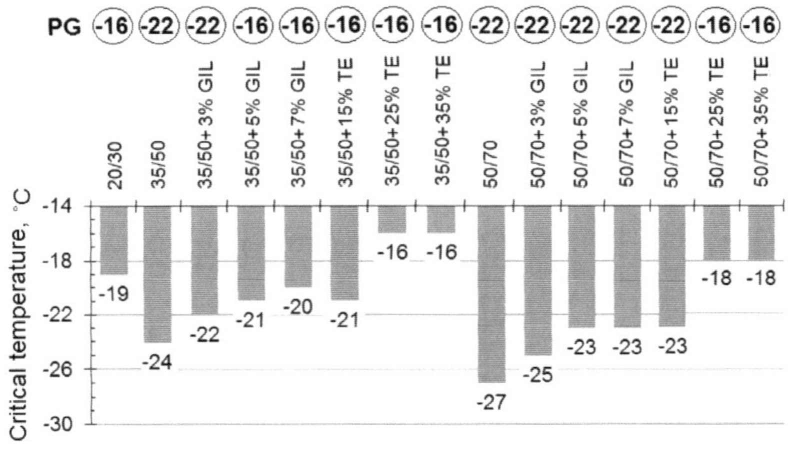

Figure 3. The values of critical temperatures of road bitumens and modified binders as well as the Performance Grade values determined on the basis of T_{cr}, as per Table 2.

The values of the equivalent temperature corresponding to pavement resistance to rutting ($|G^*|/\sin(\delta)$ = 1.0 kPa values in the case of unaged asphalt binders and $|G^*|/\sin(\delta)$ = 2.2 kPa in the case of asphalt binders aged by use of the RTFOT method) and fatigue cracking ($G^*\cdot\sin(\delta)$ = 5000 kPa in the case of binders aged by short-term and then long-term ageing) have been presented in Figures 4 and 5.

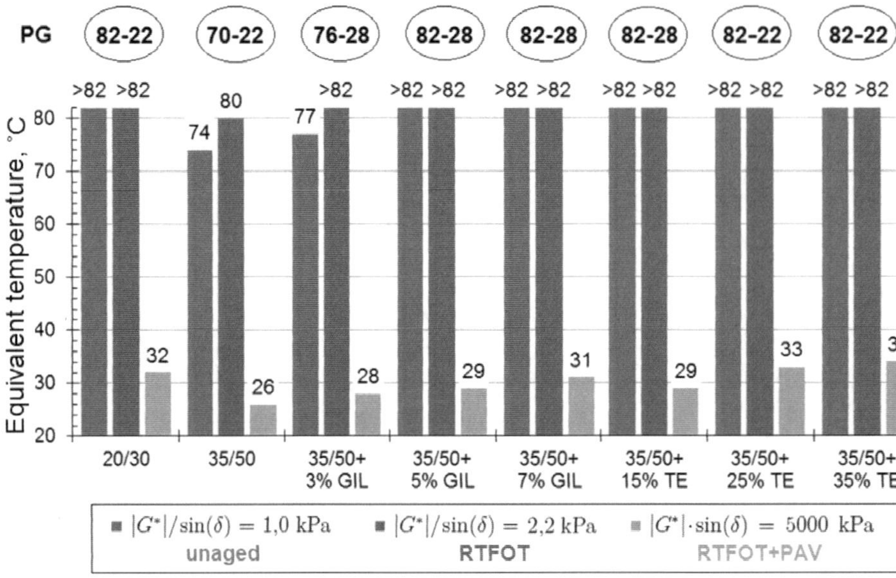

Figure 4. The values of equivalent temperature determined in the case of 20/30 bitumen and modified binders containing 35/50 bitumen, as well as the Performance Grade values determined on the basis of ET_R and ET_F as per Table 2.

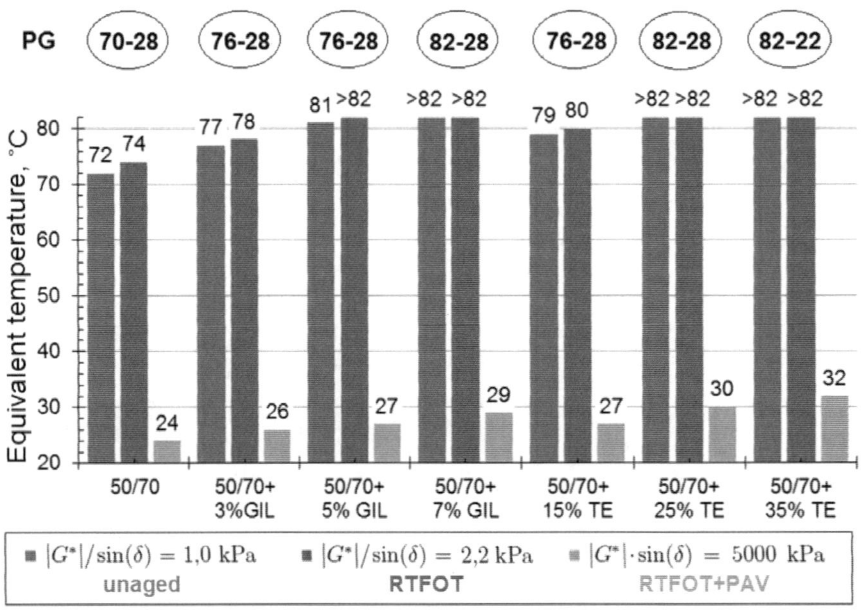

Figure 5. The values of equivalent temperature determined in the case of 50/70 bitumen and modified binders containing 50/70 bitumen, as well as the Performance Grade values determined on the basis of ET_R and ET_F as per Table 2.

It has been observed that 20/30 penetration grade bitumen is characterised by a higher equivalent temperature in relation to resistance to fatigue cracking ($|G^*|\cdot\sin(\delta)$ = 5000 kPa) than binders containing additives of natural asphalts with comparable resistance to rutting ($|G^*|/\sin(\delta)$ = 1.0 kPa and $|G^*|/\sin(\delta)$ = 2.2 kPa). The exception to this is 35/50 penetration grade bitumen with at least 25% Trinidad Epuré additive content, which shows greater susceptibility to fatigue cracking than 20/30 penetration grade bitumen (higher values of equivalent temperature).

By use of the data presented in Figures 3, 4, and 5, as well as Table 2, values of the Performance Grade have been determined for all asphalt binders studied in this paper, and shown in Table 4.

Table 4. The PG values of asphalt binders studied in the paper.

Asphalt binder	PG	Asphalt binder	PG
20/30	82-16		
35/50	70-22	**50/70**	70-22
35/50+3% GIL	76-22	**50/70+3% GIL**	76-22
35/50+5% GIL	82-16	**50/70+5% GIL**	76-22
35/50+7% GIL	82-16	**50/70+7% GIL**	82-22
35/50+15% TE	82-16	**50/70+15% TE**	76-22
35/50+25% TE	82-16	**50/70+25% TE**	82-16
35/50+35% TE	82-16	**50/70+35% TE**	82-16

The analysis of the Performance Grade values presented in Table 4 has yielded conclusions that binders with a percentage of the natural Gilsonite or Trinidad Epuré asphalt discussed in this paper in most cases bear identical PG values to 20/30 penetration grade bitumen. Applying an even softer 50/70 penetration grade bitumen produces binders characterised by an even greater resistance to low-temperature cracking than the 20/30 penetration grade bitumen.

6. Discussion

Based on the analysis of the achieved results it has been concluded that the 20/30 penetration grade bitumen can be effectively replaced with 35/50 and 50/70 penetration grade bitumens containing from 3% up to 7% of the natural Gilsonite asphalt, as binders modified in this manner are characterised by greater resistance to low-temperature cracking - the values of the critical temperatures (Figure 3) are lower by 1°C to up to 6°C, with unimpaired resistance to rutting (Figure 4 and 5) in the case of climate conditions prevailing on the territory of Poland. Moreover, asphalt binders including Gilsonite in their composition have greater resistance to fatigue cracking (Figures 4 and 5) compared to 20/30 penetration grade bitumen. The effect of modifying 35/50 and 50/70 penetration grade bitumens with the Trinidad Epuré additive is not as favourable as in the case of Gilsonite. The only asphalt binders with properties even better than 20/30 penetration grade bitumen are exclusively the ones with the 15% percentage of additive in the examined samples. In the case of 25% or 35% content of the Trinidad Epuré additive, the asphalt binders discussed in this paper have higher values of the critical temperature, ranging from 1°C to 3°C (Figure 3), and lower resistance to fatigue cracking (Figure 4) in the case of modifications of 35/50 penetration grade bitumen.

The analysis of the determined values of the Performance Grade (Table 4) demonstrates that asphalt binders containing the natural Gisonite or Trinidad Epuré additive are characterised in most cases by and identical PG to that of 20/30 penetration grade bitumen, whereas the application of the softer 50/70 penetration grade bitumen produces binders with even higher resistance to low-temperature cracking than the 20/30 penetration grade bitumen, with unimpaired resistance to rutting in the climate conditions prevalent on the territory of Poland.

In papers [15, 16], the authors have attempted to apply the Superpave requirements regarding classifying bitumens according to the Performance Grade to Polish conditions. The results for the binder course and base course are presented in Figures 6 and 7. The analysis of Figures 6 and 7 has demonstrated that the 20/30 penetration grade bitumen does not fulfil the Performance Grade requirements at all, when used in the binder course or base course of roads constructed on the territory of Poland. In conformity with the PG values presented in Figures 6 and 7, these requirements are met by 35/50 penetration grade bitumen containing 3% of Gilsonite and 50/70 penetration grade bitumen with the Gilsonite additive of 3% and 5% or Trinidad Epuré additive of 15%. On account of an extremely high PG value connected with the average 7-day maximum air temperature for the 20/30 penetration grade bitumen and asphalt binders containing natural asphalts (in most cases these are higher than the maximum PG value of 82°C).

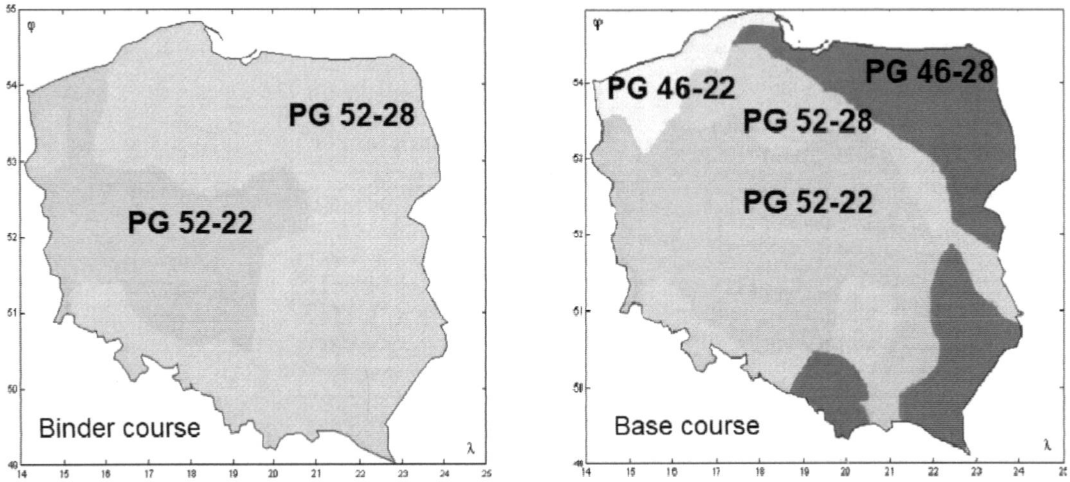

Figure 6. The values of the Performance Grade of the asphalt binders recommended to use in Poland for the binder course and base course as per [15].

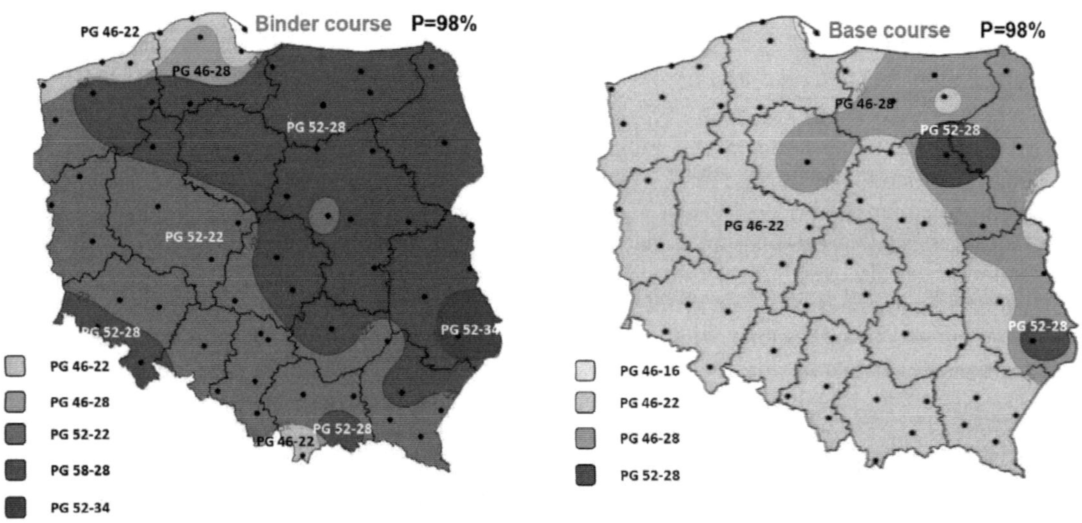

Figure 7. The values of the Performance Grade of the asphalt binders recommended to use in Poland for the binder course and base course of motorways and expressways as per [16].

In reference to the results from papers [15, 16] it is reasonable to use softer bitumens as binders for the basic substance for modifications using natural asphalts (assuming most unfavourable conditions, the maximum PG value connected with the average 7-day maximum air temperature is 64°C) due to better low-temperature properties. Notably, the requirements conforming with the Performance Grade are fulfilled by the 35/50 and 50/70 penetration grade road bitumens not submitted to modifications, which are not included in technical requirements for asphalt pavements on national roads [5] for high stiffness modulus asphalt concrete.

7. Conclusions

- The 20/30 penetration grade road bitumen used in mixtures such as high stiffness modulus asphalt concrete can be effectively replaced with 35/50 and 50/70 penetration grade bitumens with the natural Gilsonite asphalt additive of 3% to 7% and Trinidad Epuré additive of 15%.

- Asphalt binders containing the natural Gilsonite or Trinidad Epuré additive are characterised in most cases by identical PG to that of 20/30 penetration grade bitumen, whereas the application of the softer 50/70 penetration grade bitumen produces binders with even higher resistance to low-temperature cracking than the 20/30 bitumen, with unimpaired resistance to rutting in the climate conditions prevalent on the territory of Poland.

References

[1] Sybilski D (ed) 2013 Katalog Przebudów i Remontów Nawierzchni Podatnych i Półsztywnych (Warsaw: GDDKiA) *on-line:* www.gddkia.gov.pl/userfiles/articles/p/prace-naukowo-badawcze-w-trakcie_3434/KPRNPP%20i%20Zalaczniki%202013.pdf

[2] Quintana H A R, Noguera J A H and Bonells C F U 2016 Behavior of Gilsonite - modified hot mix asphalt by wet and dry processes *J. Mater. Civ. Eng.* **28** 2 DOI 10.1061/(ASCE)MT.1943-5533.0001339

[3] Jiao S, Jiang Z and Bu J 2010 Research on design and pavement performance of TLA modified asphalt mixture *Adv. Mat. Res.* **146-147** pp 217-220 DOI 10.4028/www.scientific.net/AMR.146-147.217

[4] Zhu Y, Dave E, Rahbar-Rastegar R, Daniel J and Zofka A 2017 Comprehensive evaluation of low-temperature fracture indices for asphalt mixtures *Road Mater. Pavement* **18** S4 pp. 467–490 DOI 10.1080/14680629.2017.1389085

[5] *Wymagania Techniczne WT-2* 2014 *Część I Mieszanki mineralno-asfaltowe Wymagania* Techniczne (Warsaw: GDDKiA) *on-line:* www.gddkia.gov.pl/userfiles/articles/z/zarzadzenia-generalnego-dyrektor_13901/zalacznik%20do%20zarz%2047.pdf

[6] Iwański M, Cholewińska M and Mazurek G 2014 Właściwości asfaltu z dodatkami modyfikującymi po procesie starzenia krótkoterminowego *Budownictwo i Architektura* **13** 1 pp 15-27

[7] Słowik M and Bilski M 2017 An experimental study of the impact of aging on Gilsonite and Trinidad Epuré modified asphalt binders properties *Balt. J. Road Bridge E.* **12** 2 pp. 71-81 DOI 10.3846/bjrbe.2017.09

[8] *Superpave mixture design guide* 2001 (Washington DC. WA: Federal Highway Administration – Department of Transportation) *on-line:* www.fhwa.dot.gov/publications/research/infrastructure/pavements/ppub/01052/1052.pdf

[9] Styk S and Błażejowski K 2014 *Technologia warstw asfaltowych* (Warsaw: WKŁ)

[10] AASHTO M 320-16 *Performance-Graded Asphalt Binder*

[11] Słowik M, Mielczarek M, Bilski M and Wiśniewski D 2016 Wpływu zawartości elastomeru SBS na wartości parametrów charakteryzujących odporność na odkształcenia trwałe asfaltów modyfikowanych *Czasopismo Inżynierii Lądowej, Środowiska i Architektury Journal of Civil Engineering, Environment and Architecture* **23** 63 1/II/16 pp. 307-314 DOI: 10.7862/rb.2016.89

[12] Nciri N, Song S, Kim N and Cho N 2014 Chemical characterization of Gilsonite bitumen *Journal of Petroleum & Environmental Biotechnology* **5** 193

[13] Boden T and Tripp B 2012 *Gilsonite veins of the Uinta Basin* (Salt Lake City. UT: Utah Geological Survey)

[14] Laborhandbuch *für Trinidad Naturasphalt* (Bremen: Carl Ungewitter Trinidad Lake Asphalt) *on-line:*www.trinidad-lake-asphalt.eu/wp-content/uploads/Publikationen/Labor/Laborhandbuch_fuer_Trinidad_Naturasphalt_270313.pdf

[15] Sybilski D and Mirski K 2000 Zalecane lepiszcza asfaltowe w warstwach nawierzchni w Polsce z uwzględnieniem warunków klimatycznych i obciążenia ruchem *Prace IBDiM* **1-2** pp. 103-157

[16] Pszczoła M, Ryś D and Jaskuła P 2017 Analysis of climatic zones in Poland with regard to asphalt performance grading. Analiza stref klimatycznych w Polsce z uwzględnieniem klasyfikacji funkcjonalnej asfaltów Performance Grade *Roads and Bridges - Drogi i Mosty* **16** 4 pp 245-264

IOP Conf. Series: Materials Science and Engineering **356** (2018) 012008 doi:10.1088/1757-899X/356/1/012008

The use of lithium compounds for inhibiting alkali-aggregate reaction effects in pavement structures

J. Zapała-Sławeta, Z. Owsiak

Kielce University of Technology, al. Tysiąclecia Państwa Polskiego 7, 25-314 Kielce, Poland

jzapala@tu.kielce.pl

Abstract. Internal corrosion of concrete caused by the reaction of reactive aggregate with sodium and potassium hydroxides from cement is a threat to the durability of concrete pavements. Traditional methods for reducing the negative effects of the reaction include the use of unreactive aggregates, low alkali cements, mineral additives or chemical admixtures, incorporated during mixing. Lowering the relative humidity of the concrete below 80% is another measure for limiting the destructive reaction.

The incorporation of lithium compounds, in particular lithium nitrate and lithium hydroxide, to the concrete mix is a method of limiting alkali-silica reaction effects. The challenge is to reduce the negative effects of aggregate reactivity in members in which the reaction has occurred because the aggregate happened to be reactive. The paper presents ways of limiting the deterioration of ASR-affected concrete in road pavements and other forms of transportation infrastructure, mainly through the use of lithium compounds, i.e. lithium nitrate. Impregnation methods that allow the penetration of lithium ions into the concrete structure were characterized, as was the effectiveness of the solutions applied.

1. Introduction

Alkali-aggregate reaction (AAR) in concrete is an example of internal corrosion that occurs in the entire volume of the material. Two currently recognized types of AAR are reactions between alkalis and aggregates containing reactive silica (ASR) and reactions between alkalis and carbonate aggregates (ACR). Due to more frequent presence of reactive forms of silica such as opal, tridymite, chalcedony, and strained quartz, in aggregates used in concrete, ASR is usually the main process responsible for the deterioration of concrete structures [1]. The reaction between aggregates and sodium and potassium hydroxides, which originate mainly from the cement, produces a sodium potassium-calcium silicate gels, which absorbs water and swells. Alkali-aggregate causes expansion and cracking in the concrete.

Experience in dealing with the devastating ASR has been reported in many countries, including United States, Canada, England, Denmark, Germany, India, China, Japan, or Korea [2-4]. In some cases the destruction of concrete was incorrectly attributed to other deterioration processes, excluding the reaction with alkalis [4]. In Poland, no cases of destructive alkaline reaction have been reported so far, however, the presence of reactive aggregates has been found in the regions of north-eastern and south-west Poland [5-7]. It is further necessary to take into account that concretes in road structures are highly exposed to AAR, which results from the limited possibility of protecting them from moisture. Also, the use of de-icing agents (sodium chloride) as an external source of alkalis increases their concentration in the pore solution of the concrete. In the ASR risk classification system introduced by the British Cement Association UK, concrete pavements are classified in the high-risk category [8].

The ASR preventive measures focus on eliminating the factors contributing to the destructive processes by using unreactive aggregates, low alkali cements, or protection against moisture. Cement with mineral additives may also be used, where it is possible to increase the permissible content of alkali in cement [9]. The methods of preventing the effects of ASR include the use of lithium compounds, mainly lithium nitrate and lithium hydroxide [10-12]. The risk of alkali attack is only possible here at the mixing stage of concrete production.

In ASR-affected elements, further protection of the material against damage can be provided by adding lithium compounds and reducing the relative humidity of the concrete. The present study aims at outlining the ASR preventive methods for use on concrete, especially on concrete pavements.

2. Alkali-silica reaction in concrete pavements

Cracking of road surfaces due to AAR starts in sites exposed to high moisture. Cracks are observed near free edges and joints as parallel to the free longitudinal edges of the road and perpendicular to the transverse joints. A model of crack development due to AAR is shown in Fig. 1:

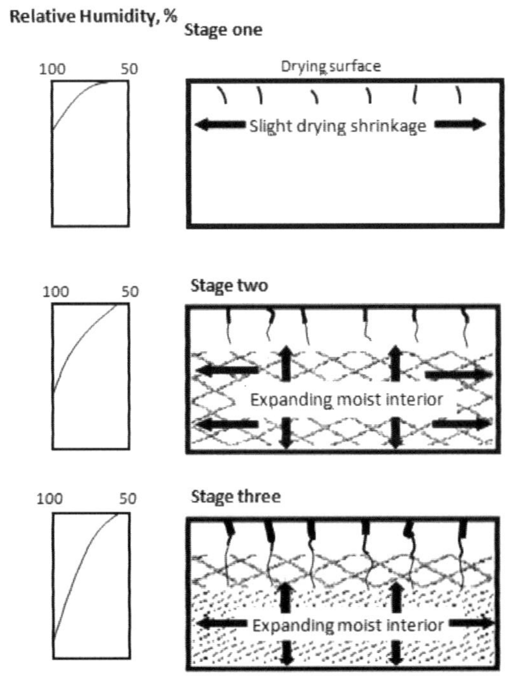

Figure 1. Model of AAR crack extension in concrete pavement [13].

The development of cracks in concrete pavements can be represented in the form of a three-stage model, depending on the relative humidity of the concrete. In the first stage, very small cracks are observed; these are caused by drying shrinkage. The effects of the alkaline reaction are not visible at this stage. In the second stage, at a relative humidity above 80%, a swellable silicate gel forms and migrates towards the surface of the concrete. Surface cracks expand as a result of concrete expansion. In the third stage, when the top layer dries, the reaction in the surface layer is reduced. A high humidity level inside the concrete ensures the progress of deterioration. The gel that exerts tensile stress on the surface of the concrete causes the growth and widening of the surface cracks. In zones with high moisture, the reaction continues until depletion or reduction in the amount of one of the substrates. The presence of reactive silica and the right amount of alkalis in the pore solution is insufficient when the concrete dries. The process is then suppressed until the moisture content in the concrete increases.

Concrete road surfaces are exposed primarily to the deterioration caused by cyclic freezing and thawing. ASR cracks affect the durability of concrete exposed to negative temperatures. The absorption of water by alkali silicate gels increases its quantity in the element, thus lowering the freeze-thaw resistance of the concrete [14]. De-icing agents used in winter, such as NaCl, increase alkali content in concrete, which results in the acceleration of their reaction with aggregate. Sodium chloride increases the pH of the pore solution, and the solubility of silica and calcium hydroxide also increases, contributing to a faster progress of the AAR and the formation of Na+K-S-H and C-N + K-S-H gels [13]. Rangaraju et al. stated that deicers such as sodium acetate or potassium acetate can trigger ASR [15].

3. ASR symptoms

The types of distress in non-reinforced and reinforced concrete pavements are summarized in Table 1.

Table 1. Distress in concrete pavements [16].

Jointed concrete pavements	Continuoustly reinforced concrete pevements
Corner break	Longitudinal cracking
D-cracking	Transverse cracking
Longitudinal cracking	D-cracking
Transverse cracking	Map cracking
Joint deficiencies	Scaling
Map cracking	Popouts
Scaling	Blowups
Popouts	Joint deterioration
Joint Faulting	
Blowups	

ASR in road pavements is manifested mainly in the form of a network of cracks, mainly on the whole concrete surface, joint deficiencies/deterioration, silicate gel exudation, local concrete chipping and ravelling (Fig. 2).

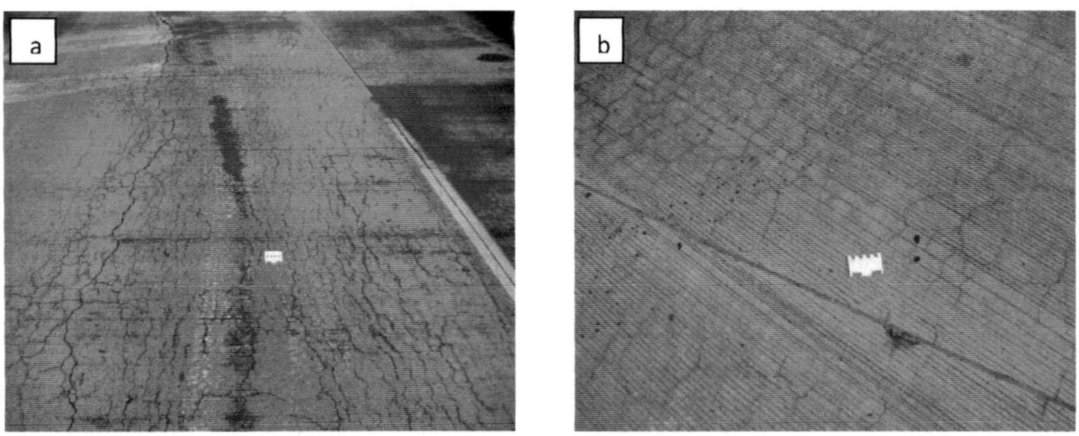

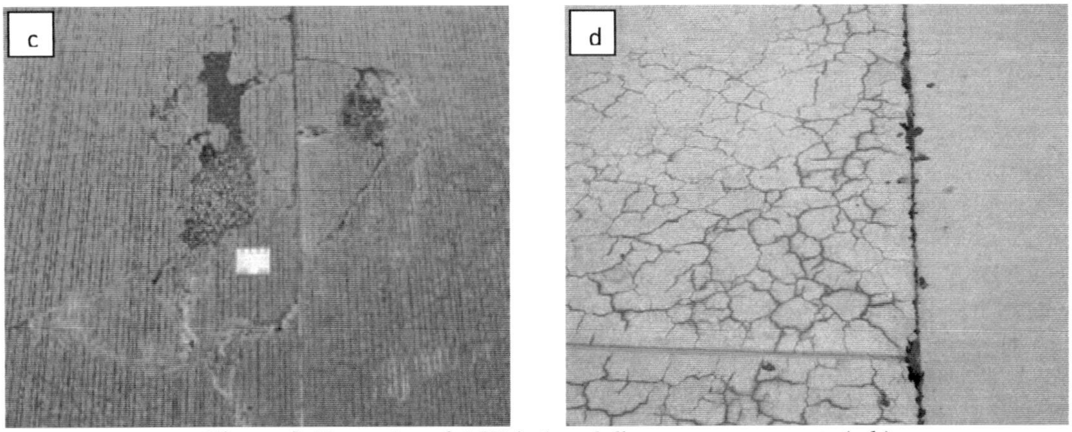

Figure 2. Examples of ASR induced distress to pavements (a-b), gel exudation and delamination (c), cracking and joint spall (d) [14,17-19].

Damage occurs mainly in the vicinity of joints and is similar in appearance to D cracking, also characteristic of frost attack. The crack pattern is closely related to the geometry of concrete, reinforcement layout, stress fields and environmental impacts. Mesh (map) cracks typically have a random orientation. Restraining the shape in one or more directions makes the cracks be oriented in the direction in which the stresses are lower. In concrete surfaces, cracks are usually parallel to the side edges restrained by curbs [20].

4. Methods of limiting ASR inducing factors

The reaction of alkali with aggregate takes place when the following three factors occur simultaneously:

- sufficient concentration of sodium and potassium hydroxides in the pore solution. Alkalis derive mainly from Portland cement, although the source may also be aggregate, mineral additives and chemical admixtures used in concrete or de-icing salts,
- the presence of reactive silica in the aggregate,
- sufficient moisture of around 80%. Water, introduced into the concrete mix as mixing water and that coming from external sources, transports the ions and when absorbed by silicate gels leads to concrete cracking.

If any of these factors is eliminated, ASR will not take place. It should be noted that the elimination of the first two factors is only possible at the mixing stage. Aggregates that do not have alkali reactivity and low alkali cements should be used, along with pozzolans or hydraulic mineral additives as a partial cement replacement. Reducing the amount of water by using a lower w/c ratio (w/s) can increase the alkali concentration in the pores in the concrete, thereby increasing the rate of silica dissolution. Another way is to incorporate a chemical admixture in the form of lithium compounds, mainly lithium nitrate and lithium hydroxide, into the concrete mix. It is believed that lithium ions protect the reactive silica grain against reaction with sodium and potassium hydroxides, forming a protective layer on its surface. This layer consists of nonexpansive lithium silicates or less expansive silicates with a low sodium and potassium contents enriched with lithium ions [21, 22].

In a concrete member, in which the aggregate happened to be reactive, limiting further destruction is possible in several ways classified into two categories [23]:

Treat the cause			Treat the symptoms		
CO_2	Lithium compounds	Drying	Crack filling	Restrain	Relieve stress

In the case of concrete pavements, recommendations include the use lithium compounds, reduction of humidity (treat the cause) and slow cutting/saw cutting (treat the symptoms). These methods will be discussed in more detail in the following sections.

4.1 Treat the cause

Lithium compounds applied to ASR-affected concrete is the promising method. Although the mechanism of lithium ions behaviour is not fully understood, it is believed that lithium modifies the properties of ASR gels changing their expansive character into nonexpansive [22]. The first studies conducted in 1993 -2004 indicated high efficiency of lithium compounds in limiting the expansion of AAR affected concrete samples analyzed in laboratory conditions. Promising research results became the driving force of further experiments, mainly in the United States as part of the Federal Highway Administration (FHWA) projects. The main application method for lithium compounds is placing them as a surface layer on pavement slabs, airport plates, road barriers, bridges. In order to increase the depth of lithium penetration, other techniques such as vacuum and electrochemical methods are used [23].

4.1.1. Topical application

The application of lithium compounds, primarily lithium nitrate, that has become quite common in the recent years is the surface overlay used on road and airfield pavements and on highway dividers. Lithium nitrate in the form of 30% $LiNO_3$ solution is applied to vertical surfaces using a handheld pressure spray applicator and a truck-mounted spray bar for depositing the solution on horizontal surfaces (Fig. 3) [24]. The $LiNO_3$ solution is most often used at a rate of 0.06 L/m^2 to 0.40 L/m^2 in one or several applications. It has been observed that better penetration of lithium ions is achieved by several applications of smaller amounts [25]. Considering the depth of penetration, the surface treatment with lithium nitrate is not as effective as vacuum or electrochemical impregnation, but in the case of large areas it is more economical.

The efficacy of lithium treatment on ASR affected surfaces of concrete, pointed out by the Federal Highway Administration U.S. Department of Transportation is affected by:

- the type of lithium compound. Lithium nitrate is the only recommended compound because it does not increase the pH of pore solution or, unlike lithium hydroxide, induce the pessimum effect;
- the quantity of lithium nitrate used and the number of repetitions. The surface needs to be covered evenly. Too large amounts of $LiNO_3$ solution may lead to runoff from the treated surface or through evaporation, to $LiNO_3$ precipitation in the form of a salt visible as a white residue on the concrete. Minimum two applications at least 30 minutes apart;
- adequate surface preparation – clean, debris-free, wet surface;
- temperature and wetness of concrete during treatment. The treatment cannot proceed during rainfall or if the rain is forecast within 6h;
- properties of concrete, i.e., the degree of damage due to ASR, permeability,
- additional environmental factors that reduce the durability of concrete, e.g., freeze-thaw cycles. If in addition to ASR, frost attack occurs, other protective measured should be applied such as crack sealing or overlays.

If the precipitated lithium nitrate covers more than 5% of the surface area, water should be applied and until the surface is deemed safe for vehicles [25].

Research on the effectiveness of lithium compounds introduced into hardened concrete, in which the reaction proceeds, was conducted in the 90s of the 20th century. Sakaguchi et al. studied mortars and concrete with reactive aggregates (Pyrex glass) impregnated with the $LiNO_2$ solution after reaching the sufficient level of expansion [26]. The impregnation was found to limit or inhibit further expansion in the specimens. Stokes et al. added a surfactant to the $LiNO_3$ solution to enhance the penetration of lithium ions deep into the concrete structure. They also demonstrated that the higher the extent of expansion, the extent of cracking and the interconnection of microcracks, the easier the path of the lithium solution penetration to a depth of 50 mm (Fig. 4). The effectiveness of concrete surface impregnation with a 30% lithium nitrate solution was investigated on State Route 1 in Delaware. Lithium nitrate was applied two times a year at a rate of 0.24 L m^2 for three years. Four years after treatment, untreated area revealed serious distress in the form of numerous cracks, chipping of concrete at transverse and longitudinal joints. The area of concrete subjected to six cycles of surface treatment with lithium nitrate showed less damage, observed in the form of smaller spalling of concrete on joints (Fig. 5) [27].

Hence the degree of damage to concrete has an impact on lithium ions penetration. Johnston pointed out that ASR cracks facilitate the penetration of the lithium nitrate solution but if the amount of deterioration is too large, it is no longer possible to protect the member against further corrosion by applying lithium nitrate [28].

Figure 3. Surface treatment with lithium compounds [24].

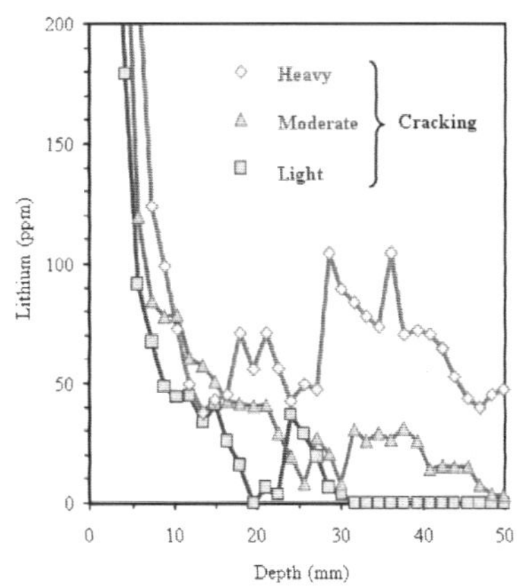

Figure 4. Lithium concentration profiles for concrete pavement after six treatments [27].

Figure 5. ASR affected pavement in Delaware a) control (untreated) concrete, b) after topical treatment with 30% lithium nitrate solution [27].

Numerous studies carried out as part of FHWA projects have shown limited penetration of lithium applied by spraying and limited effectiveness of the treatment at the long term protection against expansion and cracking. Folliard and co-workers demonstrated that after three lithium applications on the surface of concrete in Idaho, the $LiNO_3$ concentration exceeded 100 ppm, the dosage considered sufficient to limit further deterioration of concrete only in the surface layer, at a depth of 3-4 mm [29]. Similar results were reported when $LiNO_3$ was applied to a 26-kilometre concrete road in Geogetown, DE, at a dosage of 0.061L/m^2. The lithium levels were determined form the cores extracted immediately after treatment and several months later. The lithium concentration above 100 ppm was found at a depth of 3-10mm. The probability of ASR inhibition was in that case very low [30].

4.1.2. Vacuum impregnation

The lithium nitrate solution can be introduced into the concrete using the vacuum impregnation method as an alternative to pressure injection. The method consists in creating a suitable underpressure to remove air and moisture from the cracks in concrete, thus creating more space for impregnation. After a vacuum of about 0.5 atmosphere was achieved, lithium is drawn (sucked) into the evacuated area in the concrete. The solution fills the air voids, microcracks and cracks [24].

Vacuum impregnation was used to repair ASR-affected structures in South California and Pennsylvania. The Pennsylvania Department of Transportation applied this method to bridge elements – abutments, sidewalks and wing walls. However, due to the difficulty in performing vacuum impregnation on larger members, this method is not used in large road sections. Vacuum impregnation, used mainly in the repair of bridge pillars and road barriers, provided a deeper penetration of lithium ions compared to the topical application. The concentration of lithium above 100 ppm was recorded at a depth of 2-4mm in road barriers and 8-10mm in bridge pillars [29]. FHWA Project's research on the effectiveness of vacuum impregnation indicated that the penetration of lithium ions is limited only to the surface layer. Drimalas and colleagues used, among others vacuum impregnation method for the application of $LiNO_3$ in columns of road bridges in Houston, which have been damaged to a different degree as a result of alkaline corrosion. A 30% solution of lithium nitrate was used at 1.63 L/m^2. The effectiveness of the vacuum impregnation was also investigated in the ASR-affected road barriers located to the west of Boston in Massachusetts. Monitoring of the columns of the bridge in terms of expansion and surface cracks indicated low effectiveness of the impregnation method. Similar results were obtained in the case of road barriers, where no reduction in expansion levels was observed. Improvement was achieved after applying hydrophobic agents [31]. High costs and complexity of the process are the main drawbacks and limitations of this method, with a small difference in the depth of lithium ion penetration compared to the topical method.

4.1.3. Electrochemical methods

Electrochemical methods have been developed for the electrochemical chloride extraction (ECE) to protect reinforcement against corrosion. By applying a low-voltage direct current, the chlorine ions located close to the reinforcement (cathode) migrate towards the surface anode. After several modifications, this method has been implemented for the application of lithium ions in the concrete structure [32, 33]. Reinforcement is usually the cathode and the anode is attached to the surface of the member. Following the application of the current of adequate voltage, usually up to 40V (bridge deck), positively charged lithium ions travel towards the cathode. The duration of the entire process is from several days up to 8 weeks [24].

Due to the greatest depth of lithium penetration achieved, this method is the most favourable of all the methods. The electrochemical impregnation with lithium ions is not used on road pavements, but offers promising results in the case of other forms of transportation infrastructure. Studies on the depth of lithium penetration in ASR-affected bridge elements showed that the concentration of lithium sufficient to limit the expansion of concrete (above 100ppm), occurred at a depth of up to approximately 50 mm [34]. One side effect of the electrochemical lithium treatment is its ability to accelerate corrosion processes. In the electrochemical process hydroxyl ions are formed on the reinforcement (a cathode). To offset the negative charge, the sodium and potassium ions from the concrete migrate towards the cathode, increasing the susceptibility of concrete in the reinforcement area to AAR. Researchers recommend proper monitoring of the structure over time for cracks and changes in the microstructure around the reinforcement to see whether the benefits from the increased depth of lithium penetration outweigh the potential effects of the pH increase in the concrete pore solution in the vicinity of the reinforcement [23, 35].

4.1.4. Reducing humidity

As mentioned before, high relative humidity is one of the factors necessary for ASR to initiate. A drainage system can be improved, concrete can be sealed using suitable materials or external cladding can be applied to reduce water penetration into concrete. However, the mixing water used for concrete mix production is often sufficient for the ASR processes to start and proceed. As indicated by Diamond, concrete degradation is possible without an external source of water, which can just be the capillary pore solution [36]. A more adequate way to protect a member against deterioration is the use of protective coatings (sealers). The coatings retard external water penetration while allowing the water vapour exchange - vapour-permeable coatings. The most commonly used agents include silanes and siloxanes. In concrete pavements, where water is available through capillary action from the base and subbase layers, the use of silanes may be ineffective and the reduction of internal humidity is only possible in the surface layer. One advantage of using agents based on silanes and siloxanes is the reduction of external chloride penetration.

Silane was used to treat a 19 km long section of pavement in Pine Bluff, Arkansans where the signs of ASR were observed [30]. Petrographic analysis revealed that the cracks and exudations were caused by alkali reactive chert. A 100% and 40% siloxane-based agent was mechanically sprayed at a dosage of 3.1L/m^2 on the sections where the distress was a) mild, b) moderate. Seven months after the treatment, the temperature, the extent of expansion and the relative humidity of the concrete at a depth of 25 mm and 75 mm were analyzed. No significant differences were found compared to the reference samples, but a need to extend the monitoring period was pointed out to be able to determine if the siloxane agent was effective in reducing the humidity and ASR-induced expansion. Pavements in Arkansas, USA were subjected to topical treatment with silane to observe the effects of both the reaction between alkalis and moderately- reactive fine aggregate and the frost attack [37]. The silane sealer was sprayed at a rate of 150 L/m^2 (40% water-based silane) and 250 L/m^2 (100% water-based silane). During the three-year long monitoring period, strains (pavement panels were instrumented with a grid for measuring strain) and the internal RH were measured. Results showed a reduction in the relative humidity and a significant reduction in strain levels following the 100% silane application. A need for further monitoring was reported.

4.2 Treat the symptoms

4.2.1. Crack filling

Filling cracks is particularly important in concrete pavements as the cracks allow water and aggressive ions, including chlorides to easily penetrate the concrete. Flexible grouts or sealers can be used as the filling material. Rigid materials, including cement-polymer mortarsepoxy resins may cause cracks around the bonded area [23].

4.2.2. Saw cutting/slow cutting, concrete replacement

Cutting slots or expansion joints can be an effective, but mostly temporary, method of relieving ASR induced stresses in the concrete. This does not, however, eliminate the causes of ASR. These methods are temporary solutions for use in the members in which the concrete expansion proceeds and the treatment needs to be applied repeatedly.

A frequent method of extending the service life of concrete pavements is the removal of some of the concrete around the joints damaged by the expansion, where ASR is most frequent. The replacement of concrete in the vicinity of joints by saw cutting is recommended when ASR is accompanied by freeze-thaw action. Soundproof concrete can be used as a complementary material. Joint details (opening, dowel bars) have to be restored [17, 19].

5. Summary

Alkali-aggregate reaction reduces the durability of concrete in road infrastructure facilities. Numerous cracks become the migration paths for corrosive agents, such as chlorides and sulphates. The increased humidity of concrete members accelerates their deterioration due to freezing and thawing cycles. AAR control or mitigation is thus important.

Minimizing the ASR effects when reactive aggregates were used in the mix is difficult. Multiple examples of internal corrosion of concrete in road infrastructure assets indicate that the problem of aggregate reactivity needs to be addressed. One of the methods for minimizing the progress of ASR is the use of lithium compounds, which have the ability to change the properties of reaction products - sodium-potassium-calcium silicate gels - from swelling to non-swelling. Road infrastructure at many sites was treated with a solution of lithium nitrate introduced into the concrete structure by surface treatment or by vacuum or electrochemical impregnation, leading to the reduction in the number of cracks and spalls. The key problem is obtaining the appropriate depth of lithium ion penetration, which depends on the specificity of the reaction occurring in the entire volume of concrete. In road pavements, the application of $LiNO_3$ is confined mainly to surface and vacuum treatments (small area) providing insufficient lithium penetration depth to ensure the efficacy of the methods. Further research and monitoring of the ASR-affected members treated with lithium compounds is thus necessary.

References

[1] Owsiak Z, Zapała-Sławeta J and Czapik P 2015 *Bull. Pol. Ac.: Tech.* **63 (1)** pp 23-29
[2] Stark J, Freyburg E, Seyfarth K and Giebson C 2006 AKR-Prüfverfahren zur Beurteilung von Gesteinskörnungen und projektspezifischen Betonen Beton – Die Fachzeitschrift für Bau+Technik, Verlag Bau+Technik GmbH 56/12 pp 574-581
[3] Freyburg E and Schliffkowitz D 2006 Bewertung der Alkali-Reaktivität von Gesteinskörnungen nach petrographischen und mikrostrukturellen Kriterien. Tagungsberichte der 16. Internationale Baustofftagung ibausil Weimar pp 355–372
[4] Hong SH, Han SH and Yun KK 2007 *International Journal of Concrete Structures and Materials* **1(1)** pp75-81
[5] Zapała-Sławeta J and Owsiak Z 2016 *Constr. Build. Mater.* **115** pp 299-303
[6] Owsiak Z, Zapała J and Czapik P 2012 *Cement Lime Concrete* **17 (3)** pp 149-154
[7] Owsiak Z 2004 *Cem. Concr. Res.* **34** pp 7–10
[8] West G 1996 *Alkali-Aggregate Reaction in concrete roads and bridges* (London: Thomas Telford Publications)
[9] Kurdowski W 2010 *The chemistry of cement and concrete* (Krakow: The Association of Cement Producers) chapter 6 pp 427
[10] McCoy WJ and Caldwell AG 1951 *ACI Mater. J.* **22 (9)** pp 693– 706
[11] Owsiak Z and Zapała-Sławeta J 2015 *Cement Lime Concrete* **20 (1)** pp 25-31
[12] Owsiak Z and Zapała-Sławeta J 2013 *Ceram.-Silik.* **57 (2)** pp 138-145
[13] Helmuth R, Stark D, Diamond S and Moranville-Regourd M 1993 *Alkali silica reactivity. An overview of Research Strategic Highway Research Program* National Research Council Washington DC, USA
[14] Thomas MDA, Fournier B, Folliard KJ and Resendez YA 2011 Alkali-Silica Reactivity Field Identification Handbook Publication no. FHWA-HIF-12-022
[15] Rangaraju PR, Sompura KR and Olek J 2006, *J Trans Res Board*, pp 69-79
[16] Thomas MDA, Fournier B, Folliard KJ and Resendez YA 2012 *Alkali Silica Reactivity surveying and tracking guidelines* Publication no. FHWA-HIF-12-046

[17] Federal Highway Administration 2013 *Alkali–Aggregate Reactivity (AAR) Workshops for engineers and practitioners reference manual* U S Department of Transportation
[18] Folliard KJ, Thomas MDA, Fournier B, KE Kurtis and Ideker JH 2006 *Interim recommendations for the use of lithium to mitigate or prevent alkali-silica reaction (ASR)* Publication no. FHWA-HRT-06-073
[19] MDA Thomas, Folliard KJ, Fournier B, Rivard P and Drimalas T 2013 *Methods for evaluating and treating ASR-affected structures: results of field application and demonstration Projects. Volume I: Summary of findings and recommendations* Publication no. FHWA-HIF-14-0002
[20] Owsiak Z 2015 Internal Corrosion of Concrete (Kielce, Kielce University of Technology)
[21] Feng X 2008 *Effects and mechanism of lithium nitrate on controlling alkali-silica reaction*, (Canada: University of New Brunswick)
[22] Leemann A, Lörtscher L, Bernard L, Le Saout G, Lothenbach B and Espinosa-Marzal RM 2014 *Cem.Concr.Res.* **59** pp 73-86
[23] Fournier B, Berube MA, Folliard KJ and Thomas MDA 2010 *Report on the Diagnosis, Prognosis, and Mitigation of Alkali-Silica Reaction (ASR) in Transportation Structures* Publication no. FHWA-HIF-09-004
[24] Thomas MDA, Fournier B, Folliard KJ, Ideker JH and Resendez Y 2007 *The use of lithium to prevent or mitigate alkali-silica reaction in concrete pavements and structures* Publication no. FHWA-HRT-06-133
[25] Folliard KJ, Thomas MDA and Kurtis KE 2003 *Guidelines for the use of lithium to mitigate or prevent ASR* Publication no. FHWA-RD-03-047, chapter 5
[26] Sakaguchi Y, Takamura M, Kitagawa A, Hori T, Tomosawa F and Abe M 1989 *Proc. 8th ICAAR* (Kyoto) pp 229-234
[27] Stokes DB, Pappas J, Thomas MDA and Folliard KJ 2002 *Proc. of 6th CANMET/ACI International Conference on Durability of Concrete* (Thessaloniki, Greece) Supplementary Papers, pp 631–642
[28] Johnston DP 2001 *Mitigation of potential alkali-silica reactivity using lithium* SHRP Strategic Highway Research Program Concrete and Structures ASR Showcase Test and Evaluation Project 34
[29] Folliard KJ, Thomas MDA, Ideker JH, East B and Fournier B 2008 *Proc. 13th ICAAR* Trondheim, Norway
[30] Thomas MDA, Folliard KJ, Fournier B, Rivard P, Drimalas T and Garber SI 2013 *Methods for evaluating and treating ASR-affected structures: results of field application and demonstration projects. Volume II: Details of field applications and analysis* Publication no. FHWA-HIF-14-0003
[31] Drimalas T, Folliard K, Thomas M, Fournier B and Bentivegna A 2012 *Proc. 14th ICAAR* (Austin, Texas, USA)
[32] Whitmore D and Abbot S. 2000 *Inter. Conf. on Alkali-Aggregate Reaction*, ed. M-A. Bérubé et al. (Quebec City, Canada) pp 1089–1098
[33] Silva De Souza LM 2016 *Electrochemical lithium migration to mitigate alkali-silica reaction in existing concrete structures* (The Netherlands: Delft University of Technology)
[34] Folliard KJ, Thomas MDA, Ideker JH, East B and Fournier B *Proc. 88th Annual Transportation Research Board Meeting* (Washington D.C., USA) Paper#09-2685
[35] Johnson DK, Warnock RL, Schnidler AK and Barnes RW 2014 *Efectiveness of silane in mitigating alkali-silica reaction in the BIBB Graves bridge* Report FHWA/ALDOT 930-802 (Alabama Department of Transportation) p 243
[36] Diamond S, Barneyback RS and Struble L J 1981 *Proc. 5th Int. Conf. on Alkali-Aggregate Reaction in Concrete* (Cape Town, South Africa) p 252/22
[37] Deschenes Jr RA 2017 *Mitigation and evaluation of alkali-silica reaction (ASR) and freezing and thawing in concrete transportation structures* (University of Arkansas, Fayetteville)

Design and verification of bituminous mixtures with the increased content of reclaimed asphalt pavement

Wojciech Bańkowski[1], Jan Król[2], Karol Gałązka[3], Adam Liphardt[2], Renata Horodecka[1]

[1] Road and Bridge Research Institute
[2] Warsaw University of Technology
[3] Budimex S.A.

wbankowski@ibdim.edu.pl

Abstract. Recycling of bituminous pavements is an issue increasingly being discussed in Poland. The analysis of domestic and foreign experience indicates a need to develop this technology in our country, in particular the hot feeding and production technologies. Various steps are being taken in this direction, including research projects. One of them is the InnGA project entitled: "Reclaimed asphalt pavement: Innovative technology of bituminous mixtures using material from reclaimed asphalt pavement". The paper presents the results of research involving the design of bituminous mixtures in accordance with the required properties and in excess of the content of reclaimed asphalt permitted by the technical guidelines. It presents selected bituminous mixtures with the content of RAP of up to 50% and the results of tests from verification of industrial production of those mixtures. The article discusses the details of the design process of mixtures with a high content of reclaimed asphalt, the carried out production tests and discusses the results of tests under the verification of industrial production. Testing included basic tests according to the Polish technical requirements of WT-2 and the extended functional testing. The conducted tests and analyses helped to determine the usefulness of the developed bituminous mixtures for use in experimental sections and confirmed the possibility of using an increased amount of reclaimed asphalt up to 50% in mixtures intended for construction of national roads.

Keywords: hot recycling, reclaimed asphalt pavement, RAP, bituminous mixture, asphalt mixing plant

1. Introduction

Nowadays, recycling is an extremely important and common issue. It covers various areas of life and human activity. Protection of the environment, waste reduction and saving of natural resources are the main objectives of recycling. Recycling of asphalt pavements fits very well in such established goals. This technology is increasingly used. The reclaimed asphalt pavement is a valuable material and its composition contains mainly mineral aggregate and asphalt binder. It is the material obtained from milling of bituminous pavements, removal of the old pavement or originating from surplus production [1,2].

Various methods of asphalt pavement recycling are known: both in hot and cold technology [3, 4, 5, 6]. Among the many known methods, the hot recycling technology in the mixing plant is considered to be the most effective and using the value of the recovered raw materials. The hot recycling technology has been known and used for years worldwide, as well as in Poland [7]. However, it is necessary to differentiate between two variations of this technology. In the first of them the reclaimed asphalt is cold-fed directly to the mixer. In the second, more advanced and efficient one, the reclaimed asphalt is preheated in a special system. Hot recycling is not widely used in Poland, and our plants are dominated by the cold feed method. However, the use of this method is associated with some drawbacks, among which it is necessary to mention restrictions in the amount of reclaimed asphalt added to the bituminous mixture. In many countries, it is quite commonly used and striving for the maximum use of RAP [8, 9, 10].

Application of the recycling technology with hot feeding in Poland is very limited. The causes of this situation are numerous, including a lack of appropriate technical guidelines and recommendations, experience, equipment park, consents from the project owner, availability of a good quality of RAP and other [11, 12]. Taking into account the properties of bituminous mixture with RAP, it must be ensured that it met the technical requirements, and the application of RAP would not deteriorate properties of the mixture. Certainly, these are the assumptions to ensure that the made bituminous layers will be characterised by the appropriate durability. The first factor which has a significant impact is the uniformity of the RAP [13], and in particular the contents and properties of asphalt binder. The larger the contents of RAP in the bituminous mixture, the more important is the factor. Another very important factor is the ageing process covering the bitumen layers, and the asphalt binder [14, 15]. Therefore, the application of reclaimed asphalt, i.e. the material after long-term and short-term ageing may raise concerns as to the characteristics of the bituminous mixture with addition of that material. A potentially increased rigidity and better resistance to permanent deformations may be expected [16]. On the other hand, those properties, where greater stiffness, and thus the resistance to cracking and fatigue resistance is not favourable, may be deteriorated. In order to verify these characteristics, it is necessary to conduct performance tests. If necessary, regenerative (rejuvenator) additives should be used, the mission of which is to improve the viscoelastic properties [17, 18]. The fatigue life is the primary characteristic determining the service life of the pavement [19]. A change of rheological properties of asphalt binder due to ageing can cause deterioration of the fatigue life of mixtures with RAP [20, 21]. There are also studies showing an opposite effect [22].

The results presented in this publication are part of the research work carried out within the framework of the "InnGA" project entitled: "Reclaimed asphalt pavement: Innovative technology of bituminous mixtures using material from reclaimed asphalt pavement". The project is carried out by the scientific and industrial consortium composed of: The Road and Bridge Research Institute (leader), the Warsaw University of Technology and BUDIMEX S.A. The project is subsidised from the funds of the Operational Programme Innovative Economy under the "INNOTECH" programme in the IN-TECH programme path. The primary goal of the project is the development of bituminous mixtures with a possibly maximum content of reclaimed asphalt, while maintaining the corresponding properties of the mixture. Due to limitations of the volume of the text, the paper discusses designing of the selected bituminous mixtures (SMA11 and AC16W) with the content of reclaimed asphalt of up to 50% and the results of tests from verification of industrial production. Testing included basic tests according to the technical requirements of WT-2 [23] and the extended functional testing. The conducted tests and analyses helped to determine the usefulness of the developed bituminous mixtures for use in experimental sections and confirmed the possibility of using an increased amount of RAP up to 50% in mixtures intended for construction of national roads.

2. Design of bituminous mixtures with reclaimed asphalt

The research works carried out within the framework of the project included asphalt concrete for all courses, asphalt concrete type ACWMS and the SMA mixture. For the purpose of designing the mixtures of asphalt concrete (AC) and stone mastic asphalt (SMA), two materials were used with an established quality, recovered from the reclaimed asphalt pavements (RAP) using selective milling. RAP-AC from milling of the binder course was used for asphalt concrete mixes, while for the mixture of stone mastic asphalt RAP-SMA from milling of the SMA wearing course was used. Properties of the recovered materials and their composition are shown in table 1.

Table 1. Properties of the materials recovered from the RAP.

Property		Unit	RAP-SMA (SMA<15 years)	RAP-AC (AC-W< 20 years)
Binder properties	Binder content	% (m/m)	5.9	4.3
	Penetration	0.1 mm	51	31
	Softening point acc. to ring and ball method	°C	55.4	62.4
	Fraass breaking point	°C	-16	-12
	Elastic recovery	%	75	43
Grading	Aggregate particles below 0.063 mm	% (m/m)	13.7	10.1
	Aggregate particles from 0.063 to 2 mm	% (m/m)	14.9	22.4
	Aggregate particles above 2 mm	% (m/m)	71.5	67.5
Properties of aggregates	Mineral mix density	[Mg/m^3]	2.703	2.635
	Percentage of crushed and broken surfaces in fine aggregate — flow indicator	seconds	33	28
	Percentage of crushed and broken surfaces in large aggregate	-	$C_{100/0}$	$C_{100/0}$

Although the materials recovered from the RAP have not been subjected to granulation, an increased content of the filler and fine aggregate was found in the particle size distribution of mineral materials. The conducted petrographic studies have shown that about 5% of the filler and fine aggregates comes from fragmentation of coarse-grain fractions. The sand fraction of the SMA mixture showed the presence of basalts and biotites from the stone mastic fraction, while the sand fraction of the AC mixture contained flakes of carbonate aggregate. The evaluation of properties of asphalt binders recovered from RAP showed that they were not stiffened as a result of ageing. The binder recovered from the SMA mixture after many years of trafficking in the road pavement shows elastic recovery at the level of 75%, which indicates that the polymer did not undergo degradation. The elastic recovery of the binder recovered from the asphalt concrete at the level of 43% proves partial presence of polymer-modified bitumens in that material, most probably originating from the wearing course.

In order to determine the maximum content of material recovered from the reclaimed asphalt pavement (RAP) in the new mixtures, five mixtures of asphalt concrete and stone mastic asphalt with a variable content of RAP were designed. The grading of the designed mixtures containing RAP is shown in figure 1, while figure 2 presents the breakdown of air voids in the tested mixtures. The samples were compacted in the Marshall hammer: asphalt concrete 2 x 75 blows, SMA 2 x 50 blows.

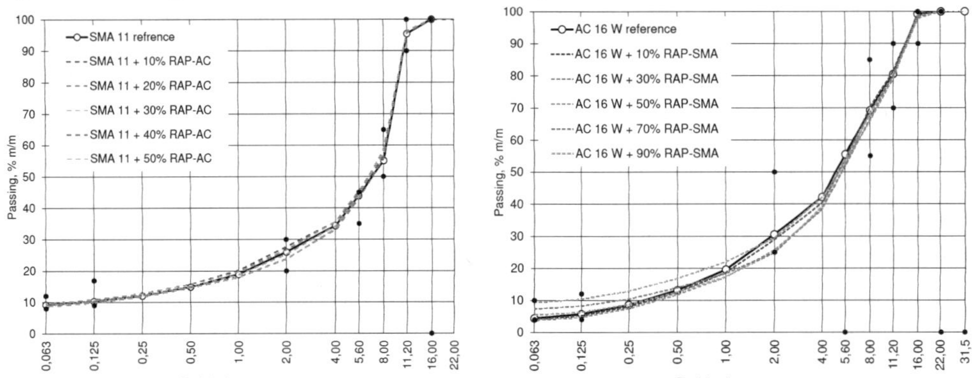

a) b)

Figure 1. Grading curves of bituminous mixtures with various RAP content: a) SMA 11 + 0–50% RAP mixtures, b) AC 16 W + 0–90% RAP mixtures.

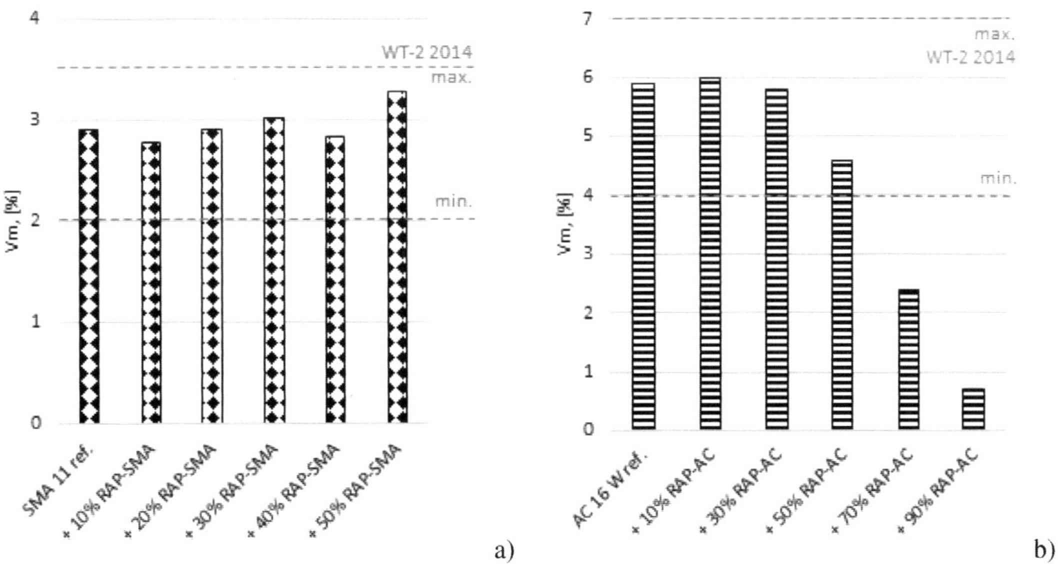

Figure 2. Comparison of air voids content of bituminous mixtures with various RAP content: a) SMA11 mixtures, b) AC16W mixtures.

The mixtures were designed so grading curves of mixtures containing RAP were possibly similar to the reference mixture. With the increase in the amount of RAP in the mixtures the amounts of filler and fine aggregates were gradually reduced. The necessity of their limitation was caused by the grading of the mineral mixture of the material recovered from the bituminous pavement containing relatively large amounts of fine fractions <2 mm. In the case of addition of RAP-AC to the AC16W at the level of 30%, it was necessary to completely eliminate the filler and crushed and not washed sand constituting 50% of the sand fraction in the reference mixture. In AC16W mixture containing 90% of RAP-AC, only one virgin aggregate in the form of 11/16 stone chippings was used. In the SMA11 mixture containing 50% of RAP-SMA, it was also necessary to reduce the amount of filler to the level of 1% and the amount of sands by half in relation to the reference mixture. In the case of mixtures containing 40% and 50% of the reclaimed asphalt admixture of 4/8 melaphyre rock was abandoned due to a high content of that fraction in the reclaimed asphalt pavement (the content of 4/8 fraction was about 42%).

The obtained results of air voids content in the AC16W mixture have shown that with the increase in the share of RAP-AC, voids content in the mixtures decreases. This phenomenon is associated with a high proportion of fine fractions in the mineral mixture of the reclaimed asphalt pavement (RAP) that fills the air voids in the mixture. Along with the increase of RAP, an increasing amount of fine fractions is added to the mixture; however, at the same time it is not possible to correct the grading using large aggregates. Figure 2 marks the range of the air voids content required according to WT-2 2014. It was found that the maximum share of the analysed reclaimed material (RAP) in the AC16W mixture allowing for meeting the requirements for the air voids content is 50% and with the favourable grading of RAP this amount could be potentially increased even up to 70%. The results of testing of SMA11 revealed that as opposed to the AC16W mixture, the use of RAP in the amount of 0–50% does not have to result in decrease in the air voids content. This is due to the fact that the material from milling of the wearing course SMA (RAP-SMA) used in this mixture is characterised by a much smaller (by about 18%) content of the fine fractions <2 mm in the mineral mixture in comparison to RAP-AC used in the AC16W mixture. All of the designed SMA mixtures with polymer-modified bitumen PMB 45/80-55 are characterised by similar contents of air voids within the limits of 2.8–3.3%.

When assessing the designed mixtures, it can be concluded that there is a possibility of designing bituminous mixtures with an increased content of RAP in the amount of 20% to 50% that meet the current technical requirements. Grading of RAP is the main factor limiting the possibility of introducing greater amounts (above 50–70%) of the pavement to mixtures. Too much content of the filler fraction and fine aggregate significantly limits the possibility to design mixtures with the adequate air voids content, which in turn determines properties of the mixtures.

3. Production technology

The mixtures were produced in the asphalt mixing plant of the company Budimex S.A. near Jędrzejów (fig.3). The plant is one of the four plants in Poland equipped with the so-called black mixing drum for feeding hot reclaimed asphalt [21].

Figure 3. Asphalt plant with RAP drum on the top and view of RAP pile.

The reclaimed asphalt is subjected to the production process in a separate process line. A vibrating feeder conveys RAP to the elevator, which in turn transports the material to the black drum. There it is heated in temperature above 100°C (figure 4).

Figure 4. View of the inside of the black drum before and during production.

In the heating process the reclaimed asphalt is deprived of moisture and fragmented. Then RAP obtains the temperature within the range of 120–140°C. As a result of the carried out process, reclaimed asphalt changes its structure to become the material reminiscent of a loose bituminous

mixture (figure 5). Further the heated material is transported to the bunker, where after weighing the proper amount is fed directly to the mixer.

Figure 5. View of the reclaimed asphalt after passing the black drum.

The increased share of RAP in the mixtures due to the applied "hot method" did not require excessive heating of aggregates in order to balance the temperatures allowing for obtaining a bituminous mixture with a minimum production temperature resulting from the type of binder used. Figure 6 shows process temperatures at the production stage with the share of 30 and 50% of RAP, respectively.

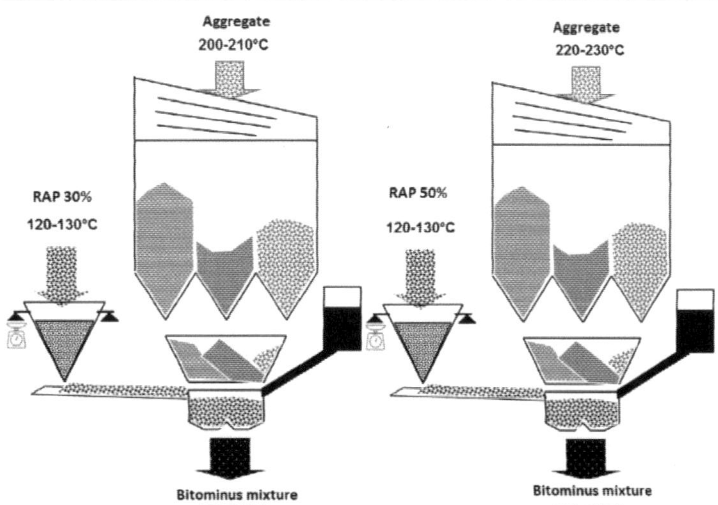

Figure 6. Process temperature in production of MMA with 30% and 50% share of RAP.

The production process used two types of reclaimed asphalt pavement: 45 RAP 0/16 — AC from the binder course and 45 RAP 0/11 — SMA from the wearing course. The batches of reclaimed asphalt were tested for homogeneity. The requirements of WT-2 2014 [23] were adopted as the criterion for assessment of homogeneity, whereas the current requirements do not allow for the use of recycled materials for the wearing course. Therefore, there are no requirements for mixtures for that course. Therefore, the requirements as for the binder course were assumed as the criterion for evaluation of the reclaimed asphalt pavement intended for wearing courses. The results of homogeneity showed compliance with the requirements. Table 2 presents the results of the homogeneity of the reclaimed asphalt pavement RAP-SMA.

Table 2. Evaluation of homogeneity 45 RAP 0/11 — SMA.

Property a_i	Unit	Span of results from n samples			Result range a_i	Requirements of WT-2 2014
Softening point of recovered binder;	[^{o}C]	59.4	-	63.8	4.4	8.0
Contents of soluble binder (m/m);	[%]	5.8	-	6.2	0.4	1.0
Aggregate with grading of < 0.063 mm, (m/m);	[%]	13.4	-	15.0	1.6	10.0
Aggregate with grading from 0.063 to 2.0 mm, (m/m);	[%]	23.5	-	30.0	6.5	16.0
Aggregate with grading of > 2.0 mm, (m/m);	[%]	55.8	-	61.5	5.7	16.0

4. Verification of production of bituminous mixtures with reclaimed asphalt

The compositions of bituminous mixtures with admixture of RAP developed under laboratory conditions were used to produce trial mixes — production verification in the bituminous mixture mixing plant equipped with a parallel drum system. The aim of this work was to select the parameters and settings of the production process, to determine the working recipe and determine the properties of the mixtures in the range of basic and performance properties. The process of production of mixtures with reclaimed asphalt under laboratory conditions is in principle similar to heating and feeding of ingredients. However, it is obvious that both the production technology and the scale are different. It was therefore considered that it is important to confirm the properties of the mixtures with the increased content of RAP based on testing of mixtures taken from production trials. It was assumed that this will be the basis for the selection of the optimum compositions of mixtures to be used in experimental sections. The production trials and tests were conducted on five types of bituminous mixtures listed in table 3.

Table 3. Bituminous mixtures with reclaimed asphalt subjected to production tests.

Mixture	Reclaimed asphalt	Contents of RAP
AC22P 35/50 KR 3-7	45 RAP 0/16 AC	50%
AC16W PMB 25/55-60 KR 3-4	45 RAP 0/16 AC	50%
ACWMS16W PMB 25/55-60 KR 3-7	45 RAP 0/16 AC	50%
AC11S PMB 45/80-55 KR 3-4	45 RAP 0/11 SMA	15%
SMA11 PMB 45/80-55 KR 3-7	45 RAP 0/11 SMA	30%

The optimum setting of production parameters and the working recipe required several repetitions for each mixture. The task included production of 5 mixes from the SMA11 mixture, 1 mix of AC11S, 5 mixes of AC16W for binder course, 7 mixes of ACWMS16 (high modulus asphalt concrete) and 6 mixes of AC22P for base course. After each production trial the mixture was preliminarily evaluated in the field laboratory in order to determine compliance with the composition of the developed recipes, as well as in terms of compliance with the requirements of WT-2 2014. Mixtures with positive results were submitted in the laboratories of research units in order to perform further tests.

Table 4 summarises the results of testing of SMA11 mixture, which fulfilled the requirements of WT-2 in terms of basic properties and composition. The exception is the result of the wheel tracking test in a small apparatus, where the value of the proportional depth of the rut was exceeded. The results within the range of 8–8.9% were obtained for the permissible value of 7%. However, performance tests confirmed a good resistance of the mixture to permanent deformations using a large rutting tester. Technical requirements do not provide the required values for this method. However, previous technical documents that take this test method into account, provided the maximum value of the rut at the level of 10% after 10,000 cycles. It is worth stressing the very good the result of resistance to low-temperature cracking. The cracking temperature at the level of -29°C is a typical result for mixtures without reclaimed asphalt with this type of polymer-modified binder. Despite the application of the reclaimed asphalt, a high resistance to water and frost was also obtained. Both features are particularly relevant for the durability of the wearing course.

Table 4. Basic and performance properties of the SMA 11 45/80-55 mixture.

Property		Unit	Laboratory: Road and Bridge Research Institute	Laboratory: Warsaw University of Technology	WT-2 2014 KR5-7
Mix bulk density		g/cm^3	2.379	2.357	-
Mix density		g/cm^3	2.437	2.429	-
Air voids		%, v/v	2.4	3.0	2.0–3.5
Resistance to water	ITSR	%	-	97	≥ 90
Resistance to permanent deformations, method B	WTS_{AIR}	mm/1000 cycles	0.09	0.08	≤ 0.1
	PRD_{AIR}	%	8.0	8.9	≤ 7.0
Rutting resistance, large apparatus, 30 thousand cycles	P	%	7.1	-	-
Stiffness (10 Hz, 10°C), MPa	E	MPa	7,853	-	-
Low-temperature cracking resistance (TSRST)	T	°C	-29.0	-	-
	σ	MPa	3.8	-	-

Table 5. Basic and performance properties of the AC16W PMB 25/55-60 mixture.

Property		Unit	Laboratory: Road and Bridge Research Institute	Laboratory: Warsaw University of Technology	WT-2 2014 KR5-7
Mix bulk density		g/cm^3	2.417	2.390	-
Mix density		g/cm^3	2.518	2.509	-
Air voids		%, v/v	4.0	4.7	4.0–7.0
Resistance to water	ITSR	%	-	77	≥ 80
Resistance to permanent deformations, method B	WTS_{AIR}	mm/1000 cycles	0.18	0.14	≤ 0.15
	PRD_{AIR}	%	7.7	7.1	≤ 7.0
Rutting resistance, large apparatus, 30 thousand cycles	P	%	3.4	-	-
Fatigue life, 10°C, 10 Hz	ε_6	µm/m	133	-	-
Stiffness (10 Hz, 10°C), MPa	E	MPa	14 669	-	-
Low-temperature cracking resistance (TSRST)	T	°C	-20.4	-	-
	σ	MPa	3.5	-	-

The AC16W 25/55-60 50%RAP mixture met the requirements in terms of basic properties and composition. The exception is the result of the wheel tracking test in a small apparatus, where the value of the proportional depth of the rut was exceeded. The results within the range of 7.1–7.7% were obtained for the permissible value of 7%. However, performance tests indicated a good resistance of the mixture to permanent deformations using a large rutting tester. The result of the proportional rut depth was 3.4%. The previous technical documents that take this test method into account, provided the maximum value of the rut at the level of 5% after 30,000 cycles. The technical requirements do not provide requirements for other performance characteristics. By evaluating the fatigue resistance, stiffness and resistance to low-temperature cracking, in general it may be concluded that the mixture does not demonstrate properties which could be expected from this type of mixture with polymer-modified bitumen. This is related to the ratio between the "old" asphalt binder and the "fresh" asphalt binder in the bituminous mixture. In this case, the reclaimed asphalt is approximately 75% of the total asphalt in the mix. A small part of the fresh polymer-modified binder is not decisive in terms of resistance to low-temperature cracking or fatigue. The obtained values are at the characteristic level for mixtures with 20/30 or 35/50 unmodified asphalt binder.

5. Evaluation of the results and a selection for test sections

By using experience from the laboratory designing of bituminous mixtures with the reclaimed asphalt pavement (RAP), bituminous mixtures for the test sections were designed.

The main criterion for evaluation of the suitability of bituminous mixtures in the wearing course was the air voids content and resistance to water (ITSR), completed with resistance to low temperature (TSRST). The limit (maximum) content of RAP was assumed at the level of:

- 30% of RAP-SMA in SMA mixtures and the recommended use of polymer-modified bitumens,
- 15% RAP-SMA in AC mixtures and the recommended use of polymer-modified bitumens.

The main criteria for evaluation of the suitability of bituminous mixtures in the structural pavement courses, i.e. the binder course and the base course were the resistance to permanent deformations (rutting) and fatigue life in the case of mixtures intended for the traffic category KR5-7. The limit (maximum) content of RAP was assumed at the level of 50% of RAP-AC and the recommended use of polymer-modified bitumens. Table 6 presents the basic properties of the designed bituminous mixtures for the test sections. The purpose of the test sections is reevaluation of the produced mixtures and the constructed courses, which in turn has to allow for selection of solutions to be used in the section of the national road.

Table 6. Properties of the designed bituminous mixtures for the test sections.

Layer	Type of bituminous mixture and traffic category	Air voids %	Resistance to water (ITSR), %	Rutting resistance small apparatus	
				WTS_{AIR} mm/1000 cycles	PRD_{AIR} %
Wearing course	AC11S KR 3÷4	3.1	96	0.05	5.8
	SMA11 KR 5÷7	2.7	95	0.08	7.7
Binding course	AC16W KR 3÷4	5.0	99	0.14	6.2
	ACWMS16 KR 5÷7	3.4	90	0.10	5.0
Binder course	AC 22 KR 3÷7	5.5	86	0.10	5.4

6. Summary

The paper indicates that it is possible to design bituminous mixtures meeting the current national technical requirements (WT-2 2014) with the simultaneous exceeding of the maximum permissible content of RAP. Grading of RAP is the main factor limiting the possibility of introducing greater amounts (above 50–70%) of the pavement to mixtures. Too much content of the filler fraction and fine aggregate significantly limits the possibility to design mixtures with the adequate air voids content, which in turn determines properties of the mixtures. It was found that it is possible to use RAP in mixtures to be used in the wearing course. In the case of SMA, it is necessary to use the material recovered from the selective milling of the SMA wearing course. In the case of mixtures with continuous grading for the binder course and the base course, it is possible to use up to 50% of RAP from milling of the lower courses of bituminous pavements and the main limitation in increase of RAP content is its grading. Successful production trials were conducted in the mixing plant equipped with an additional drum for heating of reclaimed asphalt. The further stage of the project provided for construction of test sections and experimental sections, which will be the subject of subsequent publications.

Acknowledgements
The article was created as part of the project entitled: "Reclaimed asphalt pavement: The innovative technology of bituminous mixtures using material from recycling of asphalt pavement" under the INNOTECH activity, In-Tech programme path and is co-funded from the European Regional Development Fund within the framework the Operational Programme Innovative Economy under the banner of "European Funds — for the development of innovative economy".

References
[1] EN-13108:2016, „Bituminous mixtures – Material specifications – Part 8: Reclaimed asphalt"
[2] Alenowicz J 2017 *Czym jest obecnie destrukt asfaltowy? Zmiany w normie EN-13108-8*, (Drogownictwo 9/2017), pp 281-284
[3] Iwański M, Buczyński P, Mazurek G 2016 *Optimization of the road binder used in the base layer in the road construction* (Construction and Building Materials vol **125**) pp 1044-1054.
[4] Dołżycki B, Jaczewski M, Szydłowski C 2017 *The long-term properties of mineral-cement-emulsion mixtures* (Construction and Building Materials vol **156**)
[5] Kukiełka J 2014 *Recykling głeboki na zimno nawierzchni asfaltowych dróg samorządowych*, (MATERIAŁY BUDOWLANE – vol **12**)
[6] Sybilski D 2011 *O potrzebie stosowania destruktu asfaltowego w Polsce* (Drogownictwo 1/2011)
[7] Szyller A, Król J, Bańkowski W 2017 *Współczesne doświadczenia ze stosowania recyklingu na gorąco w wytwórni mieszanek mineralno-asfaltowych w Polsce* (Nawierzchnie asfaltowe 2/2017)
[8] Al-Qadi I L, Elseifi M A, Carpenter S H 2007 *Reclaimed Asphalt Pavement – A Literature Review* (Research Report FHWA-ICT-07-001), Illinois Center for Transportation, University of Illinois at Urbana-Champaign
[9] Zaumanis R, Mallick R, Frank E 2016 *100% hot mix asphalt recycling: challenges and benefits*, (Transportation Research Procedia vol.**14**), pp 3493–3502
[10] Izaks R., Haritonovs V., Klasa I., Zaumanis M. 2015 *Hot mix asphalt with high RAP content,* (Procedia Engineering **114**, pp 676-684
[11] Bańkowski W, Sybilski D, Król J, Kowalski K, Radziszewski P, Skorek P 2016 *Wykorzystanie destruktu asfaltowego – konieczność i innowacja*, (Budownictwo i Architektura vol.**15** (1)), pp 157-167
[12] Michalski W 2014 *Sposoby dozowania granulatu* (Nawierzchnie asfaltowe 2/2014)
[13] Król J, Włodarski P, Jackowski Ł 2014 *Właściwości mieszanek mineralno-asfaltowych ze zwiększoną ilością granulatu asfaltowego*, (Drogownictwo 11/2014)
[14] Radziszewski P 2007 *Zmiany właściwości lepkosprężystych lepiszczy modyfikowanych i mieszanek mineralno-asfaltowych w wyniku procesu starzenia* Wydawnictwo Politechniki Białostockiej, 2007.
[15] Woo W J, Chowdhury A, Glover C 2008 *Field aging of unmodified asphalt binder in three texas long-term performance pavements* (Transportation Research Record: Journal of the Transportation Research Board) pp 15–22
[16] Mogawer W, Bennert T, Daniel J, Bonaquist R, Austerman A, Booshehrian A 2014 *Performance characteristics of plant produced high RAP mixtures* (Road Materials and Pavement Design)
[17] Kowalski K, Król J, Bańkowski W, Radziszewski P, Sarnowski M 2017 *Thermal and Fatigue Evaluation of Asphalt Mixtures Containing RAP Treated with a Bio-Agent*, (Applied Sciences, vol.**7**)
[18] Tran N, Taylor A, Willis R 2012 *Effect of rejuvenator on performance properties of HMA mixtures with high RAP and RAS contents* (NCAT Report 12-05 Alabama)
[19] Mackiewicz P 2016 *Trwałość zmęczeniowa mieszanek mineralno-asfaltowych stosowanych w nawierzchniach drogowych*", Oficyna Wydawnicza Politechniki Wrocławskiej, Wrocław
[20] Walubita I, Epps M, Jung S, Glover C, Park E, Chowdhury A, Lytton R 2008 *Comparison of fatigue analysis approaches for two hot mix asphalt concrete mixtures* (FHWA/TX-05/0-4468-2) Texas Transportation Institute.
[21] Daniel J, Gibson N, Tarbox S, Copeland A, Andriescu A 2013 *Effect of long-term ageing on RAP mixtures: laboratory evaluation of plant-produced mixtures* (Road Materials and Pavement Design)
[22] Tapsoba N, Benedetto H, Sauzeat C, Baaj H, Ech M 2013 *Behaviour of asphalt mixtures containing reclaimed asphalt pavement and asphalt shingle* (Road Materials and Pavement Design)
[23] WT-2 2014 część 1 – „Mieszanki mineralno-asfaltowe - wymania techniczne", GDDKiA 2014

Influence of bitumen type on cracking resistance of asphalt mixtures used in pavement overlays

P Jaskula[1], C Szydlowski[1] and M Stienss[1]

[1] Gdansk University of Technology, Faculty of Civil and Environmental Engineering, Poland

E-mail: piotr.jaskula@pg.edu.pl

Abstract. Cracking is one of the predominant distresses occurring in flexible pavements, especially in old pavements that were rehabilitated with an asphalt overlay. In such cases asphalt mixtures should be designed to ensure high resistance to reflective cracking because new asphalt layers are exposed to existing cracks of the old pavement. The nature of these cracks can be various (transverse, longitudinal as well as crazy cracking). One factor that minimizes this type of distress is the proper mix design process, which should involve selection of specific bitumen binder and mineral mix gradation. However, still there is no universally adopted laboratory test method that would allow to clearly assess resistance of asphalt mixtures to reflective cracking. This paper describes the usage of one of the devices developed to test asphalt mixtures in terms of such distress – Texas Overlay Tester. For this test, samples prepared in laboratory conditions (i.e. compacted with the use of Superpave Gyratory Compactor) as well as obtained in the field (by core drilling) can be used. The results are obtained not only quickly and easily, but also with sufficient repeatability. The described method characterizes both crack initiation and crack propagation properties of asphalt mixtures. In this work one type of mineral mixture was tested with 4 different types of bitumen (one neat bitumen, two ordinary polymer-modified and one polymer-modified with high polymer content). For selected cases extra additives (rubber and loose fibres) were also tested. In total, six asphalt mixtures were tested. A ranking of the used binders was created on the basis of the results in order to conclude which bitumen would ensure the best performance characteristics in terms of reflective cracking. The results have clearly shown that deliberate choice of the binder used in the asphalt mixture for the overlay will significantly improve its reflective cracking resistance or even fatigue resistance.

1. Introduction

The risk of cracking is one of the most important issues that can affect pavement condition and performance during its operation life. This problem is often considered as significant mainly for old pavements (especially rigid ones, with concrete wearing courses) that are due to be rehabilitated by placing new asphalt overlay, but undesirable cracking can also occur in relatively new pavements that contain stiff, rigid base courses, e.g. aggregates or soils bound with hydraulic binders. It should be emphasised that the issue of cracking susceptibility of asphalt mixtures also influences the fundamental pavement characteristics, such as fatigue life and ageing-related performance. Even new pavements, which otherwise would be structurally sound and long lasting, when constructed with asphalt mixtures that are stiff and highly susceptible to cracking will show premature distress, often much earlier than anticipated.

There are several approaches to mitigating the problem of pavement cracking. In situations when there is a risk that cracks from the existing old pavement or hydraulically bound base course will reflect in the new asphalt layers placed over the top, the following techniques are often considered and used: (1) reinforcing of asphalt layers with geosynthetics of different nature (geogrids, geonets, geocomposites, glass grids, fabrics) or metal mesh, (2) introducing stress relief layers (SAMI, chip sealing, very open asphalt mixture), (3) controlling of reflective cracking by sawing the asphalt overlay [28]. More radical technologies, but used only when concrete pavement is rehabilitated, involve elimination of movements in the concrete layers by fracturing concrete slabs into smaller fragments or rubblizing them into coarse aggregate fraction.

Another approach to increasing general pavement resistance to cracking is to enhance anti-cracking characteristics of the asphalt mixture. This can be achieved by various means: by selecting a certain type of the binder and its amount, type and gradation size of the aggregate mixture and also by incorporating different additives into the binder or asphalt mixture as a whole, e.g. polymers, recycled rubber, fibres or natural bitumen [3, 10, 11, 15, 17, 27].

Obtaining asphalt mixture that would possess higher resistance to cracking depends not only on the mix design process alone, but must also incorporate proper test method [27], allowing to differentiate asphalt mixtures in terms of cracking characteristics. For many years efforts were mainly put into development of test methods concerning resistance to permanent deformation (Wheel Tracking Test, Hamburg Wheel Tracking Test, Asphalt Pavement Analyzer), fatigue life (4-point bending test, 2-point bending test) or low-temperature properties (TSRST, ATCA). Development of test methods dedicated specifically for cracking characteristics was initiated only in recent years, when it transpired that the current asphalt mixtures, which are optimized for good rutting performance and have a high RAP content, can be more prone to cracking than the mixtures used previously. Nowadays, there are several methods of measuring asphalt mixture cracking susceptibility: (1) Semi Circular Bending test at different temperatures (SCB), (2) Disc Shape Compact Tension test (DCT), (3) Bending Beam Fatigue (BBF), (4) Indirect Tensile test (IDT) and (5) Texas Overlay Tester (TxOT). Each of the mentioned methods has its own specific characteristics and advantages, while none of them is widely adopted. Nevertheless, taking into consideration their different specific criteria of assessing cracking resistance of asphalt mixtures, TxOT method seems to be quite promising.

The first version of the Texas Overlay Tester was developed by Germann and Lytton [9]. The method at that stage used rectangular prismatic beams and modelled horizontal displacements resulting from thermal stresses that occur in overlaid old pavements. It was developed to assess resistance of asphalt mixtures to reflective cracking that can occur in new layers placed on existing pavements. The influence of fabric was also investigated. In later years the method was modified by Zhou and Scullion [24] to accommodate other shapes of specimens, so that it would be possible to prepare them more easily in the laboratory or core them in the field. At the moment, overlay tester is routinely used by Texas DOT which has developed its own testing procedure Tex248-F [19, 20], but a similar international standard ASTM WK26816 [2] does also exist. In Poland there are still no publications or research reports available concerning the use of the TxOT in Polish conditions (Polish types of mixtures, common bitumens etc.).

2. Objectives and scope of the research

The objective of the research was to verify the feasibility of the TxOT method in assessing cracking characteristics of the asphalt mixture prepared with binders routinely used in Poland, especially in terms of different bitumen types and special additives incorporated in the mixture composition.

The scope of the research included testing of one type of asphalt mixture with different bitumens and selected additives. The asphalt mixture chosen for the research was specifically prepared for overlay applications (such as levelling course placed on old concrete asphalt pavement before placing final wearing course) and was based on typical asphalt mixture AC 11W for traditional binder course. Four different bitumens included one neat bitumen, two polymer-modified bitumens (soft and hard) and one highly polymer-modified bitumen. All tested bitumens are normally used in Poland in asphalt mixtures for binder course. In case of the mixture with neat bitumen two different additives were also used: crumb rubber and polymer fibres.

3. Tested materials

3.1. Bitumens

Four different bitumens were selected for this study: neat bitumen 35/50, two polymer SBS-modified bitumens: 10/40-65, 25/55-60 and highly SBS-modified bitumen 25/55-80. All bitumens were produced in a Polish refinery. Properties and performance grade (PG) of bitumens (acc. to AASHTO M320 [1]) used in this research are shown in Table 1.

Table 1. Properties of bitumens.

Property		Type of bitumen				
		35/50	**35/50 R**	**10/40-65**	**25/55-60**	**25/55-80**
Penetration at 25°C, 0.1 mm, acc. to PN-EN 1426	Original	45	39	26	34	50
	RTFOT	28	24	21	25	41
R&B Temperature, °C, acc. to PN-EN 1427	Original	53.0	60.7	68.0	62.6	87.5
	RTFOT	60.1	69.9	71.4	68.2	89.1
Performance Grade, acc. to AASHTO M 320		70-16	82-16	82-16	76-22	82-22

R – rubber modification

3.2. Asphalt mixture

Asphalt concrete used in this study was designed according to EN 13108-1 standard [6] and Polish Technical Guidelines WT-2 2014 [14]. The composition and volumetric properties of basic mixture AC 11 W with 35/50 bitumen compacted by Marshall hammer are presented in Table 2. For other bitumens the same aggregate composition and binder content (set to 5.6%) was used and only the type of bitumen was changed.

Table 2. Composition and volumetric properties of basic asphalt mixture AC 11 W 35/50.

Type of layer	binder/levelling course							
Type of traffic	medium, design life from 0.50x10^6 to 7.3x10^6 of 100 kN standard axle loads							
Aggregate type	crushed granite							
Filler type	limestone							
Binder content (% by mass)	5.6							
Air voids in Marshall samples (2 x 75 blows) [%]	3.3							
Voids filled with bitumen VFB [%]	80.4							
Voids in the mineral aggregate VMA [%]	16.6							
Sieve size (mm)	16	11.2	8	5.6	4	2	0.125	0.063
% Passing (by mass)	100	99	83	65	54	43	12	7.4

Asphalt mixtures were prepared with the use of laboratory mixer in accordance with EN 12697-35 standard [5]. Before compaction, loose portion of each asphalt mixture was aged in accordance with the procedure given in the Appendix 2 of the WT-2:2014 [14] to simulate short-term aging of bitumen during production, transport and paving of asphalt mixture.

3.2.1. Additives to asphalt mixture
For asphalt mixture with neat 35/50 bitumen, two different methods of asphalt mixture modification were also tested: addition of polymer fibres and rubber modification.

The first method consisted of adding 19 mm long polymer aramid-polyalphaolefin fibres directly to asphalt mixture. The fibres that were used are specially designed for use as reinforcement in asphalt mixtures for pavement structures. Fibres were added during mixing process with a dosage rate of 0.05% by weight of asphalt mixture. The addition of fibres improves asphalt mixture performance properties [3, 11, 12].

The second method consisted of adding recycled tyre rubber. This was achieved in the laboratory by mixing the proportion of 191 grams of ground tyre rubber (0.2/0.8mm), 9 grams of a specific polymer (polyoctynamer) and 2000 grams of asphalt binder. The mixing process was performed by using laboratory mixer with 200 RPM and 120 minutes of mixing time. The temperature during mixing process was held between of 170 – 180°C. The process of mixing described above is used for the specific additive only during laboratory phase because during full scale production it is added directly to the mixing chamber of the asphalt plant and mixed with dry aggregate for 8-10 seconds, before liquid bitumen is added. Any kind of rubber modification improves asphalt mixture properties, especially in the field of low-temperature cracking [4, 13, 15, 16, 18, 23].

3.2.2. Basic properties of asphalt mixtures
To obtain basic information about the influence of binder and additives on asphalt mixture properties, wheel tracking test acc. to EN 12697-22 was conducted. The results of this test are shown in table 3.

Table 3. Wheel tracking test results.

Binder type	35/50	35/50 R	35/50 F	10/40-65	25/55-60	25/55-80
Rutting resistance, method B in air, 60°C, 10 000 cycles						
WTS_{AIR}, [mm/1000 cycles]	0.28	0.09	0.25	0.19	0.15	0.12
PRD_{AIR}, [%]	6.7	3.3	6.2	5.3	4.5	3.8

R – rubber modification, F – fibre modification

4. Test methodology
The methodology of the test that was used in this study was similar to Texas Standard Procedure Tex-248-F [20].

4.1. Sample preparation (compaction, trimming, mounting in the test device)
The overlay test specimen consists of a 150 mm long, 75 mm wide and 37.5 mm thick sample that is trimmed from a typical gyratory compacted sample with a diameter of 150 mm and height of 115 mm. Gyratory samples were compacted in accordance with EN 12697-31 standard [4], to obtain target density of 98-100% of Marshall density in the further trimmed test specimens. Trimmed specimen dimensions are shown in Fig. 1. Table 4 shows air voids content in trimmed test samples for each asphalt mixture. For each asphalt mixture three specimens were prepared and tested.

The compaction temperature for asphalt mixtures prepared with neat bitumen 35/50 was 135°C ± 5°C, whereas for asphalt mixtures with polymer-modified bitumens it was 145°C ± 5°C.

Table 4. Air voids in final trimmed test specimens.

Binder type	35/50	35/50 R	35/50 F	10/40-65	25/55-60	25/55-80
Air voids content [%]	5.3	5.4	5.2	5.3	5.2	5.1

R – rubber modification, F – fibre modification

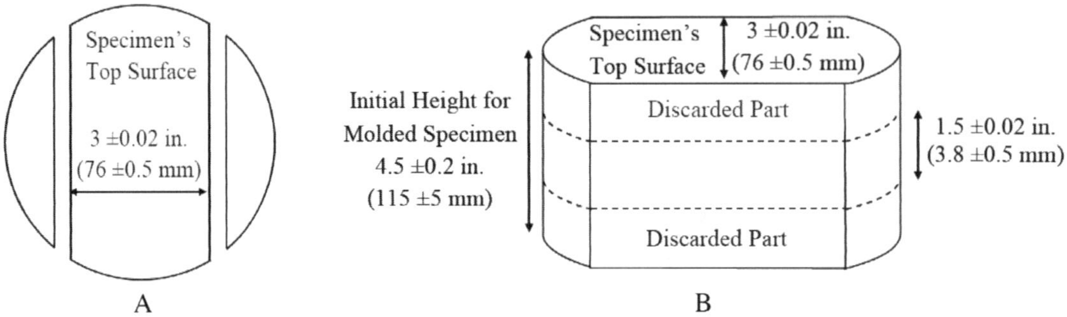

Figure 1. Test specimen dimension according to Tex-248-F standard [20].

Once prepared, each sample was glued with epoxy glue on two metalllic plates, which were securely fixed to a wide mounting plate, as shown in Fig. 2.

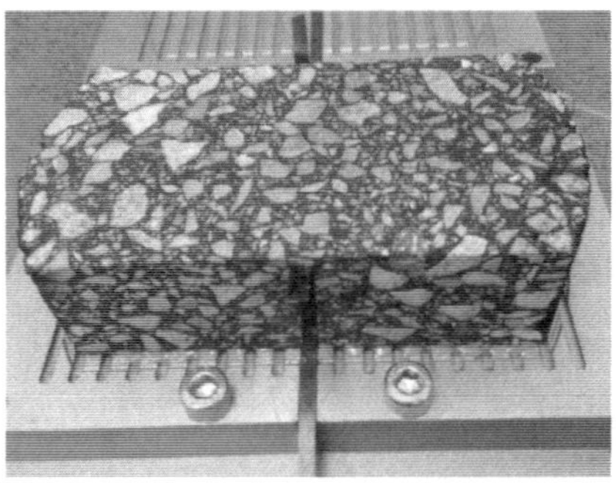

Figure 2. Trimmed specimen glued to the mounting plates.

4.2. Load mode and measurements

Once dried, the samples glued to the plates were mounted in the UTM 130 Texas Overlay Tester jig (see Fig. 3) making the setup ready for the test. The test was conducted in a controlled displacement mode, at a loading rate of one cycle per 10 seconds with a maximum displacement of 0.63 mm, at temperature +25 ± 0.5°C. Each cycle consisted of 5 seconds of loading and 5 seconds of unloading (see Fig. 4). The test was conducted until the failure of the specimen occurred.

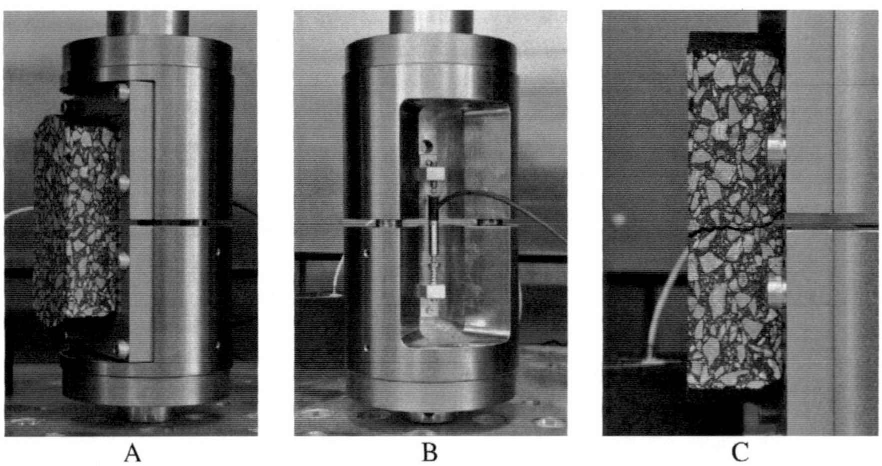

Figure 3. A – specimen mounted in the UTM 130 Overlay test jig, B – external LVDT transducer, C – cracked samples at the end of the test.

The number of cycles to failure was defined as the number of cycles to reach a 93% drop in initial load which is measured during the opening cycle. If a 93% reduction in initial load was not reached within a certain specified maximum number of cycles, the test was stopped automatically. Initially, a total number of 1,000 loading cycles was selected as a maximum number of cycles for stopping the test; however, later it was increased to 2,000 cycles, since no failures were observed after 1,000 loading cycles for the evaluated overlay mixtures, especially with highly modified polymer bitumen. At the end of the test, initial and final values of load, percent reduction in load and number of cycles to failure were reported.

Typical plot of load and displacement during the test is shown in Figure 4. As it can be seen, during the first stage of load cycle (first 5 seconds of cycle) sample is subjected to tension load, while during the second stage (from 5 to 10 seconds of cycle) sample is subjected to compression. Load needed to obtain a constant deformation amplitude is changing during the test. At the first cycle tension load is higher than compression load needed to set sample deformation to zero position, while during the rest of the test the compressive force begins to exceed the tensile force. Compression during cycles can be considered as a form of material's healing.

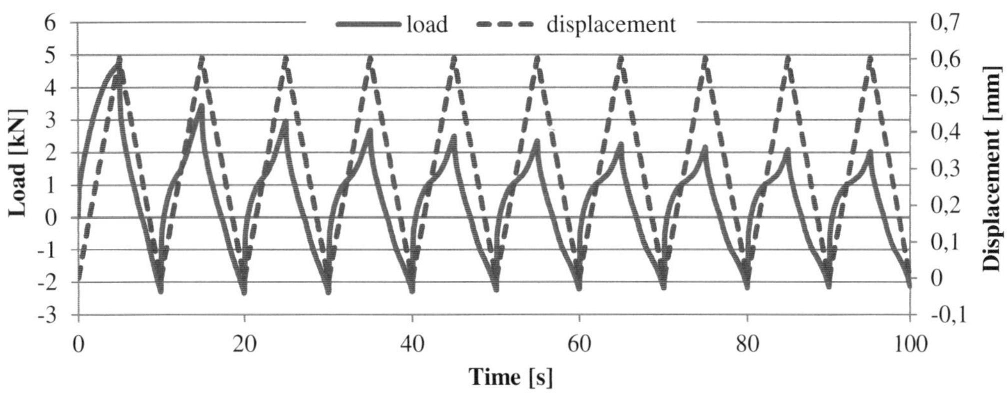

Figure 4. Typical plot of load and displacement during the test (first 10 cycles).

4.3. Cracking resistance criteria

4.3.1. The maximum number of cycles to failure

Based on the current TxOT failure criterion, asphalt mixtures that last over 300 load cycles to failure, defined as load reduction of 93% in the initial load, are judged as acceptable in terms of resistance to cracking [25]. The mixture that reach over 750 load cycles to failure can be treated as an anti-cracking layer [8, 21]. In the latest version of TxOT standard [20] the maximum number of cycles to failure is treated only as an informational value.

4.3.2. The critical fracture energy

The idea of critical fracture energy computation is shown in Figure 5. Critical fracture energy is represented by the area under the load vs. displacement chart (from zero to maximum value) during the first cycle of loading [20].

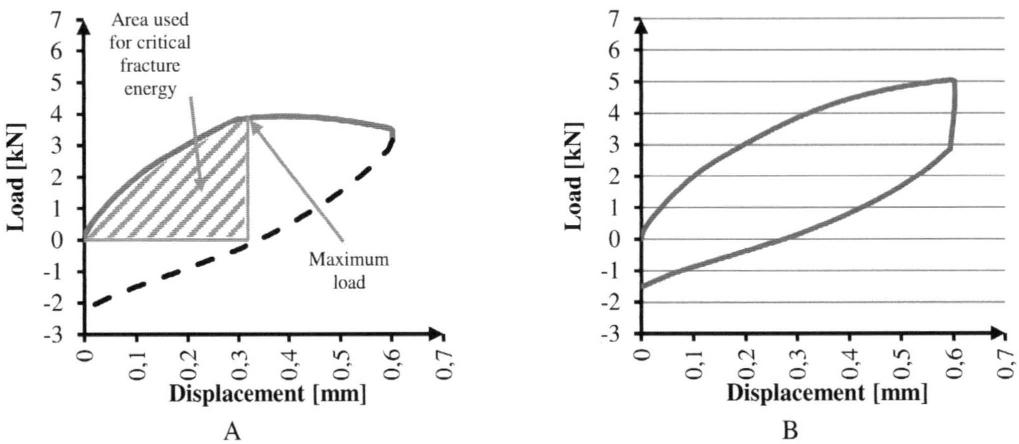

Figure 5. Graphical representation of the critical fracture energy of the first cycle, A – the idea of the method, B – shape of hysteresis loop achieved for tested mixtures.

4.3.3. Crack propagation (Crack Resistance Index, Crack Progression Rate)

Crack propagation stage of asphalt mixture behaviour can be described by Crack Resistance Index [20] or by Crack Progression Rate [8], both calculated from load reduction curve by fitting a power equation. This idea is shown in Figure 6.

According to Garcia et al. [8], the cracking resistance of asphalt mixtures can be subjectively divided into four categories: I) Tough-Crack Resistant: good resistance during crack initiation (tough) and propagation (flexible). Asphalt mixtures with acceptable cracking resistance should be in this quadrant; II) Tough-Crack Susceptible: asphalt mixtures with good resistance to crack initiation (tough) but susceptible to crack propagation (brittle); III) Soft-Crack Resistant: easy crack initiation (soft) but good abilities to attenuate the propagation of the crack (flexible); IV) Soft-Crack Susceptible: asphalt mixtures with significantly poor resistance to crack initiation and propagation. This idea, along with the obtained test results, is shown in Figure 8.

5. Results and analysis

Figure 6 shows crack driving force during loading cycles for all tested mixtures. Fig. 7 presents hysteresis loops from the first loading cycles. Overlay test results are summarized in Table 5.

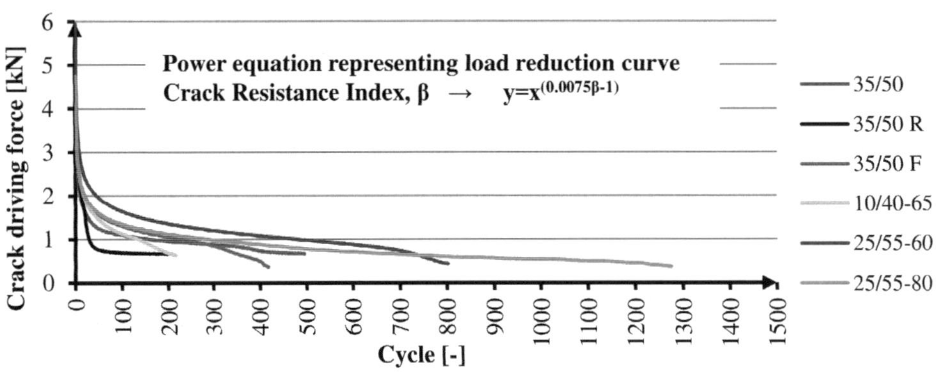

Figure 6. Crack driving force diagram, one line for each asphalt mix (first specimen from the series, not a mean value).

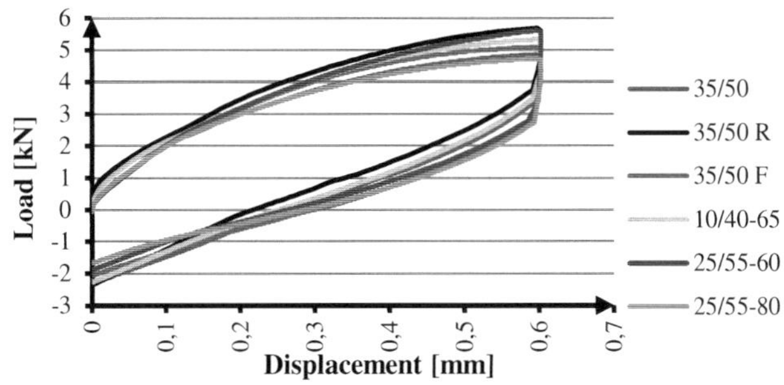

Figure 7. Hysteresis loops from the first load cycles, mean values for three specimens.

Table 5. Summary of TxOT results, mean values for three specimens.

Binder type	35/50	35/50 R	35/50 F	10/40-65	25/55-60	25/55-80
Initial load,						
mean value, [kN]	4.862	5.684	5.094	5.374	5.607	4.726
[lbs]	1093.0	1277.8	1145.3	1199.2	1260.5	1062.4
st. deviation [kN]	0.155	0.038	0.132	0.115	0.314	0.324
[lbs]	34.8	8.6	29.8	36.1	70.5	72.9
COV, [%]	3.2	0.7	2.6	2.1	5.6	6.9
Cycles to failure,						
mean value, [-]	552	120	433	241	854	1069
st. deviation	131	69	102	24	470	203
COV, [%]	26.7	57.5	23.7	9.8	55.0	19.0
Critical fracture energy, G_C						
mean value, [kN-mm/mm^2]	0.000705	0.000815	0.000727	0.000764	0.000765	0.000668
[lbs-in./in.^{2}]	4.03	4.65	4.15	4.36	4.37	3.82
st, deviation [kN-mm/mm^2]	0.000029	0.000016	0.000016	0.000038	0.000022	0.000038
[lbs-in./in.^{2}]	0.16	0.09	0.09	0.22	0.12	0.22
COV, [%]	4.1	2.0	2.1	5.0	2.9	5.7
Crack resistance index, β, (Crack progression rate)						
mean value, [-]	90.68	70.36	91.05	84.60	94.87	88.44
	(0.32)	(0.47)	(0.32)	(0.37)	(0.29)	(0.34)
st. deviation	2.96	2.85	2.18	2.93	1.68	6.16
	(0.02)	(0.02)	(0.02)	(0.02)	(0.01)	(0.05)
COV, [%]	3.26	4.05	2.40	3.46	1.77	6.97

In terms of the first basic criterion (the number of cycles to failure), only mixture with neat bitumen 35/50 modified with rubber and mixture with polymer-modified bitumen 10/40-65 (which in fact is regarded as a hard polymer-modified bitumen) did not meet the requirements, because for both of these mixtures the numbers of cycles to failure was below 300. The addition of rubber to neat bitumen 35/50 caused a significant decrease in number of cycles to failure in comparison with pure neat bitumen 35/50, whereas in case of addition of polymer fibres to the mixture with the same neat bitumen 35/50, the decrease in this value was only minor.

By using the criterion based on critical fracture energy calculated after the first cycle it can be noted that the lowest resistance to reflective cracking was obtained for the mixture with high modified bitumen 25/55-80, and the highest resistance was observed in case of the mixtures that also contained neat bitumen 35/50, but with the addition of rubber.

In terms of the third criterion that characterizes asphalt mixtures in respect to abilities to attenuate the crack which has already occurred, best performance was achieved with asphalt mixture with polymer-modified bitumen 25/55-60. The lowest results in this case were obtained for mixture with neat bitumen 35/50 with the addition of rubber. But it can be also concluded that the overall results for this criterion (combining both cracking toughness and abilities to attenuate the crack) were good for all tested asphalt mixtures and all of them can be qualified as having good anti-cracking properties. It should be also noted that the coefficient of variation was low and at the acceptable level for values calculated for second and third criterion, whereas the coefficient of variation for number of cycles to failure (which is used in the first, basic criterion) was much higher (see Tab. 5).

Graphical interpretation of cracking initiation and cracking propagation properties of tested mixes is presented in Fig. 8.

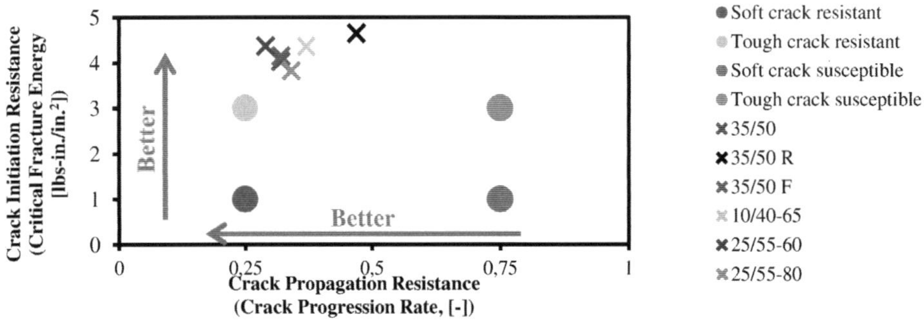

Figure 8. Graphical interpretation of cracking resistance.

6. Conclusions

The type of bitumen that was used in tested asphalt mixtures had significant influence on their cracking resistance.

Polymer-modified bitumen 25/55-60 and highly polymer-modified bitumen 25/55-80 increased resistance of asphalt mixture to cracking. Mixtures with these binders achieved the best results. Polymer-modified bitumen 10/40-65, which was the hardest one in comparison with 25/55-60 and 25/55-80, achieved worse results, even when compared with neat bitumen 35/50. In this case, the overall stiffness of the bitumen was responsible for a decrease in anti-cracking properties.

Asphalt mixture with polymer-modified bitumen 25/55-60 had higher critical fracture energy and better resistance to crack propagation than the mixture with highly polymer-modified bitumen 25/55-80. On the other hand, the latter mixture had better performance in terms of number of cycles to failure.

The addition of polymer fibres did not affect cracking properties in a positive way. Also, the addition of rubber in fact decreased the anti-cracking properties. The mixture with neat bitumen 35/50 modified with ground rubber in most cases performed worse than even the neat bitumen 35/50.

On the basis of the obtained results a following ranking of asphalt mixtures and their cracking resistance can be presented: 1 (the best performance) – 25/55-80 and 25/55-60, 2 – 35/50 and 35/50 F, 3 – 10/40-65, 4 (the worst performance) – 35/50 R.

References

[1] AASHTO M 320 2010 *Standard Specification for Performance-Graded Asphalt Binder* American Association of State Highway and Transportation Officials

[2] ASTM WK26816 2009 *New Test Method for Determining the Susceptibility of Asphalt Mixtures to Cracking Using the Overlay Teste*

[3] Chomicz-Kowalska A 2017 Laboratory testing of low temperature asphalt concrete produced in foamed bitumen technology with fiber reinforcement *Bulletin of the Polish Academy of Sciences Technical Sciences* 65.6 (2017): 779-790

[4] EN 12697-31 2007 *Bituminous mixtures. Test methods for hot mix asphalt. Specimen preparation by gyratory compactor*

[5] EN 12697-35 2016 *Bituminous mixtures. Test methods. Laboratory mixing*

[6] EN 13108-1 2006 *Bituminous mixtures. Material specifications. Asphalt Concrete*

[7] Garcia V M et al. 2016 Evaluation of Operational Variables That Can Improve Overlay Tester Results *Transportation Research Board 95th Annual Meeting* Washington D.C. p 1-16
[8] Garcia V M et al. 2017 *Improved Overlay Tester for Fatigue Cracking Resistance of Asphalt Mixtures* Research Center for Transportation Infrastructure Systems The University of Texas at El Paso p 176
[9] Germann F P and Lytton R L 1979 *Methodology for predicting the reflection cracking life of asphalt concrete overlays* Research Report Number 207-5 Texas Transportation Institute Texas A&M University p. 149
[10] Iwanski M, Buczynski P and Mazurek G 2016 Optimization of the road binder used in the base layer in the road construction *Construction and Building Materials* Vol. 125 30 p. 1044-1054 http://dx.doi.org/10.1016/j.conbuildmat.2016.08.112
[11] Jaskula P, Stienss M and Szydlowski C 2017 Effect of Polymer Fibres Reinforcement on Selected Properties of Asphalt Mixtures *Procedia Engineering*, Volume 172, p 441–448
[12] Kaloush K E et al. 2010 Evaluation of Fibre-Reinforced Asphalt Mixtures Using Advanced Material Characterization Tests *J. Test. Eval.* vol. 38 no. 4 p. 102442
[13] Król J B et al. 2014 Laboratory and Field Investigations of Polymer and Crumb Rubber Modified Bitumen *Journal of Civil Engineering and Architecture* V8, 10 (83) pp. 1327-1334
[14] Polish Technical Guidelines WT-2 2014 General Directorate for National Roads and Motorways
[15] Pszczola M et al. 2017 Evaluation of Low Temperature Properties of Rubberized Asphalt Mixtures *Procedia Engineering* Vol. 172 p 897–90
[16] Radziszewski P. et al. 2012 Asphalt rubber as an alternative of polymer modified bitumen *Asphalt Rubber Conference 2012* Munich
[17] Swieczko-Zurek B et al. 2017 Rolling resistance and tyre/road noise on rubberised asphalt pavement in Poland *Road Materials and Pavement Design* Vol. 18 Issue 1 p. 151-167 http://dx.doi.org/10.1080/14680629.2016.1159245
[18] Sybilski D et al. 2011 Rubber-bitumen granulate for asphalt pavements – laboratory comparative analysis *5th International Conference on Bituminous Mixtures and Pavements* Thessaloniki pp. 1413-1421
[19] TxDOT Designation 2014 Tex-248-F *Test Procedure for Overlay Test Texas*
[20] TxDOT Designation 2017 Tex-248-F *Test Procedure for Overlay Test* Texas
[21] Walubita L F et al. 2011 Search for a Laboratory Test to Evaluate Crack Resistance of Hot-Mix Asphalt *Transp. Res. Rec. J. Transp. Res. Board* p 73-80 DOI: 10.3141/2210-08
[22] Walubita L F, Simate G S, and Oh J H 2010 Characterising the ductility and fatigue crack resistance potential of asphalt mixes based on the laboratory direct tensile strength test *J. SOUTH AFRICAN Inst. Civ. Eng.* vol. 52 no. 2 p 31-40
[23] Zborowski A. et al. 2004 Material Characteristics of Asphalt Rubber Mixtures *Annual Meeting of the Transportation Research Board*
[24] Zhou F and Scullion T 2003 *Upgraded overlay tester and its application to characterize reflection cracking resistance of asphalt mixtures* Report 0-4467-1 Texas Transportation Institute Texas A&M University p 44
[25] Zhou F and Scullion T 2005 *Overlay tester: a rapid performance related crack resistance test* Report 0-4467-2 Texas Transportation Institute Texas A&M University p 86
[26] Zhou F et al. 2016 *Experimental Design for Field Validation of Laboratory Tests to Assess Cracking Resistance of Asphalt Mixtures* Texas Transportation Institute Texas A&M University Project no 9-87 p 262
[27] Zhu Y et al. 2017 Comprehensive evaluation of low-temperature fracture indices for asphalt mixtures *Road Materials and Pavement Design* 18:sup4 pp. 467-490 DOI: 10.1080/14680629.2017.1389085
[28] Zofka A. et al. 2017 Glass and carbon geogrid reinforcement of asphalt mixtures *Road Materials and Pavement Design* Vol. 18(1), pp: 471-490, 10.1080/14680629.2016.1266775

Structural strength assessment of the reconstructed road structure in terms of the loading time and yield criterion

Grzegorz Mazurek[1] and Marek Iwański[1]

[1] Assistant Professor, Kielce University of Technology, PL

[2] Full Professor, Kielce University of Technology, PL

e-mail: gmazurek@tu.kielce.pl

Abstract. This article reports the results of numerical simulations of the stress-strain states in the rebuilt road structure compared to the solutions contained in the Polish Catalogue, with the true characteristics of the layer materials taken into account. In the case analysed, a cold-recycled base layer with foamed bitumen as a recycling agent was used. The presented analysis is complementary to the mandatory in Poland procedure of mechanistic pavement design based on a linear elastic model. The temperature distribution in the road structure was analysed at the reference temperature of 40°C on the asphalt layer surface. The loading time was included in the computer simulations through the use of the classic generalized Maxwell model and thus the stiffness-time history of the layers had to be determined. For this purpose, the dynamic modulus E* tests of the loading time frequency from 0.1 Hz to 20 Hz were carried out, and the yield point was modelled using the Coulomb-Mohr failure criterion calculated on the basis of triaxial compression tests. The analytical solution to the problem was found with ABAQUS. The results demonstrate that the high temperature of asphalt layers and long loading time noticeably reduces the stiffness modulus in those layers. That reduction changes the principal stress levels, which significantly influences the shear stress both in the recycled base layer and in the subgrade soil. Should the yield point be exceeded rapidly in the recycled layer, the horizontal stresses in the asphalt layers will increase and adversely affect the durability of the reconstructed road pavement structure, especially in the zones of slow heavy vehicle traffic.

1. Introduction

The individual design of the pavement layer system in Poland is based on the assumption that a single load from the equivalent standard axle (ESAL) causes very small deformations. Therefore, to analyse the stress-strain states in the flexible pavement the elastic model is used at a fixed loading time of 0.02s (approximately 60 km/h) and equivalent temperature of + 13°C [1]. This model allows correct design of the pavement structure for the reference conditions assuming proper drainage of the road frame, but it does not consider the loading time. The provisions of Polish legislation provide for the use of viscoelastic models that allow the implementation of loading time in the design [2]. There are numerous road sections with a speed limit where the loading time is much greater than the design value given in the Catalogue. In this case, the phenomenon of pavement layer stiffness relaxation should be introduced in the pavement rehabilitation or reconstruction design method.

Another extremely important issue, usually overlooked in structural analysis, is the assessment of the limit state of stresses in the subgrade and in recycled bases, especially in typical pavement layer systems operating under KR1-KR2 traffic where, unlike in much thicker road structures operating under the K3-KR7 traffic, exceeding the ultimate limit state at the same load level is highly probable due to the thin package of structural layers. With these factors taken into account, the diagnosis of the KR1-2 surface will indicate exceeding the ultimate limit state due to among other factors extended vehicle loading time in synergy with high temperature as the probable cause of excessive deformations. In this situation, the elastic-plastic model will complement the stress-strain state considerations [3]. The design of flexible pavements working within elastic stress has to include a very important issue of meeting several conditions according to which the limit state cannot be exceeded in any of the layers, the structure should not deform when passing the permissible elastic stresses and tensile stresses should not be greater than the permissible stress [4, 5]. Nevertheless, structural damage may occur in the case of a thin compressible layer present in the subgrade or due to the moistening of the subgrade layers.

This article focuses on the stress-strain state in the KR1-2 traffic pavement manufactured with a foamed bitumen stabilized cold recycled base course and the subgrade with the specified parameters of the Coulomb-Mohr plastic model. It should be noted that due to the properties of the foamed bitumen, the mineral mix can contain a lower quality mineral material and even quite a large amount of mineral dust up to 15% relative to other types of recycled base courses [6]. Unfortunately, engineering practice hardly ever considers the subgrade ultimate limit state and relaxation in the pavement asphalt layers and recycled base courses, assuming the new pavement layer system is based on the optimization resulting from the solutions adopted in the Catalogue [1].

2. Materials and methods

2.1. Standard based parameters of pavement structural layers

The upper layers of the pavement structure were produced from AC11S asphalt concrete (wearing course) and AC16W (binder course). The gradation curves and the physical and mechanical parameters complied with the requirements set forth in PN-EN 13108 [7] and in WT2/2010 [8] applicable in Poland for traffic load category KR1-2. The mix standard parameters and frame compositions are summarized in Table 1.

Table 1. Frame composition and standard parameters of bituminous and recycled mixtures.

Property	Symbol	u.m.	AC11S	AC16W	RAP50/VA50
Bitumen content	Bitumen/ cement	[%]	5.7/0	4.6/0	2.0/2.0
Sensitivity to moisture damage to PN-EN 12697-12,	ITSR	[%]	96.2	91.0	-
Resistance to moisture damage [12]	TSR		-	-	81.9
Void space to PN-EN 12697-8	V_m	[%]	2.5	4.6	12.6
Voids filled with bitumen in the mineral mix,% (v/v),	VFB	[%]	83.9	69.0	-
Voids in mineral aggregate, % (v/v)	VMA	[%]	15.3	15.0	-
Aggregate grain size above 2 mm,	-	% (m/m)	60.0	68.0	60.4
Aggregate grain size of 0.063/2 mm,	-	% (m/m)	29.5	26.5	34.2
Aggregate grain size below 0.063 mm	-	% (m/m)	10.5	5.5	5.4

The recycled base layer used for the pavement reconstruction was designed as a cold deep-recycled mix with foamed bitumen (MCAS) [9]. The denotation 50RAP/50VA means 50% RAP content and 50% virgin aggregate content. The primary coarse aggregate (VA) had a continuous gradation curve with the maximum grain size "D" equal to 31.5 mm and met the Polish requirements of WT - 2010 [8] based on EN 13242 [10]. In order to ensure the required size of aggregate in the Marshall sample, the components of the recycled waste material were screened through a 22.4 mm sieve. The mineral mixture was designed following the Wirtgen guidelines [11]. The results of grouping with the given recycled mixture in relation to other waste materials are discussed in [13].The final MCAS mixture was conditioned suggesting results in [14].

The first testing stagewas devoted to all layers of the structure and required the determination of the generalized Maxwell (GM) model parameters in the linear viscoelasticity (LVE) range. The tension-compression test was carried out undercontrolled strain of a low amplitude of 25με [12, 15] in accordance with PN-EN 12697-26 Annex D [16]. The test temperatures were: -7°C, 5°C, 13°C, 25°C and 40°C. The complex modulus E* and the phase angle δ were determined.

The fracture condition for the MCAS layer containing 50RAP/50VA aggregate was determined according to the Coulomb-Mohr (CM) failure criterion using elements of ASTM D2664 [17] and information provided in [18]. The 80 mm (diameter) x 160 mm (height) specimens were core drilled from a set of six compacted MCAS samples150 mm (diameter) x 200 mm (height). The angle of internal friction "φ" was 41.5°, while the cohesion quantity *c* was assumed to be 473.2 kPa at the recommended load speeds given by Jenkins [19].

2.2. Mechanical parameters of pavement layers structure

The first stage of testing consisted in determining the rheological parameters of the materials from which the structural layers were made. Since the analysis took into account the phenomenon of relaxation, a model accounting for the changes in stiffness over time had to be used. Hence the behaviour of pavement structural layers was described using the generalized Maxwell (Weichert) model in the linear viscoelasticity range (LVE) [20, 21]. The accuracy of asphalt concrete rheological parameters determined using the generalized Maxwell model greatly influences the accuracy of viscoelastic strains determined in asphalt concrete. The viscoelastic parameters were based on the results of dynamic tests with the oscillating load applied in the controlled strain conditions according to PN-EN 12697-26 Annex D [16]. This type of stiffness modulus test is much more convenient in the situation when the parameter searched for is the instantaneous stiffness modulus E_o.
Viscoelastic parameters of the GM model were obtained by determining the imaginary part (E") and the real part (E') of the complex modulus $E^* = E' + iE''$ of the frequency function [21]. The form of the complex modulus components are represented using Fourier transforms in formulas (1)(2)

$$E' = E_o - \sum_{i=1}^{N} E_i + \sum_{i=1}^{N} \frac{E_i \cdot \tau_i^2 \cdot \varpi^2}{1+\tau_i^2 \cdot \varpi^2} \quad (1)$$

$$E'' = \sum_{i=1}^{N} \frac{E_i \cdot \tau_i \cdot \varpi}{1+\tau_i^2 \cdot \varpi^2} \quad (2)$$

where: E_i–the i-th modulus of elasticity in the GM module, E_o – the modulus of instantaneous elasticity, τ_i–the i-th relaxation time in the GM model, ω_i–reduced frequency.

The temperature shift factor was determined from equation (3):

$$\alpha_T = \omega \cdot e^{(A_1 + T \cdot A_2)} \quad (3)$$

where: T – the test temperature, ω–the frequency at the test temperature, A_1, A_2–model parameters.

Goodness-of-fit measures such as coefficient of determination R^2 and normalized mean error MNE were used to determine the model fit to the experimental data [23]. The further wear diagnosis of the pavement structure under the KR1-KR2 traffic category accounted for the limit state with a certain assumption of the ground condition. The load carrying capacity parameters of the subgrade in relation to the Coulomb-Mohr model were derived from the literature data. Finally, the subgrade parameters were assumed to be G1 category, i.e., a layer of sand and gravel mix with the angle of friction "φ" of 40° and the cohesion value *c* of zero [24].

The viscoelastic character of the pavement layer system was described by means of a maximum of five Maxwell elements. Knowing that asphalt concrete in a given range of small strains is a thermorheologically simple material [25], the load time-temperature relationship was established using a horizontal temperature shift factor [23]. The final form of the master curve included the optimization of the real and imaginary parts of the stiffness modulus, linking in the analysis the relationship between the complex stiffness modulus and the phase angle. Table 2 summarizes the parameters fitted in the GM model in the form of a master curve and the Coulomb-Mohr boundary condition.

Table 2. Master curve parameters of the pavement structural layers on the basis of the Generalized Maxwell (GM) Model.

Asphalt concrete AC11S - Wearing course						
	Parameters of the generalized Maxwell model		α_T (exponential function)		Criterion CM	
	g_i [-]	τ_i [s]	A_1	A_2	φ [°]	c [kPa]
the generalized Maxwell (GM) model	g_1=0.29731	τ_1=0.00027	8.31	-0.274	-	-
	g_2=0.29731	τ_2=0.00457				
	g_3=0.24517	τ_3=0.09663				
	g_4=0.16021	τ_4=5.01952				
	E_o = 23930MPa					
	R^2 = 0.99; RMSE = 12.45%					
Asphalt concrete AC16W –Binder course						
	Parametersof the generalized Maxwell model		α_T (exponential function)		Criterion CM	
	g_i [-]	τ_i [s]	A_1	A_2	φ [°]	c [kPa]
the generalized Maxwell (GM) model	g_1=0.29808	τ_1=0,00001	12.45	-0.264	-	-
	g_2=0.23997	τ_2=0.00013				
	g_3=0.20785	τ_3=0.00157				
	g_4=0.17046	τ_4=0.02198				
	g_5=0.08365	τ_5=1.42083				
	E_o = 20690.39MPa					
	R^2 = 0.99; RMSE = 7.1%					
Recycled base with foamed bitumen (50%RAP/50%VA)						
	Parameters of the generalized Maxwell model		α_T (exponential function)		Criterion CM	
	g_i [-]	τ_i [s]	A_1	A_2	φ [°]	c [kPa]
the generalized Maxwell (GM) model	g_1=0.21416	τ_1=0.00001	11.32	-0.232	41.5	473.2
	g_2=0.21416	τ_2=0.00063				
	g_3=0.20886	τ_3=0.02993				
	g_4=0.20886	τ_4=5.72972				
	g_5=0.115396	τ_5=5.72972				
	E_o = 10784.41 MPa					
	R^2 = 0.96; RMSE =5.1%					

The results of the analysis compiled in Table 2 indicate that the model fit to the experimental data was done with a moderate mean square error of less than 13%. The correlation coefficient was greater than 0.96. The largest dispersion of results, based on the root-mean-square error RMSE, was observed for AC11S. Probably an increased amount of bitumen in the wearing course caused slight disturbances in master curve fitting at high temperature. Nevertheless, all the results of the qualitative evaluation of the master curve model fit were considered satisfactory. The set of master curves at 25°C is shown in Figure 1.

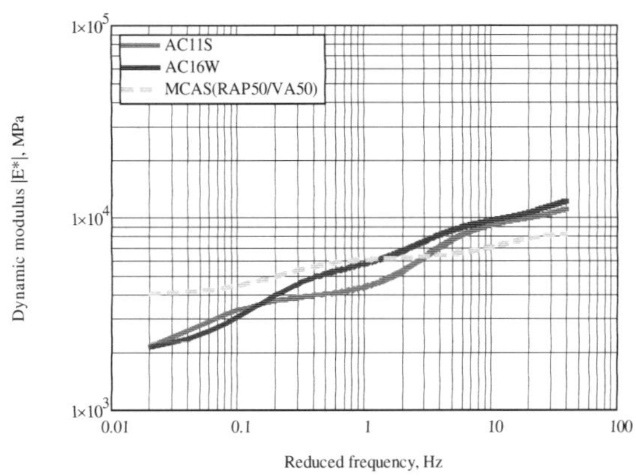

Figure 1. Dynamic modulus (master curves) at 40°C of the pavement structure.

Analysis of the fitting results for the Maxwell model parameters (Table 2) and of the master curve plots indicates that the AC11S layer containing the largest amount of bitumen attained the highest level of the instantaneous modulus. The MCAS recycled layer had the lowest relaxation rate, being the least sensitive to the vehicle loading time. The cement content in the MCAS mixture (2.0%) was responsible for this result. The parameters of the physical Maxwell model from Table 2 were directly implemented to the ABAQUS numerical model.

3. Numerical model

The numerical model assumed considering the stress-strain state for two types of structures under light traffic KR1 and KR2. It should be remembered that the standard load of a vehicle is the same as for structures operating under higher traffic categories. The numerical model included the pavement surface temperature of 40°C, whereas the temperature distribution was based on the findings and equations contained in [26]. The temperature distributions measured at the mid-thickness of analysed layers are shown in Table 3 [1, 26].

Table 3. Pavement layer system and temperature distribution in the pavement.

Traffic category	Type	thickness	depth	Mid-thickness	Temperature of the layer [23]	Graphical representation
-	-	[m]	[m]	[m]	[°C]	[-]
KR1	AC11S	0.03	0.03	0.015	37,5	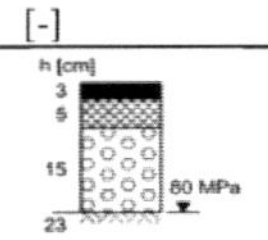
	AC16W	0.05	0.08	0.06	33	
	MCAS	0.15	0.23	0.16	30	
	subgrade	-	below 0.23		25	
KR2	AC11S	0.04	0.04	0.020	37	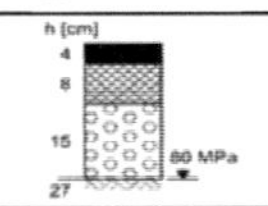
	AC16W	0.08	0.12	0.08	30	
	MCAS	0.15	0.27	0.20	31	
	subgrade	-	below 0.27		25	

The ABAQUS modelling software was used in the numerical analysis. A finite element rotationally symmetric model was chosen for the surface of the structure. The subgrade was determined in such a way that the dimensions of the model did not affect the state of stresses and strains from external loads. The cylindrical area was adopted, with dimensions 7 m in diameter and 7 m in thickness. The model required that the boundary conditions be defined. The adopted variant made the displacement of points on the side walls in directions perpendicular to them and the displacement in the base of the lower cylindrical area impossible. The finite element mesh together with the finite element dimensions was based on the results of simulations [27, 28]. The height of the finite element in the layer the base layer was set at a maximum level of 5 cm, while the asphalt layers were divided according to the principle of 4 elements for the thickness of the layer. As a result, the maximum height of the asphalt layer element was less than 5 cm. The computational model consisted of two-dimensional elements with non-linear parabolic shape functions of the CAX8R type [29]. The analysis assumed full interlayer bonding and full isotropy of the pavement structure materials.

The cyclic load was applied on the pavement structure, with two variants: 1 Hz and 10 Hz, i.e. an approximate vehicle speed of 6 km/h and 60 km/h respectively. The cyclic load corresponded to the form of the haversine function determined in ABQUS by fitting it to the recorded modulated wave function (4):

$$a = A_o + A \cdot \sin\omega_1 \cdot (t - t_o) \cdot \sin\omega_2 \cdot (t - t_o) \qquad for \quad t > t_o \tag{4}$$

where: A_o, A, ω_1, ω_2 – parameters of the function fitted to haversine form

After fitting the parameters of the modulated function, the representation of the haversine cycle of the vertical stress on the contact surface between the wheel and the pavement is shown in Figure 2.

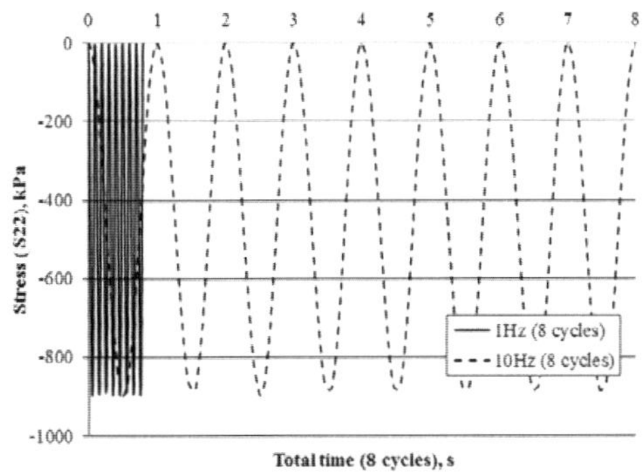

Figure 2. Load cycle haversine for 1Hz and 10 Hz..

Eight load cycles were used in the analysis to simulate the maximum number of standard vehicles that can pass within 24 hours on the KR1 traffic category road. For the KR2 traffic category, the load pattern also included 8 load cycles.

4. Analysis of road structure load bearing capacity

4.1. The influence of loading time on the stress and strain distribution

The analysis started with the simulation of the stress-strain state in the cross-section of the pavement structure, taking into account the influence of the loading time (1Hz, 10Hz) and pavement layer system (KR1, KR2). The simulation of the stress-strain state in the structure at the unusually low frequency of 1 Hz (8 cycles) was aimed at drawing attention to the fact that the level of damage to the pavement structure, for example in the intersection areas, is of great importance in terms of the durability of roads under light traffic. Three significant quantities were chosen from the large set of numerical model results. The first quantity was the total horizontal strain (E11). In the classic elastic model, the strain value at the underside of asphalt layers is used in estimating fatigue life. It should be noted that taking into account the parameters of the plastic model changes the stress field distribution in the structure. The accuracy of determining this parameter has a considerable impact on the quality of the pavement fatigue life estimation [27]. Figure 3 shows the change in the total horizontal strain (E11) and plastic vertical strain (PE22) in the cross-section of the pavement structure.

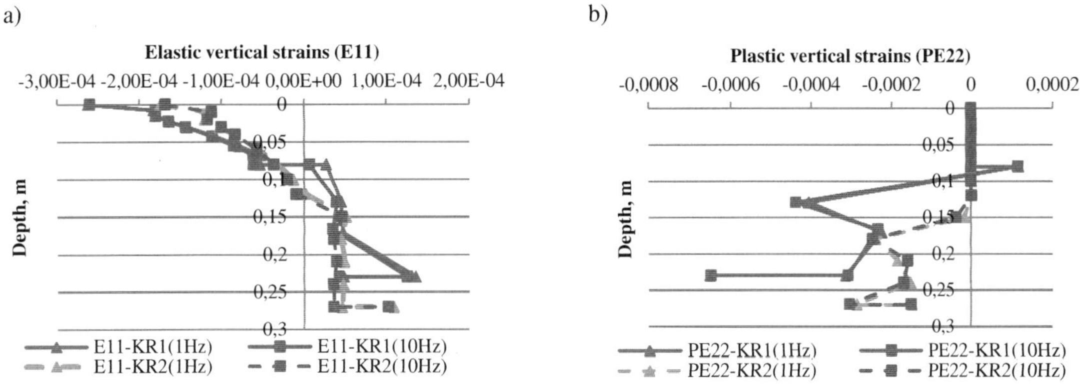

Figure 3. Strain distribution after 8 loading cycles: a) total horizontal strains (E11); b) plastic vertical strains (PE11)

The increased loading time (lower frequency) caused an increase in the maximum horizontal strain (Fig 3a) due to the stiffness modulus relaxation in the structural layers at 10 Hz being much smaller than that for 1 Hz. Therefore, in the structure under KR1traffic (h = 0.08m), the loading time change from 10Hz to 1Hz causes a14%increase in horizontal strain, while in the structure under KR2 traffic (h = 0.12m) the horizontal strain increase can be up to 76%. This suggests that ignoring the phenomenon of relaxation, especially in the area of intersections during the summer, is a gross oversimplification. The reduced asphalt layer stiffness due to relaxation also contributes to increased horizontal strains in the recycled layer. The increase in horizontal strains and the initiation of the cracking process in the base layer will make it act not as a cracked layer of large blocks but as a conventional unbound mixture. Then, the design assumption set forth in the Catalogue that the stiffness modulus E = 1500 MPa further in the in-service period will become significantly overstated. The reduced stiffness of the asphalt layers will also lead to an increase in vertical strains in the base and subgrade. Vertical plastic strains determine the scale of potential structural deformations and ultimately create a need for the reconstruction of the entire pavement structure. Analysis of the results (Fig. 3b) shows that the decrease in the stiffness of asphalt layers (low frequency) caused an increase in horizontal plastic strains in the KR1 subgrade (about 3.5%), with only 1.5%in the KR2 subgrade. This increase is slight but considering the thickness of the structure, the level of vertical plastic strains in the KR2 subgrade and base layers is noticeably lower than in the KR1 structure. This indicates high sensitivity of the KR1 road pavement structure to overloading or to an excessive number of vehicles with a standard wheel load> 50 kN. Exceeding the permissible number of vehicles may lead to the rapid destruction of the road surface. The effect of stiffness change in the structure due to the loading time was also reflected by the deflection level (vertical deformation U22 in the wheel axis of the vehicle). The changes in vertical deflection are shown in Figure 4 in the cross-section of the road structure.

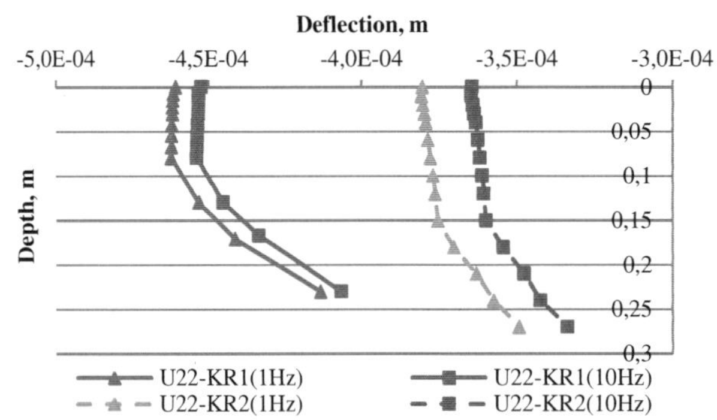

Figure 4. Deflection of the road structure after 8 loading cycles.

As expected, the decreased stiffness in asphalt layers due to relaxation increased the deflection at the contact area between the wheel and the pavement surface by about 2% in KR1 and 4.5% in KR2. However, the deflection difference between KR1 and KR2 for the frequency of 1Hz was 21%. The elastic-plastic model adopted for the analysis was a major factor in the change of the stress field distribution (plane state of strain). Hence the friction angle will be of crucial importance for the redistribution of vertical and horizontal stresses in the subgrade.

4.2. The influence of internal friction angle in the subgrade

Limit state parameters in geomaterials depend on a number of factors, especially on the compaction level and moisture content [5, 28]. The cohesion of the recycled base layer with foamed bitumen can be reduced, as in other bituminous mixtures, through the increase in bitumen temperature. According to Tan [30], the friction angle is independent of temperature and, often, a change in the moisture content (improper drainage of the road frame)reduces its value. Figure 5 shows the distribution simulations for the vertical plastic strains (PE22) and horizontal strains (E11)against the internal friction angle (40°, 30° and 20°)in the structure designed for the KR1 traffic category.

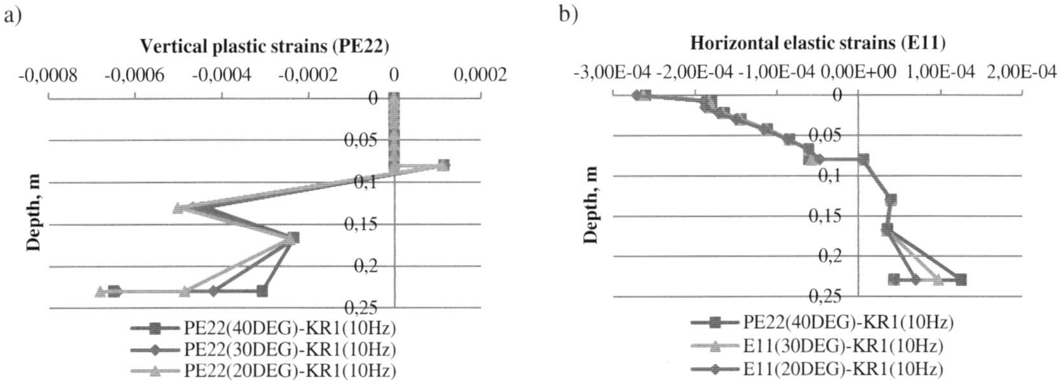

Figure 5. Strain distribution vs. friction angle: a) plastic vertical strains (PE11); b) total horizontal strains (E11).

Simulations were performed for friction angles of 40°, 30°and 20°at the load frequency of 10Hz after 8 loading cycles. The results in Fig. 5bshow that the reduction in the friction angle in the subgrade has a destructive effect on the pavement strength due to the progressive accumulation of vertical plastic strains. Increased strain values were observed both at the top and bottom surface of the recycled base. The difference of the vertical permanent deformation (PE22) at the underside of the base layer increased by 37%in relation to the friction angle reduction in the subgrade from 40° to 20°. This is a high value considering the small number of cycles. This plastic deformation gain does not include the "shakedown" effect discussed in detail in [31].

As for the horizontal deformations (E11) in Figure 5a, the reduction in the friction angle in the subgrade also increased their values in the recycled base layer. After the 8th loading cycle, the recycled base was entirely within the large tensile strain region, which may lead to the premature loss of load carrying capacity in the structure as a whole. A large tensile strain could contribute to high cracking propagation in the base coarse. Taking into consideration many imperfections during a recycled base course manufacturing it is highly possible to observe many transverse cracks in pavement in a short time. In that respect, it should be pointed out that the effect of longer loading time other than that specified in the catalogue [1], considering high temperatures in the summer, may cause permanent deformations regarded as structural. Therefore, paradoxically, the layers designed for the KR1 and KR2 traffic need a well compacted subgrade with a high friction angle. Also, the load cycles increased above 115 kN/axle will cause permanent deformations to appear in the early service period of those pavement structures.

5. Conclusions and observations

As a result of the rheological tests of the materials and numerical analyses of pavement structures made with those materials, the following conclusions were formulated:

- pavements designed for light traffic (KR1, KR2) are sensitive to exceeding the ultimate limit state in the subgrade and in the recycled base layer in the summer;
- the time of loading is an important parameter to be considered at intersectionsduring high temperature periods;
- the use of 8 loading cycles (oneday for the KR1 category) in the summer causes structural deformations to appear in the structure;

- longer time of loading leads to increased horizontal strains, which in the cases analyzed was up to 76%, in the asphalt layers of the pavement, affecting its fatigue life;
- a reduction in the friction angle in the subgrade increases plastic deformations in the recycled base layer (MACAS);
- the major friction angle reduction caused the recycled base layer to work within the large tensile strain region after the 8th loading cycle, which suggests that fast exhaustion of the load carrying capacity of the structure as a whole should be expected;
- when constructing pavements designed for the KR1-2 traffic, attention should be given to the quality of the subgrade. Considering the small overall thickness of structural layers, insufficient quality of the subgrade may lead to premature failure of the pavement.

References

[1] Douangphachanh V and Oneyama H 2014 *EURASIP Journal on Wireless Communications and Networking* **1** 114
[2] Judycki J, Jaskula P, Pszczola M, AlenowiczJ, Dolzycki B, Jaczewski M, Rys D and Stienss M, 2014*Catalogue of typical flexible and semi-rigid pavement structures.* Gdansk University of Technology, Gdansk, Poland (in Polish) GDDKiA (General Directorate for National Roads and Motorways)
[3] (Dz.U. 1999 nr 43 poz. 430) *Rozporządzenie Ministra Transportu i Gospodarki Morskiej z dnia 2 marca 1999 r. w sprawie warunków technicznych, jakim powinny odpowiadać drogi publiczne i ich usytuowanie*(in Polish)
[4] Huang C W 2008 Development and numerical implementation of nonlinear viscoelastic-viscoplastic model for asphalt materials, texas a&m University, *Doctoral Dissertation*
[5] Wiłun Z 2013 *Zarys geotechniki (in Polish) [Introduction in Geotechnic]* WKŁ Warsaw
[6] Helwany S 2007*Applied Soil Mechanics: with ABAQUS Applications* John Wiley & Sons Inc.
[7] ISBN: 978-0-471-79107-2
[8] Iwański M, Buczyński P and Mazurek G 2016 *The use of gabbro dust in the cold recycling of asphalt paving mixes with foamed bitumen. Bulletin of the Polish Academy of Scieneces Technical Sciences,* Vol. 64, No. 4, pp. 763-773.2016 DOI: 10.1515/bpasts-2016-0085
[9] EN 13108-1:2016 *Bituminous mixtures. Material specifications. Asphalt Concrete*
[10] General Directorate for National Roads and Highways (GDDKiA) 2010 *WT-2/2010—Techical Requirements—Bituminous Mixtures—Annex 2* Order Nr 102 GDDKiA: Poland
[11] Iwański M and Chomicz-Kowalska A*Evaluation of the pavement performance. Bulletin of the Polish Academy of Scieneces Technical Sciences,* Vol. 63, No. 1, pp.97-105, 2015 DOI: 10.1515/bpasts-2015-0011
[12] EN 13242:2013 *Aggregates for unbound and hydraulically bound materials for use in civil engineering work and road construction*
[13] Wirtgen Group. 2012*Wirtgen Cold Recycling Technology,*1st ed.; Wirtgen GmbH: Windhagen, Germany
[14] Buczynski P and Iwański M 2017 Inactive Mineral Filler as a Stiffness Modulus Regulator in Foamed Bitumen-Modified Recycled Base Layers *IOP Conf. Series: Materials Science and Engineering* **245** https://doi:10.1088/1757-899X/245/3/032042
[15] Mazurek G and Iwański M Multidimensional Analysis of the Effects of Waste Materials on Physical and Mechanical Properties of Recycled Mixtures with Foamed Bitumen *Appl. Sci.* 2018 *8*(**2**) pp. 282 doi:10.3390/app8020282
[16] Dolzycki B, Jaczewski M and Szydlowski C 2017 The long-term properties of mineral-cement-emulsion mixtures *Construction and Building Materials*, Volume **156** 799-808 https://doi.org/10.1016/j.conbuildmat.2017.09.032.
[17] Kim Y R 2009 *Modeling of Asphalt Concrete*, McGraw-Hill Construction, 2009.
[18] EN 12697-26 *Bituminous mixtures – Test methods for hot mix asphalt- Part 26: Stiffness, Annex D*
[19] ASTM D2664-04: *Standard Method for Triaxial Compressive Strength of Undrained Rock Cores Specimens Without Pore Pressure Measurements, American Society for Testing and Materials* Available in Internet: http://www.astm.org/Standards/D2664.htm

[20] Blanc B, Gabet T, Hornych P, Piau J M, Di Benedetto H 2015 Cyclic triaxial tests on bituminous mixtures*Road Materials and Pavement Design*, Editions Hermes 16 (**1**) pp.46-69
[21] Jenkins K J 2009 Determination of Triaxial Shear Parameters Using Simple Triaxial Test *Method 7*South Africa
[22] Marques Severino P, Creus C and Guillermo J 2012 *Computational Viscoelasticity* SpringerBriefs in Computational Mechanics
[23] Schapery R A 1999 Nonlinear Viscoelastic and Viscoplastic Constitutive Equations with Growing DamageInternational Journal of Fracture 97 pp. 33-36
[24] Park S W and Schapery R A 1999 Methods of interconversion between linear viscoelastic material functions. Part I—A numerical method based on Prony series*Int. J. Solids Struct.* **36** 1653-1675
[25] Yusoff N I, Mounierb D, Marc-Stéphanec G, Aireye M and Di Benedetto H 2013 Modelling the rheological properties of bituminous binders using the 2S2P1D Model, *Construction and Building Materials* **38** 395–406 doi.org/10.1016/j.conbuildmat.2012.08.038
[26] Obrzud R. & Truty, A THE HARDENING SOIL MODEL - A PRACTICAL GUIDEBOOK Z Soil.PC 100701 report, revised 31.01.2012
[27] Chailleux E, Ramond G, Such C and de La Roche C A 2006 mathematical-based master-curve construction method applied to complex modulus of bituminous materials *Road Materials and Pavement Design* Volume **7**
[28] Judycki J, Pszczoła M, Jaczewski M, Mejłun Ł and Ryś R, 2013 *Badanie wpływu zastosowania warstw betonu asfaltowego o wysokim module sztywności (ac-wms) w konstrukcjach nawierzchni na spękania nieskotemperaturowe i na zmniejszenie powstawania deformacji trwałych* Raport 3
[29] Nagórski R 2014 *Mechanika nawierzchni drogowych w zarysie (in Polish)* PWN
[30] Gonzalez A A 2009 An Experimental Study of the Deformational and Performance Characteristics of Foamed Bitumen Stabilised Pavements *PhD Thesis* University of Canterbury
[31] ABAQUS *Theory Manual. Information on the SIMULIA* site: http://www.simulia.com/support/documentation.html
[32] Tan S A, Lo B H and Fwa T F 1994 Behavior of Asphalt Concrete Mixtures in Triaxial Compression *Journal of Testing and Evaluation* vol. **23** N° 3 pp. 195-203
[33] Werkmeister S 2004 Permanent deformation behaviour of unbound granular materials in pavement constructions *PhD Thesis* Technical University of Dresden Germany

IOP Conf. Series: Materials Science and Engineering **356** (2018) 012012 doi:10.1088/1757-899X/356/1/012012

Influence of Excessive Moisture in the Subgrade on the Durability and Load-Bearing Capacity of Road Pavements

Mieczkowski P., Budziński B.

West Pomeranian University of Technology Szczecin

E-mail: pawel.mieczkowski@zut.edu.pl

Abstract. When well performed, pavement renewal or alteration shall ensure the desired properties of the road during the assumed period of operation. Presence of water in the subgrade can be one of the main factors affecting the structural capacity of pavement and can result in cracking of the bituminous layers, even after a very short period of trafficking. Reconstruction of one of regional roads in Poland has been chosen to serve as an example of inappropriate approach to the problem of the presence of water in the road structure. The project included construction of new layers of pavement and increasing the design life of the whole pavement structure to 4.06 million ESAL of 100 kN (as per the Standard Catalogue of Typical Flexible and Semi-rigid Road Pavement Structures, issue of 1997). After a relatively short period of trafficking (3-5 years) localised alligator cracking appeared on the surface along with structural deformations. The pavement condition assessment including FWD tests was carried out to reveal excessive deflections (over 500 μm) which classify the road for renewal. The analysis of data showed that the main cause of distress was softening of the subgrade caused by an ingress of precipitation water under the pavement layers through the roadway and shoulder edges. The deficiencies of the performed reconstruction occurred both in the roadway (including small step-outs in the cement-treated layer) and partly in the shoulders where the existing soil was in places replaced with impervious material, with the existing (cohesive) material left in place on a major part of the overall length.

1. Introduction

Pavement renewal or alteration works, when well performed, shall ensure the desired properties of the road during the assumed period of operation. Planning of such works should be preceded by comprehensive investigation of the existing condition of pavement, including the hydrogeological conditions of the soil subgrade. A well designed the road drainage system is also very important. Presence of water in the subgrade can be one of the main factors affecting its structural capacity and can result in fatigue cracking of the bituminous layers, even after a very short period of trafficking. The activities related to protecting the pavement from water must not be limited to dealing with ground water and attention must also be paid to storm water infiltrating under the pavement layers. Storm water should be evacuated, as soon as possible, beyond the area of the most severe traffic impact through open or closed (subsurface) drainage system. Quality of works and appropriate longitudinal profile of the pavement layers and subgrade are among the main factors ensuring efficient drainage and correct behaviour of the whole road structure.

Typical pavements are protected from infiltration of storm water by an impervious surface layer. This layer must have appropriate longitudinal and transverse slopes for efficient draining of water to open ditches or to a closed drainage system. In open drainage systems particular attention must be paid to pervious dirt shoulder which may hinder the flow of water in places where it overtops the roadway edge elevation. As a result water may penetrate into the pavement structure. This can be particularly dangerous if water cannot drain towards open ditches through dirt shoulders made of cohesive materials and there is no proper drainage or separation layer in place.

2. Case Study – Current Situation

Due to a poor condition of the road and increasing level of traffic to be handled (including a large percentage share of HGV's) the road operator decided to undertake reconstruction of a 3.5 km long section of the road. The reconstructed section would be a single carriageway road with two 3.5 m wide traffic lanes and 0.70-1.2 m wide shoulders (varying). Subsurface conditions were investigated and ground water was not encountered (to 4 m depth below the bottom of pavement) while the subgrade soil was rated as most frost-susceptible (G class, clay). On this basis, taking into account the vertical alignment of the road the subgrade was classified in G4 strength group ($E2 \geq 25$ MPa) according to the Polish subgrade classification system. This rating was adopted in the design of the new pavement:

- Wearing course SMA 11 PMB 45/80-55 - 4 cm,
- Binder course AC 16 W 35/50 - 8 cm,
- Base course AC 22 P 35/50 - 10 cm,
- Base course Crushed stone 0/31,5 mm - 20 cm,
- Stabilized course Rm=2,5 MPa - 25 cm.

The design included strengthening of a 50-70 cm wide strip of shoulder with 15 cm thick layer of recycled asphalt pavement (RAP).

The pavement structure satisfies the frost-resistance requirement (Hdes = 67 cm > Hreq = 60 cm).

The pavement structure has fatigue life of 4.06 million ESAL of 100 kN (calculated with the Asphalt Institute method [1]), i.e. is designed for KR4 traffic loading according to the Polish classification system. The reconstructed road was designed to ensure the properties of pavement at the required levels in the 20-year period of operation.

Taking into account the new issue of the Typical Flexible and Semi-rigid Pavement Design Catalogue [2] the fatigue life of pavement was calculated using the method of AASHTO 2004 [3]. Assuming the value of FCbottom at 10% (for bottom-up fatigue cracking) fatigue life of 8.2 million ESAL of 100 kN was obtained, meeting the requirement for KR5 traffic loading. The governing criterion was fatigue cracking of the asphalt layers.

According to the construction program the works were to be carried out in spring and in summer. The scope of work included removal of existing layers, grading and compacting of the road subgrade, placing of the cement-treated soil layer (Rm = 2.5 MPa) and finally placing the remaining layers of pavement. The soil removed from shoulders was to be reused subject to meeting the permeability requirement. However, the design did not specify the required properties of soil (neither grading or permeability). The pavement configuration showing connection with the dirt shoulder is presented in Fig. 1.

Fatigue cracks started to appear locally after 3-4 years of trafficking along with structural deformations (Fig. 2). The distress was most severe at low spots of the road (e.g., within concave vertical curves).

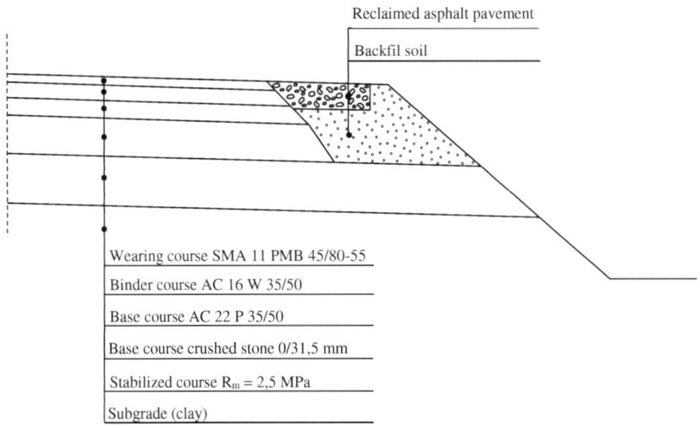

Figure 1. Layers of pavement reconstructed on voivodeship road.

Figure 2. Fatigue cracking and rutting after 3-4 years of open road into service.

3. Field and Laboratory Testing

A detailed field investigation was indispensable for correct determination of the causes of pavement distress. Non-destructive tests were performed in the first stage, including FWD tests to determine the structural capacity of the pavement and GPR tests to determine the thicknesses of the pavement layers. This stage of testing was commissioned by the road administrator. The tests were carried out in the first days of September in partly cloudy weather with no heavy rainfall and at ambient temperatures of 10-20°C. In the second stage (at the end of September) cores were cut from the pavement for more accurate thickness determinations and for testing the physical and strength properties of the pavement layers. Additionally, samples were taken from the shoulder.

Deflection tests were carried out early in the morning (for safety reasons) in the right wheel path. The test load of 50 kN (±0.5 kN) was applied at 25 m intervals. The temperature of the bituminous layers was adopted at 13-14°C. The deflections were measured with an array of geophones deployed at the plate centre and at pre-determined offsets of 200, 300, 450, 600, 900, 1200, 1500 and 1800 mm. The analysed parameters were the maximum deflections at the plate centre and surface curvature index SCI300. The maximum deflections at the plate centre on the slow lane are presented in Fig. 3. Similar results were obtained in the fast lane.

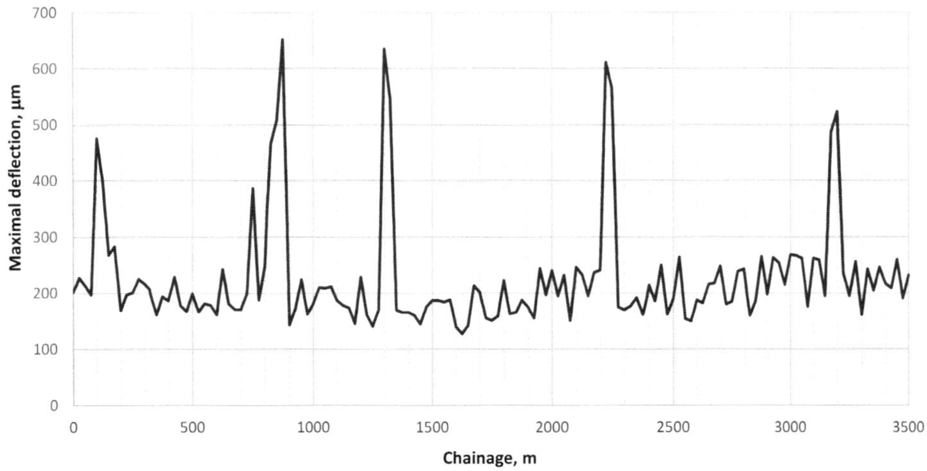

Figure 3. Maximal deflection from FWD test (nearest wheel, nearest line).

The maximum deflections at the plate centre can be classified in two groups. The first group includes results falling in the range from 125 to 270 µm making up over 90% of all the results. The second group includes results obtained at twelve test points. There the deflections exceeded 350 µm. This grouping of results was also relevant for the deflection bowl shapes. This concerns primarily much greater increase of deflection with the decreasing distance to plate centre recorded at these twelve test points where high values of maximum deflections were recorded. Fig. 4 presents the character of deformation obtained through deflection measurements on the respective geophones for the above-mentioned two groups of results.

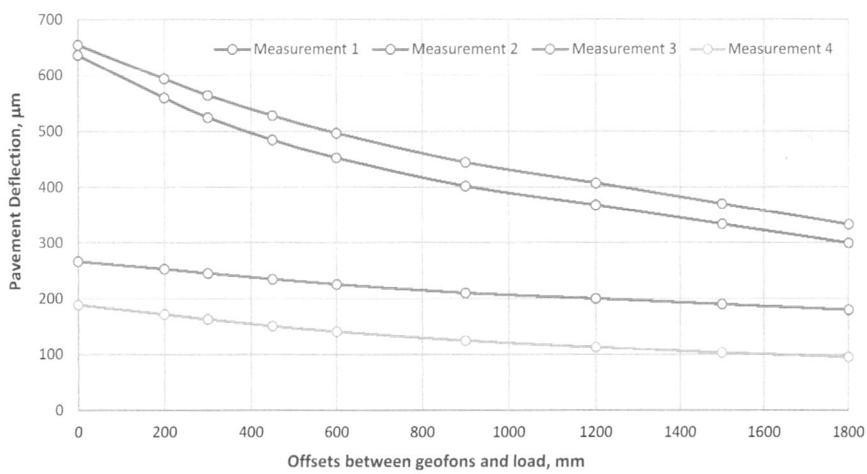

Figure 4. Pavement deflection in every geophone from FWD test (test 1 and 2 – places where are large maximal deflection, place 3 and 4 – places where are small maximal deflection).

In addition to determining the thicknesses of pavement layers the GPR system can also be used to detect changes or failures of the pavement structure, loss of bonding between pavement layers or excessive moisture content in the soil subgrade or base layers. The measurements were carried out with two antenna systems operating at different frequencies, namely 2.2 GHz and 1.0 GHz. The two antenna systems had different penetration depths: ca. 45 cm and ca. 90 cm. The results of thickness measurements on the both lanes are presented in Table 1.

IOP Conf. Series: Materials Science and Engineering **356** (2018) 012012 doi:10.1088/1757-899X/356/1/012012

Additionally, the layer thicknesses measured in the slow lane in relation to the road chainage are presented graphically in Fig. 5. The analysis of GPR results did not reveal any loss of bonding between the bituminous layers or excessive moisture content in the lower layers of the pavement structure (soil subgrade, cement-treated soil or crushed aggregate base).

Table 1. Layer thickness from ground-penetrating radar (GPR).

Statistical parameters	Asphalt layers		Base course Crushed stone		Stabilized course	
	Right lane	left Lane	Right lane	Left lane	Right lane	Left lane
Average [cm]	21.2	21.5	20.6	19.7	22.1	22.3
Standard deviation [cm]	2.5	2.2	2.8	2.7	3.6	3.2
Maximum value [cm]	25.5	24.3	25.6	24.7	28.5	28.2
Minimum value [cm]	17.0	17.3	16.0	16.7	16.8	17.3
Quantile 75% [cm]	19.1	18.9	18.2	18.4	19.7	19.4
Quantile 85% [cm]	18.2	18.4	17.2	17.7	18.4	18.5
Quantile 95% [cm]	17.4	17.7	16.3	16.6	17.4	17.3

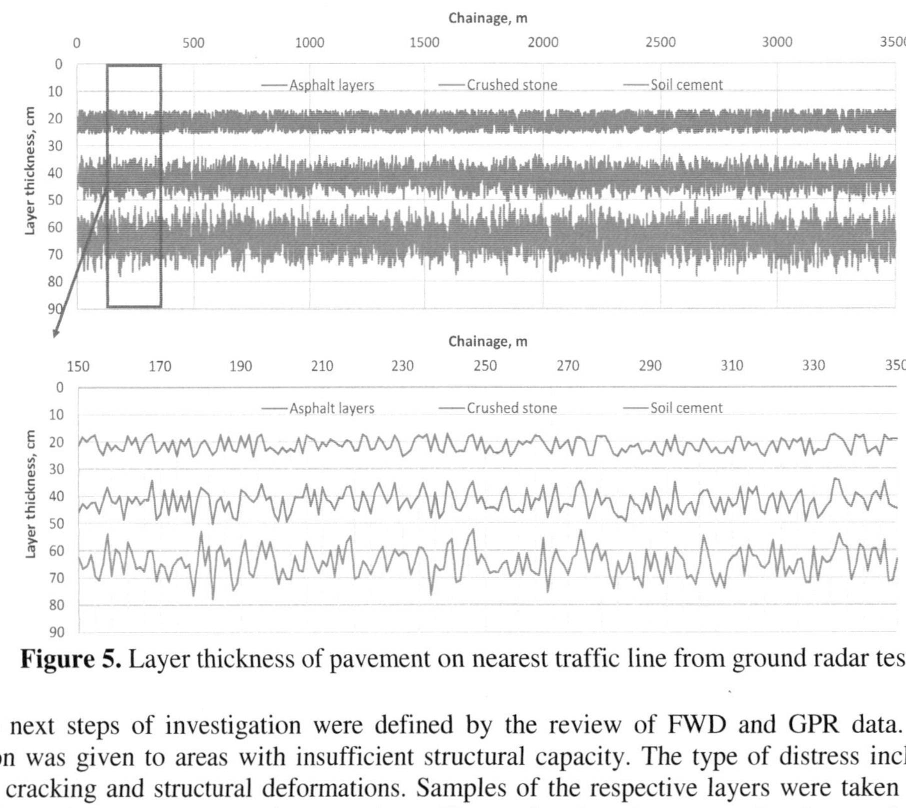

Figure 5. Layer thickness of pavement on nearest traffic line from ground radar test.

The next steps of investigation were defined by the review of FWD and GPR data. Particular attention was given to areas with insufficient structural capacity. The type of distress included both fatigue cracking and structural deformations. Samples of the respective layers were taken by cutting cores from the pavement on these sections. The coring locations were in places of maximum deflections (5 No.) and additionally at two points with small deflections for comparison. The scope of tests included determination of different parameters of the pavement layers, including their thicknesses, stiffness moduli determined with the indirect tensile test on cylindrical specimens (IT-CY), compressive strength of the cement-treated soil layer and CBR ratios of crushed aggregate base and of the subgrade soil. In test points 1-5 cores were cut in the cracked areas and their vicinity (at ca. 3-10 m distance). The samples taken from uncracked areas (bituminous mixtures and cement-treated soil designated "**") had undisturbed structure which enabled determination of the stiffness modulus of the bituminous base course by the indirect tensile tests (IT-CY), determination of the relative compaction of this layer and determination of the compressive strength of the cement-treated soil layer. In areas of apparent cracks samples were taken from all the pavement layers and cracks were found both in the asphalt layers and in the cement-treated soil layer. This definitely confirms that cracks were initiated at the bottom of the pavement structure, i.e. in the cement-treated soil and in the bituminous road base layers. The test results are presented in Table 2.

Table 2. Parameters of pavement layers based on core drilling.

Parameter	Sample						
	1	2	3	4	5	6*	7*
Chainage, m	100	875	1300	2225	3200	925	2600
Asphalt layers thickness, cm	21.7 / 21.2**	23.5 / 22.8**	22.4 / 23.1**	21.1 / 21.7**	22.6 / 22.3**	19.6	21.1
Compaction factor (AC 22 P), %	– / 97.6**	– / 99.4**	– / 98.7**	– / 100.3**	– / 98.5**	99.2	98.3
Voids in total mix (AC 22 P), %	– / 6.9**	– / 5.3**	– / 6.2**	– / 4.8**	– / 6.1**	5.5	6.4
Crushed stone layer thickness, cm	– / 23.4**	– / 17.2**	– / 22.4**	– / 19.2**	– / 17.1**	23.2	19.6
Stabilized course layer thickness, cm	– / 24.3**	– / 24.0**	– / 23.6**	– / 23.6**	– / 23.8**	21.8	23.6
Stiffness module IT-CY, 20°C, MPa	– / 7654**	– / 7023**	– / 6578**	– / 6487**	– / 7787**	6203	7436
Compressive strength of stabilized course, MPa	– / 3.26**	– / 3.51**	– / 2.96**	– / 4.12**	– / 2.67**	3.75	4.06
California bearing ratio of crushed stone, %	124	141	138	126	154	131	129
California bearing ratio of subgrade(optimum moisture), %	14	9	13	16	11	12	17
California bearing ratio of subgrade (4-day soaked), %	5	2	3	3	4	2	3

* place where pavement wasn't destroyed, ** place next to where pavement was destroyed (3 - 10 m)

The structure of the cement-treated soil in the area affected by surface cracking was disturbed, specifically it was partly mixed with both crushed aggregate and with subgrade soil.

Soil samples taken from the subgrade featured a relatively high bearing capacity when tested at optimum moisture content (70 to 108 MPa elastic modulus according to Powell). With increasing moisture content the bearing capacity decreased considerably and so did the value of the elastic modulus (25 to 50 MPa). This could be one of the causes of the observed structural failures.

The soil in the shoulder area was found to be highly diverse. It included both loose material (medium and coarse sand) and cohesive material (clay, silty clay, pre-Quaternary clay, loamy sand), the latter type of material being in majority.

4. Analysis of the Test Results

One of the main performance characteristics which is used to rate the condition of pavement is the elastic deflection. It can be determined by FWD measurements. The parameters measured during this tests include the maximum deflection at the plate centre, which is converted to normalised deflection value (according to [4, 5]) and used for determining uniform sections (for example with the cumulative sum chart). The equivalent deflections are then calculated for these uniform sections. The second parameter is the surface curvature index SCI300 calculated as a difference between the centre deflection and the deflection measured at 300 mm distance from the plate centre. These two parameters are used to determine the so-called decision level by assigning condition rating to the pavement [4, 5].

Fig. 6 presents the normalised deflection values (obtained by converting the results to standard conditions of 50 kN weight and 30 cm plate diameter and 20°C temperature, taking into account the season of the year and the road base material) as cumulative sums.

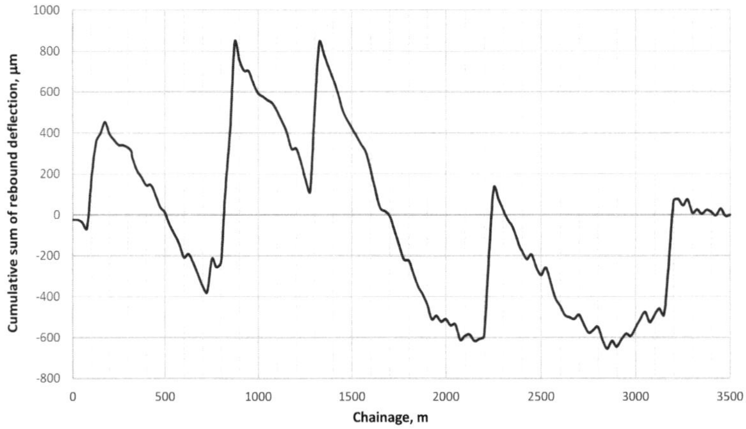

Figure 6. Cumulative sum of rebound deflection on voivodeship road.

It clearly transpires from Fig. 6 that higher values were obtained in isolated spaces influencing the shape of the curve and the number of uniform sections. In order to obtain a better illustration of the effect of places with deflections exceeding the maximum allowed value basic static calculations were conducted for the whole analysed section (before adjustment) and for the section without areas with higher deflections (after adjustment) for both lanes (Table 3).

Table 3. Statistics of deflection measurements on left and right lane.

Statistical parameters	Right lane		Left lane	
	Before correction	After correction	Before correction	After correction
Average [µm]	303.8	266.9	295.4	254.5
Standard deviation, D_U [µm]	133.8	48.4	138.7	46.2
Maximum value [µm]	875.3	380.1	891.5	376.9
Minimum value [µm]	170.8	170.8	175.3	167.8
Standardized deflection (deflection index), U [µm]	437.6	315.3	434.1	300.7
Quantile 85% [µm]	349.4	326.6	334.3	311.8
Quantile 95% [µm]	654.5	352.9	626.9	334.5

According to the results presented in Table 3 the differences between the average measured deflections before and after adjustment fall in the range of 36-40 µm. Much greater differences are obtained for the typical deflections, namely ca. 120-130 µm. This is attributed to high statistical variability indicated by the value of standard deviation. The equivalent deflections obtained after omitting weak spots correspond to the values specified for the KR4 traffic loading in the DSN pavement rating system used in Poland (except for a small exceeding in the slow lane). When the whole section is considered the equivalent deflection is close to the value critical for KR4 traffic loading (on both travel lanes) and immediate renewal is required.

The equivalent value of the curvature index SCI300 which is one of the parameters for rating the bearing capacity of pavements is calculated as the sum of the average standard deviation of SCI300 values for the whole section (before adjustment), for the section without higher deflection areas and for high deflection areas. These values, together with additional statistical parameters for SCI300 curvature index, are presented in Table 4.

Table 4. Statistical parameters, Surface Curvature Index SCI300.

Statistical parameters	Right lane			Left lane		
	before correction	after correction	high deflection	before correction	after correction	high deflection
Average [μm]	63.3	43.5	261.9	65.7	41.8	252.1
Standard deviation, D_U [μm]	66.0	11.2	55.5	62.1	10.4	53.2
Maximum value [μm]	366.6	78.4	366.6	339.2	79.2	339.2
Minimum value [μm]	28.1	28.1	197.4	32.2	32.2	169.8
Surface Curvature Index SCI [μm]	129.3	54.6	317.4	127.8	52.2	305.3
Quantile 85% [μm]	66.5	54.4	296.7	67.2	49.5	284.2
Quantile 95% [μm]	233.3	69.3	343.3	206.8	64.3	326.9

Considering the equivalent curvature index values for the whole section (including higher deflection areas) the road obtains C rating for KR4 traffic loading (at the boundary of D rating). Note that for KR5 traffic loading the condition rating would be D and the road would require renewal (according to [5]). When the areas with increased deflections are left out the road gets B rating for both KR4 and KR5 traffic loading. In higher deflection areas the values of SCI were very high, considerably exceeding the limit of D rating irrespective of the traffic loading. This clearly indicates damage to bituminous layers, resulting in distribution of stresses on a smaller area of soil subgrade.

For selected locations (two high deflection spots marked B and two low deflection spots marked L) the resilient moduli were back-calculated for the respective layers of pavement. For the crushed aggregate sub-base a constant value of the resilient modulus was adopted at 400 MPa. The thicknesses of the pavement layers were determined on the basis of GPR data. The thickness of the soil subgrade layer was taken at 5.9 m. The calculation results are presented in Table 5.

Table 5. Stiffness module from „backcalculation".

Layer	B1		B2		L1		L2	
	h, mm	*S*, MPa	*h*, mm	*S*, MPa	*h*, mm	*S*, MPa	*h*, mm	*S*, MPa
	204	2175	216	2841.4	224	15336.4	221	9431.0
Base course Crushed stone	190	400*	200	400*	236	400*	229	400*
Stabilized course	225	567.1	236	454.9	257	7097.2	240	6237.8
Subgrade	5000	45.8	5000	37.8	5000	101.8	5000	109.9

* – constant value for calculations, *h* – layers thickness, *S* – Stiffness module (modulus of elasticity)

The calculated values show much smaller resilient moduli of the bituminous and cement-treated soil layers in high deflection areas, as compared to areas where smaller deflections were obtained in measurements. This is in agreement with the structural damage of these layers due to cracking. Different resilient moduli were also obtained for the subgrade. This could be attributed to a higher moisture content affecting its bearing capacity.

Insufficient thickness of pavement layer(s) or their mechanical properties (compaction, strength, etc.) were excluded as the causes of pavement distress on the basis of GPR data and determinations on undisturbed samples taken at the damaged sites. During the site visit it was established that the defective areas are located at low spots along the road. Consequently, longitudinal profile measurements were carried out. The vertical alignment of the road, related to the maximum deflections measured with FWD is presented in Fig. 7.

IOP Conf. Series: Materials Science and Engineering **356** (2018) 012012 doi:10.1088/1757-899X/356/1/012012

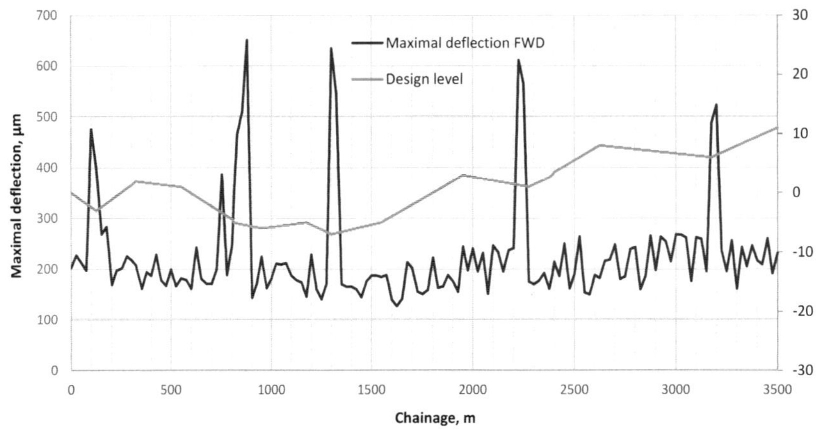

Figure 7. Road vertical alignment with maximal deflection.

From the analysis of results presented in Fig. 7 it can be seen that the greatest deflections coincide with concave vertical curves. Storm water could pond there and penetrate under the structure through the edge of dirt shoulder. This affected the bearing capacity of subgrade and, subsequently, also of the pavement courses, in particular the layers of increased stiffness, i.e. cement-treated and bituminous.

At the dips of the road profile storm water penetrates under the pavement structure when it cannot drain into open ditches. This suggests that efficient drainage of water outside the roadway prism was not ensured during the works. In order to verify this, test pits were dug at pre-determined locations at the road edge. It was found out that the cross section and the layer locations vary and conform to the design only in places (Fig. 1), deviating from it on a major part of the section, in particular where shoulder extends in width - see Fig. 8.

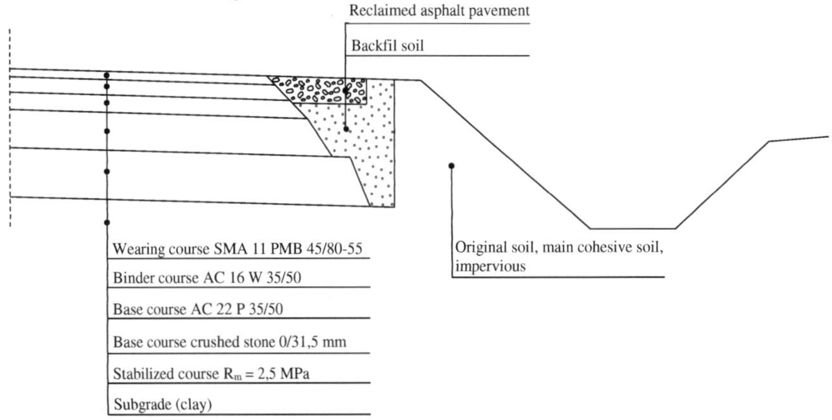

Figure 8. Cross section of road with marginal strip.

5. Summing up

The analysis of the test and measurement data and findings from condition assessment indicate that the causes of distress should not be attributed to insufficient thicknesses of pavement courses or to the values of their performance parameters. This has been confirmed by comparing the data obtained at the places of distress, in their immediate vicinity and in areas where no apparent damages were found.

Therefore, with a high degree of probability we can attribute the cause of pavement distress to bearing failure of the subgrade soil which was caused by storm water penetration. Plastic yielding of soil in the subgrade caused structural failure of the cement-treated soil layer and, subsequently, also of the bituminous layers. This is confirmed by the bearing capacity data obtained in FWD test and by the condition of the pavement courses.

It clearly transpires from Fig. 8 that accumulation of storm water under the pavement and plastic yielding of subgrade soil was caused by reduced width of the cement-treated soil layer and replacement of the native soil in the shoulder with loose material (permeable and non-swelling). The native soil which was left undisturbed in the shoulder area has varying composition in which cohesive, swelling and impervious materials prevailed. An additional factor hindering the flow of storm water were higher places in shoulder area (beyond the RAP strip) created by growth of vegetation.

The recommended treatments include immediate repair of the affected areas, replacement of the shoulder material and widening of the cement-treated soil layer to the whole width of the road prism.

References

[1] Katalog Typowych Konstrukcji Nawierzchni Podatnych i Półsztywnych. IBDiM, GDDKiA Warszawa 1997

[2] Katalog Typowych Konstrukcji nawierzchni Podatnych i Półsztywnych. Załącznik do zarządzenia nr 31 GDDKiA, Politechnika Gdańska, 2014

[3] Judycki J., Jaskuła P., Pszczoła M. i inni: Analizy i projektowanie konstrukcji nawierzchni podatnych i półsztywnych. WKiŁ, Warszawa 2014

[4] System oceny stanu nawierzchni SOSN. Wytyczne stosowania. GDDKiA. Załącznik do Zarządzenia Nr 5 z dnia 01.02.2010

[5] Diagnostyka stanu nawierzchni i jej elementów. Wytyczne stosowania. GDDKIA. Załącznik do zarządzenia Nr 34 GDDKiA z dnia 30.04.2015

Analysis of pavement structure sensitivity to passage of oversized heavy duty vehicle in terms of bearing capacity

Rys Dawid, Jaskula Piotr

Gdansk University of Technology, Faculty of Civil and Environmental Engineering, Highway and Transportation Engineering Department.

dawrys@pg.gda.pl; piotr.jaskula@pg.edu.pl

Abstract. Oversized heavy duty vehicles occur in traffic very rarely but they reach extremely high weights, even up to 800 tonne. The detrimental impact of these vehicles on pavement structure is much higher than in case of commercial vehicles that comprise typical traffic, thus it is necessary to assess the sensitivity of pavement structure to passage of oversized vehicles. The paper presents results of sample calculations of load equivalency factor of a heavy duty oversized vehicle with usage of mechanistic-empirical approach. The effects of pavement thickness, type of distress (cracking or rutting) and pavement condition (new or old with structural damage) were considered in the paper. Analysis revealed that a single pass of an 800 tonne oversized vehicle is equivalent to pass of up to 377 standard 100 kN axles. Load equivalency factor calculated for thin structures is almost 3 times lower than for thick structures, however, the damage effect caused by one pass of an oversized vehicle is higher in the case of thin structure. Bearing capacity of a pavement structure may be qualified as sufficient for passage of an oversized heavy duty vehicle when the measured deflection, for example in an FWD test, does not exceed the maximum deflections derived from mechanistic-empirical analysis. The paper presents sample calculation of maximum deflections which allow to consider passage of an oversized vehicle as safe over different pavement structures. The paper provides road administration with a practical tool which helps to decide whether to issue a permit of passage for a given oversized vehicle.

Keywords
Oversized vehicles, heavy duty vehicles, bearing capacity of pavement structure, flexible pavements, load equivalency factors, standard axle, equivalent axle.

1. Introduction

Oversized heavy duty vehicles are used to carry untypical freights which cannot be divided into smaller parts. Most commonly the oversized freights include turbines of power plants, transformers, parts of ships or industrial machines. Oversized heavy duty vehicles occur in traffic very rarely but they reach extremely high weights, even up to 800 tonnes, while the weight of a typical commercial vehicle does not exceed 40 tonnes, except for overloaded vehicles. The definitions of an oversized heavy duty vehicle and an overloaded vehicle should be clearly separated. An overloaded vehicle is one that exceeds legal axle load or gross weight, while an oversized heavy duty vehicle is an untypical vehicle constructed specially for carrying freights heavier and with a larger volume than typical freights. The problem of vehicle overloading has been widely described in other publications [1, 2, 11-18] and it does not lie within the scope of this paper. In the case of oversized heavy duty vehicles, the basic problem is the issuing of permission for passage. Officers of road authorities most often have to make a decision without any specific data to assess if the passage of a given oversized vehicle is safe or if it may cause pavement damage.

2. Objectives and scope

The main objective of this paper is to present the methodology of calculations of fatigue damage and load equivalency factor of heavy duty oversized vehicles. The methodology is supported by calculations performed for a sample oversized vehicle. Calculations include both thin and thick pavements as well as two variants of pavement technical condition: good and poor. Further part of the paper presents the determination of the maximum pavement structure deflection which allows to permit safe passage of an oversized vehicle without pavement failure.

3. Load of the chosen oversized vehicle

The load of freight in the oversized heavy duty vehicle chosen for calculations (OSV) is distributed onto a group of wheels in a multi-axle trailer. A group of ballast tractors pulls or pushes the trailer with the freight, as shown in Figure 1. Loads of particular axles usually do not exceed 80 kN, so the level of axle load is comparable to loads of typical heavy traffic. On the other hand, the distances between particular axles and wheels are much smaller than in typical vehicles. Moreover, in the case of oversized vehicles the number of axles and wheels is much higher.

In the calculation example considered in this paper the vehicle consists of two trailers and two ballast tractors. Each trailer is equipped with 16 rows of axles, 2 axles per row, and each axle consists of 4 wheels. The distance between neighboring axle rows equals 1.65 m. The wheelbase across the vehicle is 3.60 m. The tare weight is 215 tonnes, the maximum weight of the transported cargo is 500 tonnes, so the maximum total weight of the trailer is 715 tonnes. The load is evenly distributed on each axle and may reach 112 kN per axle. The suspension of each axle is hydraulically adjusted in such a way that the load of each axle remains constant during transport. The ballast tractors are equipped with two steering axles with single wheels and two drive axles with twin wheels. The total weight of one ballast tractor is 48 tonnes. The load of ballast tractor is distributed on each axle and equals 120 kN per axle. The tire-pavement contact pressure equals 850 kPa. The speed of the oversized vehicle is much lower than the speed of typical traffic and does not exceed 20 km/h, which has an impact on elastic modulus of asphalt layers and in consequence on increase of fatigue damage.

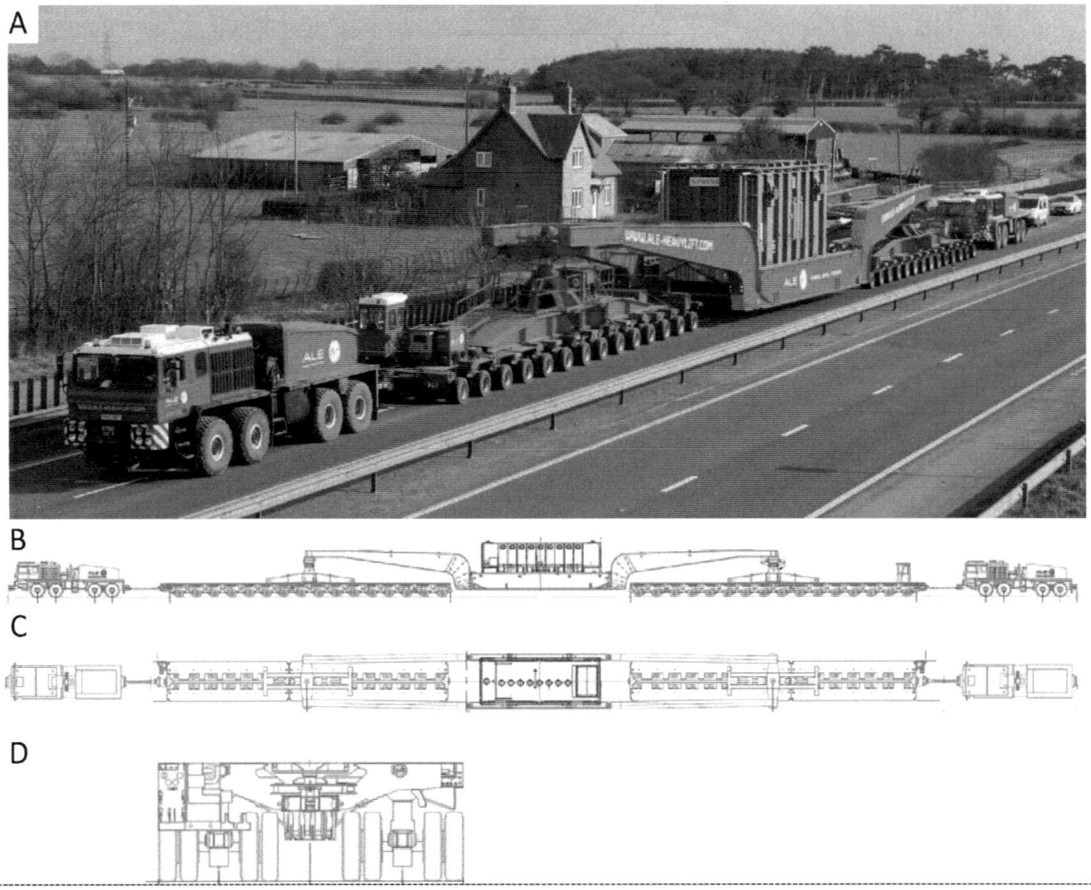

Figure 1. An example of oversized heavy duty vehicle – a group of two trailers and two ballast tractors A) general photograph [19], B) side view scheme C) top view scheme D) trailer front view scheme.

4. Pavement structure assumed for analysis

Calculations were performed for flexible pavement structures assumed on the basis of the Polish catalogue of typical flexible and semi-rigid pavements. Structures similar to those assumed for analysis are commonly used on the Polish road network. Thicknesses and mechanical parameters of structure layers are presented in Table 1. Calculations were performed for two pavements: thick (30 cm of asphalt layers) for very heavy traffic and thin (12 cm of asphalt layers) for light traffic.

Typical values of stiffness modulus of asphalt layers were assumed for calculation of fatigue damage caused by a standard axle load (variant 0). Typical values were assumed on the base of research project described in brief in [9]. However, for calculation of fatigue damage caused by the axles of the oversized vehicle, reduced values of stiffness modulus were assumed. The reduction of stiffness modulus values results from lower speed of the oversized vehicle. Two variants of pavement structure condition were considered: good condition without damage (variant 1) and bad condition near the level of terminal serviceability of pavement structure (variant 2). In the variant 2 stiffness modulus of asphalt layers was reduced to 50% and stiffness modulus of subbase and subgrade was reduced like for bad drainage conditions. All values of stiffness modulus used in the analysis are presented in Table 1.

IOP Conf. Series: Materials Science and Engineering **356** (2018) 012013 doi:10.1088/1757-899X/356/1/012013

Table 1. Thicknesses and mechanical properties of pavement structure layers.

No.	Pavement layers	Layer thickness [cm]	Volumetric properties of asphalt layers [% v/v]		Stiffness modulus [MPa]		
			Binder content	Air voids	**Variant 0** Standard load, good pavement condition v=60 km/h	**Variant 1** Oversized vehicle good pavement condition, v=20 km/h	**Variant 2** Oversized vehicle bad pavement condition v=20 km/h
			Thick structure (traffic category KR7)				
1.	Wearing course, SMA11 50/70	4	16	3	7300	5300	2650
2.	Binder course, AC16W 35/50	8	10.5	6	10300	8500	4250
3.	Asphalt base, AC22P 35/50	18	10	7	9800	8100	4050
4.	Subbase, crushed stone	20	-	-	400	400	250
5.	Subgrade	-	-	-	120	120	60
			Thin structure (traffic category KR2)				
1.	Wearing course, AC11S 50/70	4	14	2.5	9300	6800	3400
2.	Asphalt base, AC22P 35/50	8	10	8	8100	6100	3050
3.	Subbase, gravel	25	-	-	250	250	150
4.	Subgrade	-	-	-	80	80	35

5. Analysis of pavement structure response under loads of the oversized vehicle

The program BISAR from Shell company was used to perform calculations of deflections and strains in pavements caused by load of oversized vehicle. Firstly, the critical locations were determined. Critical locations are defined as points where tensile or vertical strains reach maximum values. When the criterion of fatigue cracking of asphalt layers is considered, the critical locations occur at the bottom of asphalt layers (horizontal strains ε_{xx} and ε_{yy}). When the permanent deformation of subgrade is considered, the critical locations occur at the level of subgrade surface (vertical strains ε_{zz}). The procedure of determination of critical locations and strain values is as follows:

1) Calculation of strains in the direction transverse to vehicle movement (along the X-axis) and identification of critical sections at which strains reach their extremes.
2) Calculation of strains along the direction of travel (along the Y-axis) at critical sections, from the load of one row of wheels.
3) Calculation of strains taking into account the influence of neighboring axles, in accordance with the superposition principle.
4) Calculations are performed separately for ballast tractor and for trailers.

The results of strain calculations for trailer on thick new pavement (variant 1) at a critical section is shown in Figure 2. The critical point at the bottom of asphalt layers occurred for x = 1000 mm, and at this point $\varepsilon_{yy,\ max}$ = 73.0 μstrain. Similarly, the critical point for vertical strain at the top of subgrade surface occurred for x = 1000 mm and equaled $\varepsilon_{zz,\,min}$ = - 214.0 μstrain. For the example considered the dimension x of the maximum strains at the bottom of asphalt layers and at the top of subgrade surface is the same (x = 1000 mm), however for other examples of calculations it can be different. Figure 3 presents strains along the Y-axis at the critical strain location (x = 1000 mm). In the case of horizontal strain at the bottom of asphalt layers, the difference between minimum and maximum values of ε_{yy} is higher than for ε_{xx}, thus strain ε_{yy} is assumed as critical strain for calculations of fatigue damage. The values of vertical strains at the top of subgrade surface are negative and for calculations of fatigue damage the extreme values are assumed. This approach is the same as in damage calculations in fatigue tests and it has been described in more detail in the work [16].

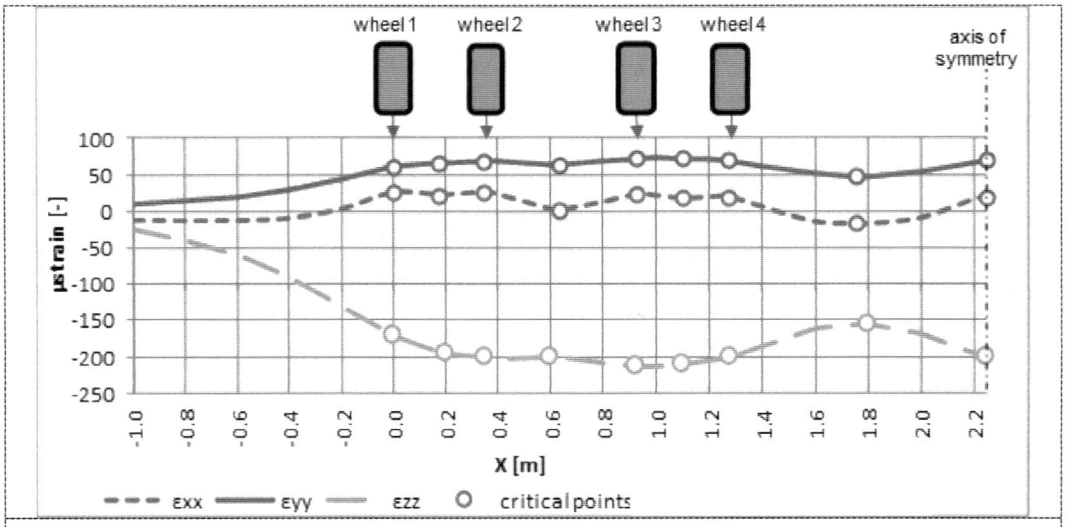

Figure 2. Strains caused along the X-axis by a row of wheels at a critical section perpendicular to the direction of vehicle movement, thin pavement, variant 1.

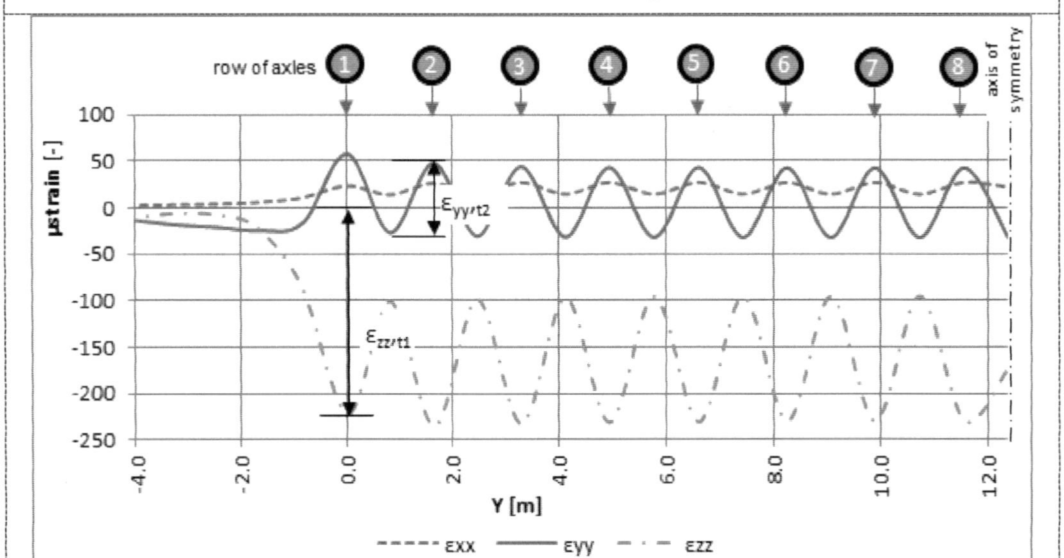

Figure 3. Strains caused by consecutive axles of the oversized vehicle trailer along the direction of vehicle movement (Y-axis), thin pavement, variant 1.

6. Analysis of the fatigue damage and load equivalency factors caused by oversized vehicles

The analysis of fatigue damage caused by oversized vehicles is based on mechanistic-empirical approach. Two variants of pavement distress were considered: bottom-up fatigue cracking of asphalt layers and permanent deformation of the subgrade. For asphalt layers the bottom-up fatigue cracking AASHTO 2004 criterion was used. It is given in equation (1). The criterion was developed in the USA and it is described in detail in the Mechanistic-Empirical Pavement Design Guide [3]. The AASHTO 2004 criterion was adapted to Polish conditions by Judycki [8].

$$N_f = D_{FC} \cdot 7.3557 \cdot 10^{-6} \cdot C \cdot k'_1 \left(\frac{1}{\varepsilon_t}\right)^{3.9492} \left(\frac{1}{E}\right)^{1.281} \quad (1)$$

where:
N_f – number of load repetitions to fatigue cracking of asphalt layers on the area FC of the lane.
D_{FC} – fatigue damage parameter,
k'_1 – correction parameter, introduced to provide a correction for different asphalt layers thicknesses,
ε_t – horizontal tensile strain at the critical location at the bottom of asphalt layers,
E – stiffness modulus of the asphalt layers [MPa],
C – laboratory-to-field adjustment factor which includes the effect of binder and air content in asphalt mixture.

Fatigue life of pavement structure due to permanent deformation of the subgrade was calculated according to the Asphalt Institute criterion:

$$N_f = k_1 \left(\frac{1}{\varepsilon_t}\right)^{n_1} \quad (2)$$

where:
N_f – number of load repetitions to occurrence of permanent structural deformation of 12.5 mm depth at the pavement surface,
ε_t – vertical compressive strain at the critical location at the subgrade surface,
$k_1 = 0.0105$ – empirical parameter,
$n_1 = 0.223$ – empirical parameter.

The result delivered from equations (1) or (2) of fatigue criteria is the number N_f of load repetitions to pavement distress. The number N_f depends on strain at critical location ε_t and, in turn, the strain ε_t depends on axle load and distance to neighboring axle, as shown in Figure 3. In the case of criterion of fatigue cracking of asphalt layers, the value of ε_t in equation (1) should be assumed as the difference between minimum and maximum strain ε_{yy}. In the case of permanent deformation, the value ε_t in equation (2) should be assumed as the absolute vale of the minimum ε_{zz}. A more detailed explanation of this approach is available in [16].

According to Miner's law, each transition of vehicle axle causes accumulation of fatigue damage. In the case of a single axle transition, the fatigue damage is equal to:

$$d_j = \frac{1}{N_{fj}} \quad (3)$$

where:
d_j – fatigue damage caused by axle with load Q_j,
N_{fj} – number of repetitions of load Q_j to occurrence of pavement distress,
j – axle number.

In the case of transition of a multiple axle, the fatigue damage is delivered from the formula:

$$d_j = \sum_{i=1}^{n} \frac{1}{N_{f\,i,j}} \tag{4}$$

where:

$N_{fi,j}$ – number of repetitions of load causing strain ε_{tj} at the critical point of pavement structure to occurrence of pavement distress,
j – axle number,
i – critical point number,
n – total number of critical points along the direction of vehicle movement (Y-axis).

The load equivalency factor of one axle in the vehicle can be calculated based on of fatigue damage according to equation (5), delivered from the work of Judycki [5, 6].

$$F_j = \frac{d_j}{d_s} \tag{5}$$

where:

F_j – load equivalency factor for axle Q_j,
d_s – fatigue damage caused by a standard axle with load Q_s = 100 kN,
d_j – fatigue damage caused by axle j, according to equations (3) or (4).

The load equivalency factor for the whole vehicle is delivered from the following formula:

$$F_v = \sum_{j=1}^{n_O} F_j \tag{6}$$

where:

F_v – load equivalency factor of the vehicle,
F_j – load equivalency factor of a given axle j (single or multiple) in the vehicle,
j – axle number,
n_o – total number of axles in the vehicle.

Calculations of fatigue damage and load equivalency factors were performed for an example of oversized vehicle OSV in accordance with equations (1-6). The results are presented in Table 2. As shown in Table 2, the load equivalency factor depends on the pavement distress criterion as well as pavement structure thickness and condition. Oversized vehicles have greater load equivalency factor for thick pavements and better pavement technical condition. Nevertheless, higher fatigue damage from the OSV is observed for thinner pavements and worse pavement technical condition. The values of fatigue damage are relatively low, even for thin pavements in poor condition, which means that one pass of the OSV should not cause serious damage to pavement structures. It may be assumed that the level of fatigue damage lower than 1% can be acceptable to permit the passage of the OSV.

According to studies of commercial vehicle loads in Poland [15, 18], an average articulated heavy vehicle with trailer or semi-trailer reaches load equivalency factor in the range from 1.60 to 1.95 100 kN ESALs. One passage of the OSV may cause the same fatigue as passage of 200 typical commercial heavy vehicles.

Table 2. Bearing capacity of pavement structures, fatigue damage and load equivalency factors of the oversized vehicle (OSV)

Pavement structure thickness	Variant of pavement conditions	Fatigue cracking of asphalt layers			Structural deformation		
		Bearing capacity [100 kN ESALs]	OSV fatigue damage d_{FC}	OSV load equivalency factor F_{FC}	Bearing capacity [100 kN ESALs]	OSV Fatigue damage d_{DEF}	OSV Load equivalency factor F_{DEF}
Thick (30 cm HMA)	Good (1)	$66 \cdot 10^6$	**0.0006%***	184	$251 \cdot 10^6$	0.0001%	321
	Poor (2)	$2 \cdot 10^6$	**0.0073%***	171	$13 \cdot 10^6$	0.0028%	377
Thin (12 cm HMA)	Good (1)	$368 \cdot 10^3$	**0.02%***	80	$405 \cdot 10^3$	0.01%	58
	Poor (2)	$31 \cdot 10^3$	0.07%	75	$17 \cdot 10^3$	**0.38%***	64

* critical value of fatigue damage

7. Determination of the required bearing capacity of pavement structure

The bearing capacity calculated from the pavement model for the assumed mechanical parameters and presented in Table 2 is theoretical. In the field the pavement thickness as well as stiffness modulus of subgrade and pavement layers may be varied. The differences can be especially high in the case of thin pavements in poor technical condition, and this variation may result in underestimation of fatigue damage caused by the OSV. In consequence, the pass of the OSV may cause local appearance of serious pavement damage. At weak points, both permanent deformation and cracks of asphalt layers can be expected. In order to avoid overestimation of bearing capacity of pavement structure it is recommended to perform a Falling Weigh Deflectometer (FWD) test, which allows to localize weak areas in the road pavement. The judgment of bearing capacity of pavement can be performed in two ways: by determining critical maximum deflections under an FWD plate or by determining stiffness modulus of pavement layers using back-calculations. The back-calculations deliver real values of stiffness modulus, which can be further used in mechanistic-empirical analysis according to the approach described above.

Critical deflections were calculated based on the mechanistic model of pavement structure and mechanical parameters of layers assumed for pavement in bad condition. The critical deflections were determined in the following steps:

1. Calculation of the maximum deflection at pavement surface under flexible plate (standard axle wheel) using the BISAR software,
2. Deflections calculated for flexible plate were multiplied by factor of 0.785 to account for differences in deflections under flexible and rigid plate [5], as well as the dynamic factor f_d=0.83 [3] and the temperature factor f_t=1.14°C [10] to standardize temperature of deflection measurement from 13°C to 20°C.

The values of critical deflections are given in Table 3.

Table 3. Critical deflections under FWD plate

Pavement structure thickness	Critical deflection calculated for standard axle [mm]	Critical deflection calculated for FWD [mm]
Thick (30 cm of HMA)	0.39	0.3
Thin (12 cm of HMA)	1.20	0.9

8. Summary

1) The methodology of calculation of fatigue damage and load equivalency factor of heavy duty oversized vehicles is based on mechanistic-empirical approach. The methodology was used to perform calculation for one example of an oversized vehicle, however, it is universal and can be adopted for any type of vehicle. The paper provides road administration with a practical tool, which helps to decide whether to issue a permit of passage of an oversized vehicle.
2) Two variants of pavement structure condition (good and poor) as well as two types of pavements structure thickness (thick and thin) were considered in the analysis. Analysis revealed that new pavements in good technical condition have enough bearing capacity to bear an oversized vehicle without any pavement failure, even if their thickness is low (12 cm). The fatigue damage caused by an 800 tonne oversized vehicle on new pavement in good condition varies from 0.0006% to 0.02%.
3) Load equivalency factor can reach a level of up to 321 ESALs (100 kN). One pass of an oversized vehicle is equivalent to pass of approximately 200 commercial heavy vehicles, occurring in typical road traffic flow. Load equivalency factor of an oversized vehicle is lower for thinner pavements and for pavements in worse condition, in contrast to fatigue damage caused by the oversized vehicle.
4) In the case of pavements in poor technical condition, fatigue damage caused by oversized vehicle is much higher. Moreover, bearing capacity of old pavements may be locally underestimated, thus it is advisable to perform an FWD test to identify and localize weak areas in pavements. The paper presents methodology for determination of critical deflections under an FWD plate. Critical deflections can be used to assess sensitivity of pavement to passage of oversized vehicles.

Literature

[1] Budzyński M., Ryś D., Kustra W., 2017: *Selected Problems of Transport in Port Towns – Tri-City as an Example*. Polish Maritime Research. Vol. 24, issue. s1 (2017), s.16-24

[2] Burnos P., Rys D., 2017: *The Effect of Flexible Pavement Mechanics on the Accuracy of Axle Load Sensors in Vehicle Weigh-in-Motion Systems*. Sensors, Vol. 17, 2053, doi: 10.3390/s17092053

[3] *Diagnostyka stanu nawierzchni i jej elementów*, (in Polish). GDDKiA, Warsaw, 2015. Available in the Internet: https://www.gddkia.gov.pl/pl/2982/Diagnostyka-Stanu-Nawierzchni

[4] *Guide for Mechanistic-Empirical Design of New and Rehabilitated Pavement Structures. Final Report, Part 3 – Design Analysis, Chapter 3, Design of new and reconstructed flexible pavements*. NCHRP, TRB, NRC, 2004.

[5] Huang Y.H. 1993: *Pavement Analysis and Design*. Prentice-Hall, New Jersey.

[6] Judycki J., 2010: *Determination of Equivalent Axle Load Factors on the Basis of Fatigue Criteria for Flexible and Semi-Rigid Pavements*. Road Materials And Pavement Design. Vol. 11(1), pp. 187-202, doi: 10.1080/14680629.2010.9690266

[7] Judycki J., 2011: *Equivalent axle load factors for design of rigid pavements derived from fatigue criteria*. Baltic Journal of Road and Bridge Engineering, Vol. 6(4), pp. 219-224, doi: 10.3846/bjrbe2011.28

[8] Judycki J., 2011: *Modele spękań zmęczeniowych warstw asfaltowych nawierzchni drogowych w mechanistyczno-empirycznej metodzie AASHTO 2004* (in Polish), Drogownictwo Vol. 11/2011, pp. 343-347.

[9] Judycki J., Jaskuła P., Pszczoła M., Ryś D., Jaczewski M., Alenowicz J., Dołżycki B., Stienss M., 2017: *New polish catalogue of typical flexible and semi-rigid pavements*. XI International Road

Safety Seminar GAMBIT 2016, MATEC Web of Conferences, Vol. 122, doi: 10.1051/matecconf/201712204002
[10] *Katalog przebudów i remontów nawierzchni podatnych i półsztywnych.* GDDKiA, IBDiM, (in Polish). Warsaw, 2013. Available in the Internet: www.gddkia.gov.pl/userfiles/articles/p/prace-naukowo-badawcze-w-trakcie_3434/KPRNPP%20i%20Zalaczniki%202013.pdf
[11] Mohammadi J., Shah N., 1992: *Statistical evaluation of truck overloads.* Journal of Transportation Engineering, Vol. 118, 651-665. ASCE, doi: 10.1061/(ASCE)0733-947X(1992)118:5(651)
[12] Pais J.C., Amorim S.I.R., Minhoto M.J.C., 2013: *Impact of Traffic Overload on Road Pavement Performance.* Journal of transportation Engineering, Vol. 139(9), pp. 873-879. ASCE. doi: 10.1061/(ASCE)TE.1943-5436.0000571
[13] Rys. D, Judycki J., Jaskula P., 2015: *Analysis of effect of overloaded vehicles on fatigue life of flexible pavements based on weigh in motion (WIM) data.* International Journal of Pavement Engineering, Vol. 17(8), pp. 716-726, doi: 10.1080/10298436.2015.1019493
[14] Rys. D, Judycki J., Jaskula P., 2017: *Impact of overloaded vehicles on load equivalency factors and service period of flexible pavements.* Bearing Capacity of Roads, Railways and Airfields. London: CRC Press, doi: 10.1201/9781315100333-66
[15] Rys D., Judycki J., Jaskula P., 2016: *Determination of vehicles load equivalency factors for polish catalogue of typical flexible and semi-rigid pavement structures.* 6th Transport Research Arena (TRA). Transportation Research Procedia, Vol. 14, pp. 2382-2391, doi: 10.1016/j.trpro.2016.05.272
[16] Rys D., 2015: *Loading of roads by heavy vehicles and their impact on fatigue life of flexible and semi-rigid pavement structures.* PhD thesis, Gdansk University of Technology.
[17] Stephens J., Carson J., Hult D.A., Bisom D., 2003: *Preservation of infrastructure by using weight-in-motion coordinated weight enforcement.* Transportation Research Record, Vol. 1855, pp. 143-150, doi: 10.3141/1855-18
[18] Zofka A., Urbaniak A., Maliszewski M., Bankowski W., Sybilski D., 2014: *Site specific traffic inputs for mechanistic-empirical pavement design guide in Poland.* Transportation Research Board Annual Meeting, Washington, USA.
[19] https://www.chroniclelive.co.uk/news/north-east-news/a19-silverlink-road-closures-cancelled-13458970

Fatigue cracking in road pavement

P Mackiewicz

Wroclaw University of Technology, Faculty of Civil Engineering,
Wybrzeże Wyspiańskiego 27, 50-370 Wroclaw, Poland

piotr.mackiewicz@pwr.edu.pl

Abstract. The article presents the problem of modelling fatigue phenomena occurring in the road pavement. The example of two selected pavements shows the changes occurring under the influence of the load in different places of the pavement layers. Attention is paid to various values of longitudinal and transverse strains generated at the moment of passing the wheel on the pavement. It was found that the key element in the crack propagation analysis is the method of transferring the load to the pavement by the tire and the strain distribution in the pavement. During the passage of the wheel in the lower layers of the pavement, a complex stress state arises. Then vertical, horizontal and tangent stresses with various values appear. The numerical analyses carried out with the use of finite element methods allowed to assess the strain and stress changes occurring in the process of cracking road pavement. It has been shown that low-thickness pavements are susceptible to fatigue cracks arising "bottom to top", while pavements thicker are susceptible to "top to bottom" cracks. The analysis of the type of stress allowed to determine the cracking mechanism.

1. Introduction

The issue of fatigue and cracking of asphalt concrete (AC) is a broad topic because it involves both material technology and fracture mechanics.

Depending on the thickness of the pavement, its upper layers undergo varying degrees of fatigue and bending [1], [2]. The asphalt base is more prone to fatigue damage than permanent deformations. The binder course is already exposed to permanent deformation, but it can also be subjected to low-temperature damage, while the initiation of fatigue cracks in this layer is likely when there is no adequate interlayer bonding of asphalt layers [3], [4], [5], [6] with base or propagation of the crack occurs in the base. The surface course is exposed to damage from the bottom in the case of loss of adhesion or initiation of propagation from the lower layers. It is also obviously vulnerable to "top to bottom" cracks due to its upper position.

Fig. 1 shows the scheme of cracks formation in asphalt layers. A crack can occur anywhere where a tensile strain is present. Three such areas can be distinguished: in the lower part of the asphalt layers under the load, on the surface on the edges of the load zone and on the surface in the contact area of the tire tread. The key area is also the edge of the tire and the surface in where shear occurs.

"Bottom-up" cracks are termed as classical fatigue cracks. These cracks start easily on the underside of thin asphalt layers and propagate upwards. Their direct cause is the concentration of repetitive tensile stresses at the bottom of the layers and low tensile strength at the bottom of the layers [7]. Another type of cracking occurs on the surface, mainly on the edges of the load zone. Field tests show that the range of cracks "from top to bottom" covers a small area deep into the surface. They usually reach a limited depth of 50-75 mm. This type of cracks was "discovered" only at the end of the twentieth century [8], [9], [10], [11]. Despite many studies and analyses [12], [13], [14] the problem still raises many discrepancies in their direct cause. It was also found that the aging and hardening of asphalt intensify these cracks. Low-temperature cracks can also be initiated, which will create "convenient" stress centres for their creation and subsequent development.

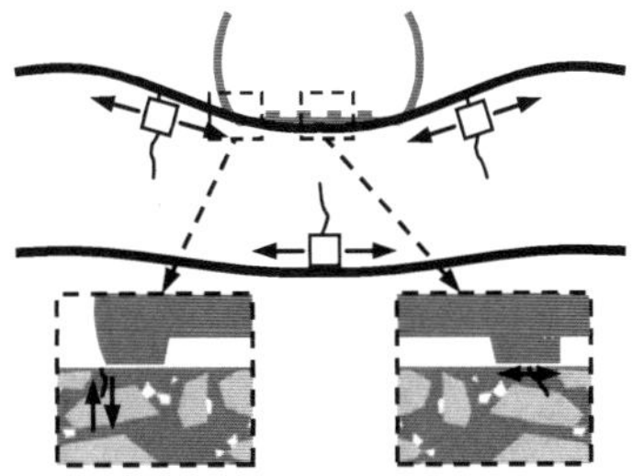

Figure 1. Scheme of crack formation in asphalt layers.

The problem of fatigue cracks in the pavement is complex. The calculation of deformations in the lower part of asphalt layers does not cause any problems, as they can be calculated unequivocally using multilayer systems. However, the designation of crack initiation sites in the upper part is not unambiguous. It is important to analyse the distribution of forces depending on the type of tire and tread. Therefore, it is not surprising that most researchers using pavement models still focus on deformations in the lower asphalt layers. This article attempts to analyse the problem of fatigue cracks, including "top-down" cracks.

2. Mechanism of formation of a complex deformation state in pavement

When the vehicle wheel passes on the pavement, the distribution of stresses in the ground changes [15], [16]. Then, a complex state of stress and strain arises. There are vertical, horizontal and tangential stresses causing mainly compression and stretching in varying degrees.

In order to accurately describe the impact of the wheel on the pavement, two cases should be analysed in two cross-sections of the pavement. The first cross-section refers to the case when the pavement is observed in the longitudinal direction of movement, in the second section, it is viewed in the transverse direction. When the wheel passes in the longitudinal direction and is in front of or behind the analysed point, i.e. it approaches a given section or passed it (Fig. 2), there are slight compressive phenomena in the lower points of the asphalt layers, while on the surface course, stretching occurs. In the transverse view, the initiation of stress and tensile strain at the bottom of the layers is observed. On the surface, however, an inhomogeneous distribution of stress and strain begins to appear, which depends on the type of tire and tread. At the moment when the wheel is over the point under consideration (Fig. 3), in the longitudinal view in the lower places of the surface layers, the direction of the interaction changes and the stretching appears at the bottom and the compression at the surface. In the cross-section view, the horizontal values of tensile stresses in the lower point of the asphalt layers increase, while stretching in the surface and shearing at the edges of the tire tracks intensify.

Based on the review of many test results and also numerical simulations [17], [18], [19], [20], [21] it was found that there is no agreement among researchers regarding the value of longitudinal deformations in relation to transverse strains. It is obvious that their relative values depend on the speed, axle load, temperature of the asphalt layers, asphalt mix type and thickness of layers. Differences in deformation depending on the type of tires and axles are also indicated. The steered axle causes greater transverse deformation and the remaining axes cause longitudinal deformations slightly exceeding the transverse values [22]. Based on the literature review, it can be concluded that the ratio of longitudinal to transverse strains can vary from 0.5 to 2.0. In practice, there is a mixed nature of initiation and subsequent development of fatigue cracks. Cracks in the lower layers can first be initiated from deformations observed longitudinally to the direction of movement and will be laid transversely in the lane. On the pavement, single, short cracks are visible. In contrast, cracks "from top to bottom" that also lay transversally, will be initiated by stretching deformations on the surface course.

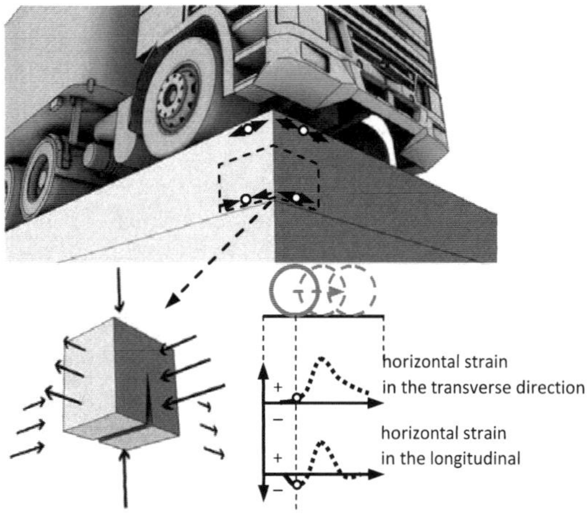

Figure 2. A typical distribution of deformations in the pavement when the vehicle approaches the analysed area.

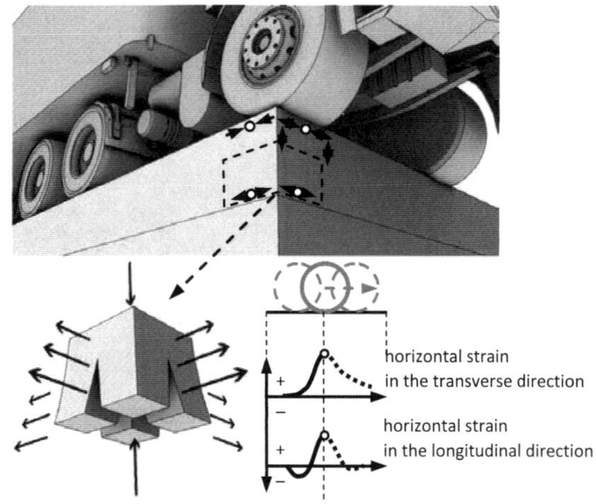

Figure 3. A typical distribution of deformations in the pavement when the vehicle is in the analysed area.

Extensive analyses of the normal and tangential stress distributions were carried out by de Beer [23]. He pointed out that in the case of heavy loads (0.9 MPa, 100 kN), the concentration of normal stresses on the edges of the radial tire footprint reaches 2 MPa, which is over twice as high as the tire air pressure. Contact stresses also concentrate on the edges of the tire. On the other hand, stretching strains alternately with squeezing arise under successive ribs. This was confirmed at work [24].

It is found that different types of tires cause different stress distributions. Therefore, the subject of contact stresses between the tire and the road pavement is still being investigated. At present, no effective analytical methods and models have been developed to predict the initiation of cracks caused by the impact of the tire on the pavement. The tire type is not used in design procedures of crack modelling, either.

3. Characteristics of the road pavement model

In the light of phenomena occurring during loading, the pavement model was developed that also allows for multi-axial conditions in the vicinity of contact between the tire and the pavement. For this purpose, the pavement model was developed in the Cartesian three-dimensional system *xyz* using the SOLIDWORKS Simulation, COSMOS/M application [25], [26], [27]. The pavement FEM model with basic geometric parameters is presented in Fig. 4.

The choice of the discretization size for SOLID volume elements was selected based on the analysis of the convergence of vertical displacements and deformation values at the characteristic points of the model. On the side surfaces, a fastening enabling only lateral displacement was used, while on the bottom surface a fastening preventing displacement and rotation was applied. In the lower part of the binding layer and on the edge of the load, geometric discontinuities were used in the form of fracturing slots. GAP contact elements were used in the cracks [28], [29], [30], [31]. They allowed to take into account the appropriate behaviour of the crack when it was closing.

A detailed mapping of the contact effect of the tire on the pavement was also carried out. The model adopted the wheel load with the value of 57.5 kN generating the vertical pressure of 0.705 MPa. Preliminary calculations were made for a radial tire often used in heavy goods vehicles, taking into account the proper pavement and tread effect. The tire trace of 26 cm × 30 cm was adopted, while the length of the tread ribs was: 5 cm, 4 cm, 4 cm, 4 cm and 5 cm with 1 cm distance between them.

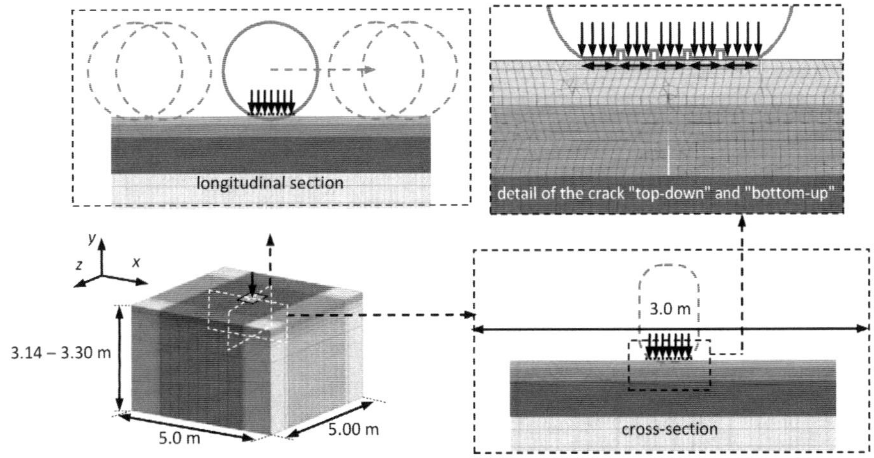

Figure 4. Pavement model.

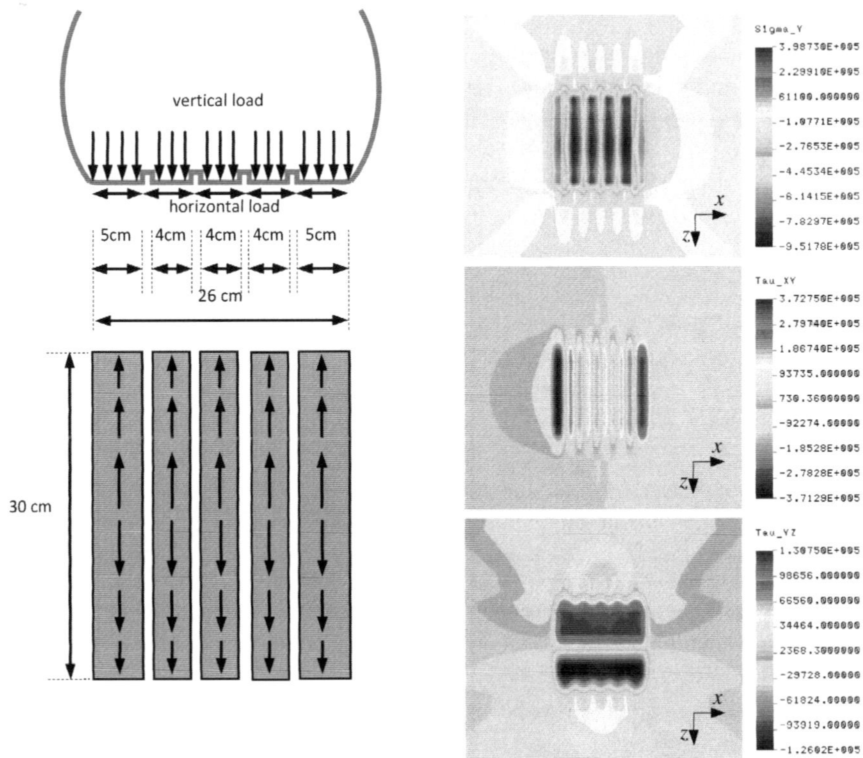

Figure 5. Load diagram and distributions of normal and tangential stresses [Pa] for a radial tire.

Based on research [23], [24], [32], [33], [34], [35], the horizontal load value 0.175 MPa was assumed in the transverse direction, while in the longitudinal direction the load was 0.035 MPa, which corresponded to the rolling wheel with the speed of 72 km/h.

Figure 5 shows the load scheme and distributions of normal and tangential stresses that have been obtained for a radial tire. High values of tangential stresses were found. These observations are consistent with the observations of other researchers who carried out field measurements. To describe the characteristics of the asphalt concrete AC layers, elastic material parameters for temperature +10°C were adopted based on the identification for different mixtures in the fatigue four bending test. This topic will be the subject of a separate publication. Two pavements significantly differing in their layer thicknesses were analysed. The characteristics of the materials in the pavement model and the pavement variants are presented in Table 1. The analyses assumed full interlayer bonding.

Table 1. Characteristics of materials in the pavement model and pavement variants.

Layer	Parameters	Pavement I	Pavement II
		Thickness [m]	
Surface course AC 11 S 50/70	E = 12 129 MPa, ν = 0,30	0.04	0.04
Binder course AC 16 W 35/50	E = 13 764 MPa, ν = 0.30	0.08	0.08
Base course AC 22 P 35/50	E = 14 603 MPa, ν = 0.30	–	0.16
Sub-Base course (aggregate)	E = 400 MPa, ν = 0.,30	0.22	0.22
Subgrade	E = 100 MPa, ν = 0.25	2.80	2.80

4. Results of numerical calculations

4.1. Effect of load direction

On the basis of calculations for the pavement model, the distribution of deformations in asphalt layers in the longitudinal and transversal direction was analysed during the vehicle wheel passing, assuming a basic load impulse of 0.015 s. Fig. 6 and 7 show an example distribution of tensile strains in the lower part of the asphalt layers of pavement I in the longitudinal (ε_z) and transversal (ε_x) directions.

The analysis of the strain distribution in both directions showed that in the transverse direction, there are 11% more deformations in relation to longitudinal strains.

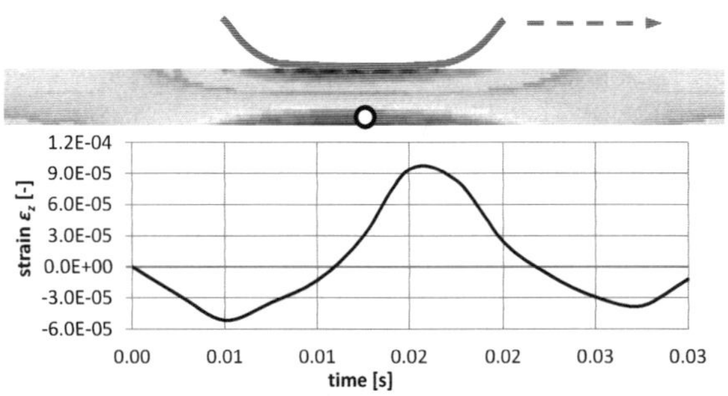

Figure 6. Distribution of strain in the longitudinal direction in pavement I.

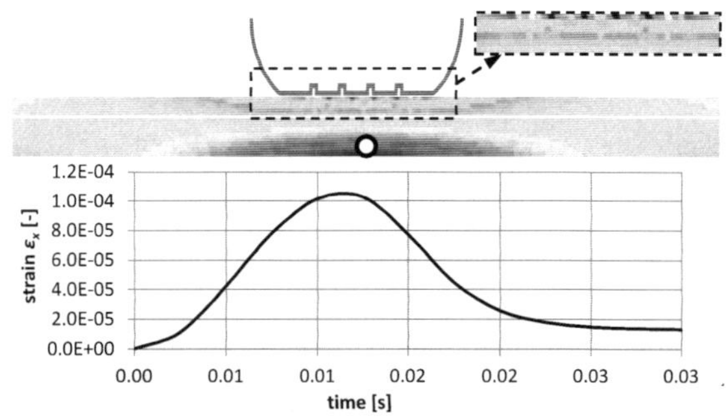

Figure 7. Distribution of deformations in the transverse direction in pavement I.

There was also found a key concentration of tensile stresses reaching 20×10^{-6} for the transverse direction in the tread area of the tire.

Analogous analyses of strain and stress distribution were performed for the thicker pavement of variant II (Figures 8 and 9). In the case of pavement II with the thickness of asphalt layers of 28 cm, there are definitely less deformations and stresses at the bottom of the layers. For the transversal and longitudinal direction, there is a similar deformation with the difference of only 2%. For this pavement in cross-section, a greater concentration of horizontal deformations is observed at the edge of the tire load than for the thinner pavement I. This can have a significant impact on the initiation of cracks "from top to bottom" and is related to the direct influence of the tire tread ribs on the pavement. For the transversal direction, the deformations are 87×10^{-6}, while in the longitudinal direction only 9×10^{-6}.

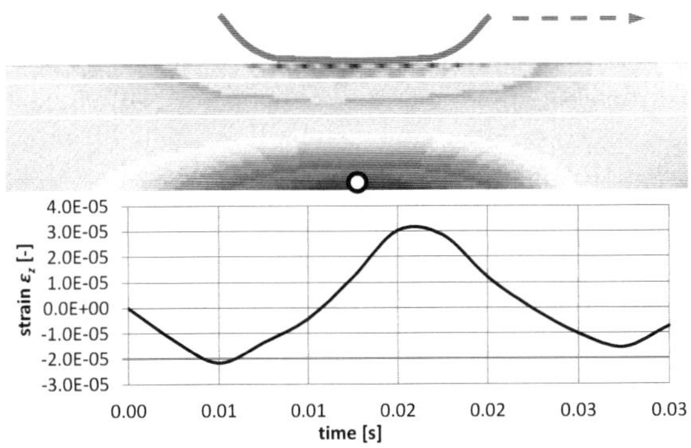

Figure 8. Distribution of strain in the longitudinal direction in pavement II.

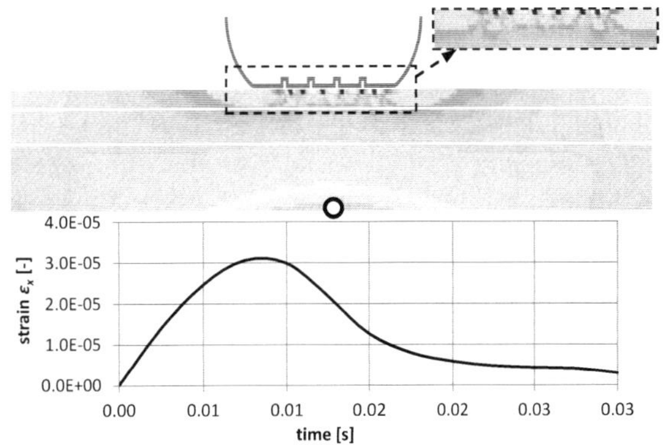

Figure 9. Distribution of deformations in the transverse direction in pavement II.

4.2. Influence of cracks

In the following, the analysis of the impact of crack length in the model of cracks propagating " from bottom-up" in the binding layer and "top-down" in the wear layer was carried out. For pavements I and II, the results of tensile stress calculations σ_x at the apex of the propagating crack "from bottom to top" is shown in Fig. 10.

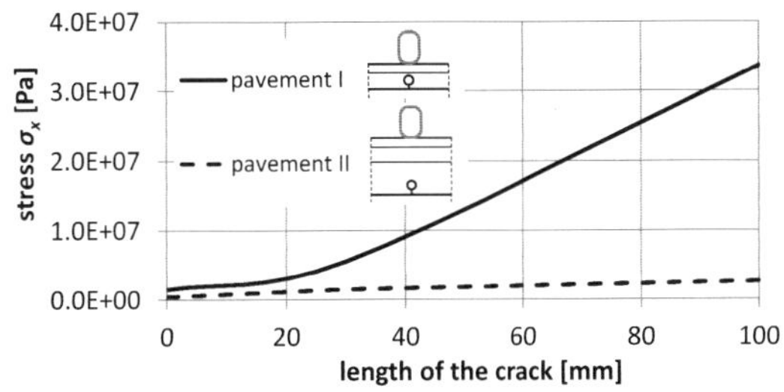

Figure 10. Influence of the "bottom-up" crack length on stress values σ_x.

For a thicker pavement, there are almost ten times less stresses than for a thinner one. In the case of the thinner pavement, above the crack with the length of 20 mm, a significant increase in stress is observed. A similar analysis was carried out on the crack located on the edge of the wheel track, propagating "from top to bottom". In this case, the value of tensile σ_x and shearing τ_{xy} stresses was analysed at the tip of the crack.

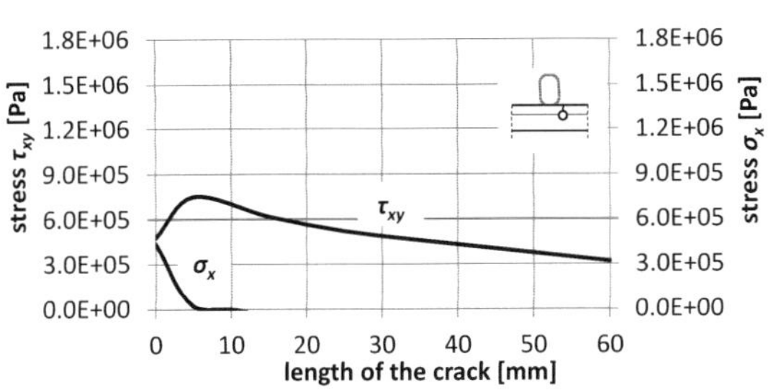

Figure 11. The influence of the crack length "from top to bottom" on the stresses σ_x and τ_{xy} for the layers in pavement I.

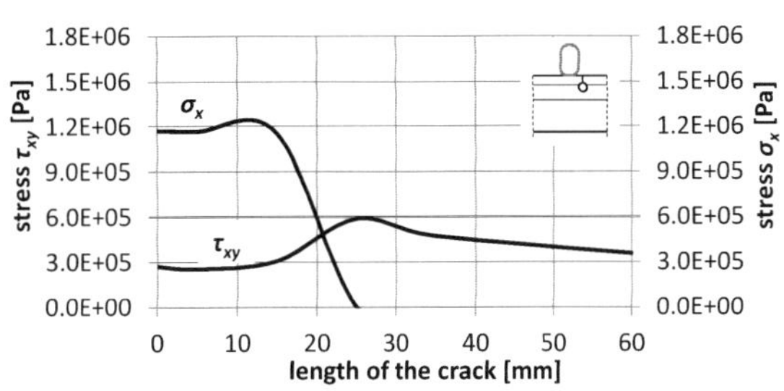

Figure 12. The influence of the crack length "from top to bottom" on the stresses σ_x and τ_{xy} for the in pavement II.

Fig. 11 shows the influence of the crack length on the stress for the pavement I, whereas Fig. 12 for the pavement II. Stresses σ_x and τ_{xy} have similar values in the thinner pavement on the surface course. In a situation, where no crack was yet formed on the surface of the wear layer, tensile stresses σ_x of the value 5.0×10^5 Pa appear. They are crucial in the initiation of this type of cracks. The tensile stresses σ_x decrease with the length of the crack. The fact of further propagation can be explained by shear stresses τ_{xy}, which increase to the maximum of 7.5×10^5 Pa for the 6 mm crack. Then, along with the crack development, these stresses decrease. The thicker pavement is initially dominated by stresses σ_x (1.2×10^6 Pa), which decreases with the length of the crack only from 15 mm. Stresses τ_{xy} reach their flat extreme 6.0×10^5 Pa for the crack with the length of approximately 25 mm. It is worth noting that the stresses reach their maximum value at a relatively short length of the crack from 6 mm to 25 mm and then their fastest propagation is possible. Then they can propagate more slowly to the length from 50 mm to 75 mm. It should be emphasized that the "top-down" crack will propagate mainly through a transverse shear. This is the second case of cracking determined in fracture mechanics, for which the crack surfaces slip in the direction perpendicular to the front of the crack (Fig. 13). For the "bottom-up" crack, shear stress is also present, but it is negligible, and the higher stress σ_x determines the vertical direction of the propagation.

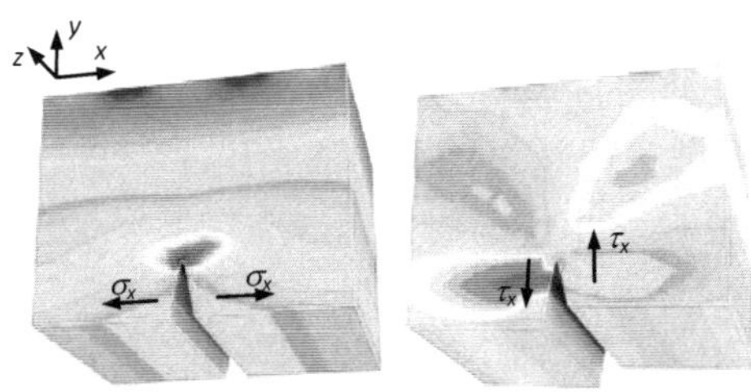

Figure 13. Considered cracking methods depending on the load.

Figures 14-17 show example distributions of stresses σ_x and τ_{xy} for pavement I and II with 30 mm lower and 15 mm upper crack. In both variants of pavements, a considerable concentration of stresses is observed at the apexes of the cracks. As previously noted, in the case of the upper cracks, the shear stress prevails in the thinner pavement, while for the pavement thicker with the upper crack, tensile stresses still prevail.

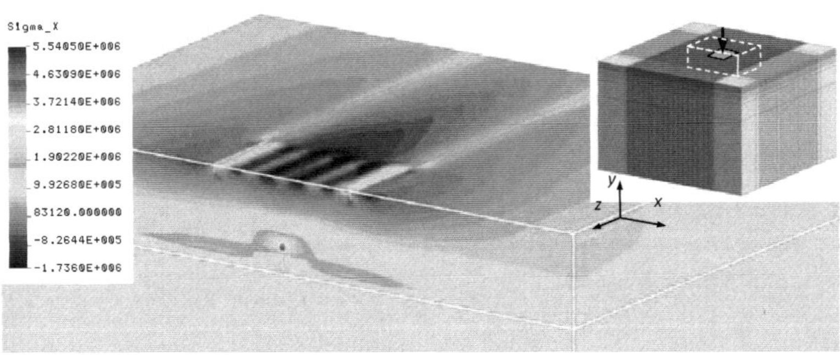

Figure 14. Distribution of stresses σ_x [Pa] for pavement I with cracks.

In the top of the bottom crack for a pavement with a thinner bundle of asphalt layers (12 cm), higher stresses σ_x were obtained relative to pavement II (28 cm). For the thinner pavement, the shear stresses are also slightly greater at the top of the upper crack. However, it should be noted that for the longer cracks (over 50 mm) the situation is reversed.

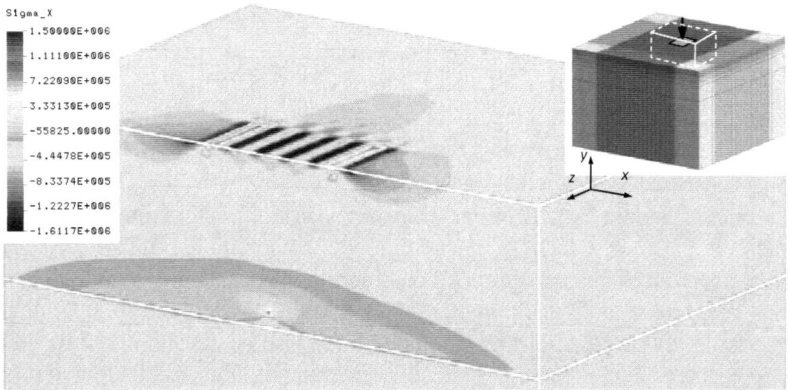

Figure 15. Distribution of stresses σ_x [Pa] for pavement II with cracks.

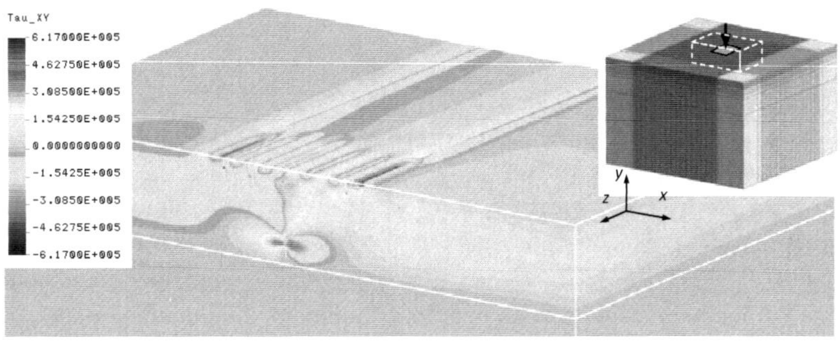

Figure 16. Distribution of stresses τ_{xy} [Pa] for pavement I with cracks.

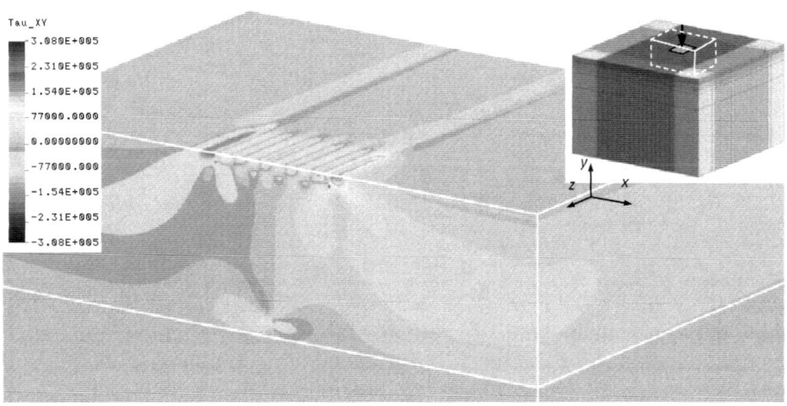

Figure 17. Distribution of stresses τxy [Pa] for pavement II with cracks.

5. Summary

The developed pavement model is suitable for the analysis of both micro- and macro-crack propagation. The examples of the two pavements show the changes occurring under the influence of load in different places of the pavement layers. In the example of the radial tire load for temperature 10°C, it was found that deformations in the transverse direction are 11% more than in the longitudinal one. Taking into account the different length and location of the crack sin the pavement allowed to analyse the critical stresses at the apex of the crack affecting the speed of crack propagation. It was shown that there is a separate way of propagation for cracks coming from "bottom to top" and "from top to bottom", which is associated with shear stresses in the further phase of propagation. In the situation where the crack has not yet been formed, tensile stresses σ_x have been identified on the surface of the wear layer. The stresses decrease as the crack length increases. Further propagation is developed by shear stresses τ_{xy}, which increase to the maximum for the 6 mm crack in the thinner pavement. In the thicker pavement, the initial strains σ_x also dominate. However, they decrease as the length of the crack increases only from 15 mm. Stresses τ_{xy} reach their extreme for a crack of about 25 mm. The stress concentration in the apex of the crack is reflected in the cracking mechanism and values of cracking parameters defined as the stress intensity factor and the Rice's integral. Analysis of these parameters taking into account different temperatures will be the subject of further research.

The model presented here, includes an exact load of the vehicle wheel and geometric discontinuities in the form of cracks. This model is useful for assessing the way of cracking in the pavement, and together with the equations describing the cracking kinetics, over which further research is underway, is suitable for fatigue analysis in a wider range of conditions. Knowledge of the mechanism and size of cracking in thin and thicker asphalt layers allows to determine future methods of repair and reinforcement of the pavement.

References

[1] MCGENNIS R B ANDERSON R M KENNEDY T W SOLAIMANIAN M 1994 Background of Superpave Asphalt Mixture Design and Analysis *Publication No.FHWA-SA-95-003*

[2] ROBERTS F KANDHAL P BROWN E LEE D KENNEDY T 1996 Hot Mix Asphalt Materials Mixture Design and Construction *NAPA Education Foundation* Second Edition 603

[3] JASKUŁA P 2014 Influence of compaction effectiveness on interlayer bonding of asphalt layers *The 9th international conference. Environmental Engineering* May

[4] KIM H ARRAIGADA M RAAB C PARTL M N 2011 Numerical and experimental analysis for the interlayer behavior of double-layered asphalt pavement specimens *Journal of Materials in Civil Engineering* **23** (1) pp 12–20

[5] MOHAMMAD L BAE A ELSEIFI M BUTTON J PATEL N 2010 Effects of pavement surface type and sample preparation method on tack coat interface shear strength *Transportation Research Record: Journal of the Transportation Research Board* **2180** pp 93–101

[6] WANG H AL-QADI I L 2010 Near-surface pavement failure under multiaxial stress state in thick asphalt pavement *Transportation Research Record: Journal of the Transportation Research Board* 2154 91–99

[7] BROWN S F 1996 Soil mechanics in pavement engineering *Geotechniqu* **46** pp 381–426

[8] GERRITSEN A H VAN GURP C A VAN DER HEIDE J P MOLENAAR A A A PRONK A C 1987 Prediction and Prevention of Surface Cracking in Asphaltic Pavements *Proceedings 6th International Conference Structural Design of Asphalt Pavements* University of Michigan Ann Arbor Michigan July pp 378–391

[9] MOLENAAR A A A 1983 Structural Performance and Design of Flexible Road Construction and asphalt Concrete Overlays *Ph.D. Thesis Delft University of Technology* Netherlands

[10] MYERS L A ROQUE R RUTH B E 1998 Mechanisms of Surface-Initiated Longitudinal Wheel Path Cracks in High-Type Bituminous Pavements *Proceedings* **67** AAPT

[11] NUNN M E 1997 Design of Long-Life Pavements for Heavy Traffic *TRL Report* 250

[12] SPŁAWIŃSKA M ZIELIŃSKI P BURNOS P 2015 Influence of traffic flow variability of heavy vehicles and temperature on pavement fatigue life *Roads and Bridges* **14** pp 117–132

[13] JUDYCKI J 2011 Budowa i kalibracja modeli spękań zmęczeniowych warstw asfaltowych nawierzchni drogowych w mechanistyczno-empirycznej metodzie AASHTO 2004 *Roads and Bridges* **4** pp 31–53

[14] GRABOWSKI W POŻARYCKI A 2006 Badania i ocena propagacji spękań zmęczeniowych w modelu nawierzchni asfaltowej *Roads and Bridges* **1** pp 5–18

[15] CHAN F K W 1990 Permanent deformation resistance of granular layers in pavements *Ph.D. Thesis* University of Nottingham United Kingdom
[16] LEKARP F RICHARDSON I R DAWSON A R 1996 Influences on permanent deformation behavior of unbound granular materials *Transportation Research Record* **1547** Washington DC TRB National Research Council pp 68–75
[17] GOPALAKRISHNAN K THOMPSON R 2006 Effect of Dynamic Aircraft Gear Loads on Asphalt Concrete Strain Responses *Journal of ASTM International* **3** 8
[18] NAGÓRSKI R BŁAŻEJOWSKI K NAGÓRSKA M 2015 Comparative analysis of deflections and strains of two road pavements with high traffic load *Roads and Bridges* **14** pp 31–46
[19] SYBILSKI D BAŃKOWSKI W MULARZUK R 2015 Ocena konstrukcji nawierzchni z BAWMS z zastosowaniem programu komputerowego VEROAD Instytut Badawczy Dróg i Mostów strona internetowa Polskiego Kongresu Drogowego www.pkd.org.pl
[20] BAŃKOWSKI W GAJEWSKI M 2012 Badania przyspieszone w skali rzeczywistej innowacyjnych nawierzchni drogowych *Roads and Bridges* **2** pp 89–121
[21] RUTTMAR I SZYDŁO A 2004 Wykorzystanie symulatora ciężkich pojazdów do weryfikacji konstrukcji nawierzchni płatnej autostrady A-2 *Drogownictwo* **34** pp 75–78
[22] GILCHRIST M D HARTMAN A M OWENDE P M O WARD S M 2001 Full scale accelerated testing of bituminous road pavement mixtures *Key Engineering Materials* pp 204–205, 443–452 Trans Tech Publications
[23] DE BEER M FISHER C JOOSTE F 1997 Determination of pneumatic tyre/Pavement interface contact Stresses under moving loads and some effects on pavements with thin asphalt surfacing layers *Eighth International Conference on Asphalt Pavements* Seattle Washington
[24] MYERS L A ROQUE R RUTH B E DRAKOS C 1999 Measurement of Contact Stresses for Different Truck Tire Types to Evaluate Their Influence on Near-Surface Cracking and Rutting *Transportation Research Record: Journal of the Transportation Research Board* **1655** Washington DC TRB National Research Council pp 175–184
[25] RUSIŃSKI E 1994 Metoda elementów skończonych System COSMOS/M *Wydawnictwa Komunikacji i Łączności* Warszawa
[26] RUSIŃSKI E 2000 Zaawansowana metoda elementów skończonych w konstrukcjach nośnych *Oficyna Wydawnicza Politechniki Wrocławskiej* Wrocław
[27] MACKIEWICZ P 2016 Fatigue life of asphalt mixtures used in pavements Oficyna Wydawnicza Politechniki Wrocławskiej Wrocław
[28] BRILL D R 2000 Field Verification of a 3D Finite Element Rigid Airport Pavement Model *FAA report* DOT/FAA/AR-00/33
[29] CARRASCO C LIMOUEE M CELAYA M ABDALLAH I NAZARIAN S 2009 NYSLAB: software for analysis of jointed pavements *Research Report FHWA* RD-07-1008-01
[30] MACKIEWICZ P 2015 Finite-Element Analysis of Stress Concentration around Dowel Bars in Jointed Plain Concrete Pavement *Journal of Transportation Engineering* **141** 6
[31] MACKIEWICZ P 2014 Thermal stress analysis of jointed plane in concrete pavements *Applied Thermal Engineering* **73** 1
[32] BLAB R 1999 Introducing Improved Loading Assumptions into Analytical Pavement Models Based on Measured Contact Stresses of Tires Paper CS5-3 *Proc. Int. Conf. on Accelerated Pavement Testing* Reno Nevada October
[33] CHOWDHURY F RALLINGS R A 1994 A Survey of Truck Tyre Pressures in Tasmania *Road and Transport Research* **3** (3) pp 80–9 Australian Road Res. Board Ltd. South Vermont Vic.
[34] SIDDHARTHAN R V SEBAALY P E 1999 Investigation of Asphalt Concrete Layer Strains from Wide-Base Tyres *Transportation Research Record* 1655 Washington DC TRB National Research Council pp 168–174
[35] WOODSIDE A R WOODWARD W D SIEGFRIED H 1999 The Determination of Dynamic Contact Stress Paper No. CS5-2 *Proc. Int. Conf. on Accelerated Pavement Testing* Reno Nevada

IOP Conf. Series: Materials Science and Engineering **356** (2018) 012015 doi:10.1088/1757-899X/356/1/012015

Influence of temperature on fatigue life or reinforced pavement by whitetopping

A Szydło, P Mackiewicz

Wroclaw University of Technology, Faculty of Civil Engineering, Wyb. Wyspiańskiego 27, 50-370 Wroclaw, Poland

antoni.szydlo@pwr.edu.pl

Abstract. The article presents the influence of temperature on the fatigue strength of concrete slabs used for reinforcing susceptible flexible pavement. In Poland, so far, there is no research on thermal interactions on concrete pavement. The article presents an analysis of various climatic conditions occurring in Poland and temperature distribution in concrete pavement. The dependence of daily temperature fluctuations on the temperatures appearing in the concrete slab was demonstrated. An analysis of thermal stresses in concrete slabs depending on their parameters was shown, and then fatigue life was determined. The applied 3DFEM model includes elements of contact, friction, and gravity in order to better approximate the behaviour of the board from temperature change. On this basis, the significant influence of cyclical daily temperature changes on the durability of the concrete pavement was indicated. The presented analyses can be applied to reinforcements of existing flexible pavements.

1. Introduction

Whitetopping is a technology based on the reconstruction of damaged pavement using a layer of cement concrete. This method was used for the first time in 1918 in the United States making reinforcements of exploited pavement of motorways, runways, and roads with smaller street loads and parking lots [1]. Until the 90s of the last century, few projects using whitetopping technology were implemented. It was not until 1992 that this technology was used on a larger scale, covering around 500 projects in the United States ([2], [3], [4]). Then this technology has come into focus, among others, in Canada, Brazil, Taiwan, and in Europe. At the time of implementation of this method, the different thickness of the concrete cover was taken into account for the renovation of existing pavement. Three methods were defined in this way: Conventional Whitetopping (slabs above 20 cm), Thin White-Topping (slabs from 10 to 20 cm) and Ultra-Thin Whitetopping (slabs from 5 to 10 cm).

According to American experiments [5], non-reinforced, reinforced, doweled, or undoweled or continuous reinforced concrete slabs can be used for whitetopping. It was found that cracks in the asphalt pavement do not transfer to the concrete cover (no reflective cracks). French experiences [6] show that whitetopping can be used with ruts smaller than 30 mm. However, according to Belgian experience [7] with ruts smaller than 40 mm. The levelling of larger uneven surfaces (e.g. ruts larger than 50 mm) can be done by milling. Another method of levelling the surface before laying the concrete overlay is to arrange the asphalt mixture as a levelling layer.

An important problem of performing whitetopping appears in summer periods. During long-lasting exposure to sunlight, the asphalt pavement heats up and the surface temperature can reach several dozen degrees Celsius. High temperature of the surface together with the temperature created during hydration of the cement may cause high stresses in the concrete during the night cooling of the cover. The result of this phenomenon may be cracking. To reduce the temperature in the US, for example, two methods are used: water spraying (water Jogging) and applying a white coating (whitewashing).

In the article, we present results of thermal analyses in the concrete layers depending on their thickness, strength and various capacities of existing pavement. Attention was focused on conventional whitetopping (from 20 to 30 cm), but the additional analysis was also carried out for thin plates with a thickness of 10 cm. The numerical calculations allowed to determine the fatigue life resulting from the cyclic impact of temperature on the concrete pavement. The work uses 3DFEM with elements of contact, friction, and gravity in order to better approximate the behaviour of the slab from temperature change.

2. The impact of temperature on concrete pavement

The first detailed stress tests on the temperature in rigid concrete pavement date back to 1900. Westergaard published two papers related to the estimation of stresses on a concrete surface. The first one concerned stresses and deformations in the pavement caused by a load of vehicles [8], the second one took into account linear changes from the temperature on the slab thickness [9]. In contrast to Westergaard, it was shown that the temperature profiles on the thickness of the concrete slab are non-linear ([10], [11]). In the next stage, modelling was carried out using the finite element method and simulations of temperature impact on rigid pavement. Initially, these were 2D models ([12], [13], [14], [15]). Along with the development of computer technology, FEM 3D models have been developed. Using the 3D FE Shoukry et al. [16] investigated the effect of linear and non-linear temperature gradients in concrete slabs. Also, many interesting computer programs have been developed. Some programs, such as the integrated climate impact model of the pavement [17] and the extended integrated climate model [18] capable of generating rainfall patterns, solar radiation, cloud cover, wind speed and air temperature. Part of the work concerns the analysis of the concentration of stresses around dowels. Riad [19] and Luoke [20] carried out analysis of the influence of load and temperature on these stresses. The same thermal analysis and the associated stress distribution in the concrete around the dowels were handled by Shoukry [21] and Mackiewicz [22].

Despite the significant number of research works accumulated in recent years in the field of concrete pavement, the assessment of the behaviour of environmental conditions on pavement still needs to be analysed, especially if it concerns different climatic conditions that occur in Central Europe, including Poland. So far, the influence of temperature on the total durability of both new concrete pavement and those used in whitetopping technology in the period of operation has not been analysed.

3. Analysis of climatic conditions in Poland

There is a temperate climate in Poland between the maritime and land climate. As a result, the climate in Poland is characterized by quite significant fluctuations in temperature and pressure. Every month it is possible to extract characteristic values of air temperature, which have a significant impact on the behaviour of concrete pavement.

Due to the considerably extended atmospheric circulation in Poland, significant climate differences are observed throughout the country depending on different periods of the year. Representative thermal conditions prevailing in a given place can be determined primarily on the basis of average values and extreme air temperatures and in the case of the need to design concrete pavement using temperature amplitude (diurnal, annual). Impacts on a concrete slab are most often analysed in a day-night cycle. In Poland, there are many days when the daily temperature goes through 0°C and a large number of days with a high thermal diurnal gradient.

Based on meteorological data, a distribution of daily temperature changes for Poland has been developed. Fig. 1 shows the distribution of the average daily temperature amplitude in the whole year from the last thirty years [23]. There is a marked reduction in the daily fluctuations of air temperature near the Baltic Sea of 6-7°C. In the prevailing area of Poland, daily fluctuations range from 7°C to

9°C. Diurnal temperature variations increase from north-west to south-east and reach the highest values (over 9°C) in the southeast of the country.

Figure 1. Diurnal air temp. amplitude in Poland during the whole year [°C], [23].

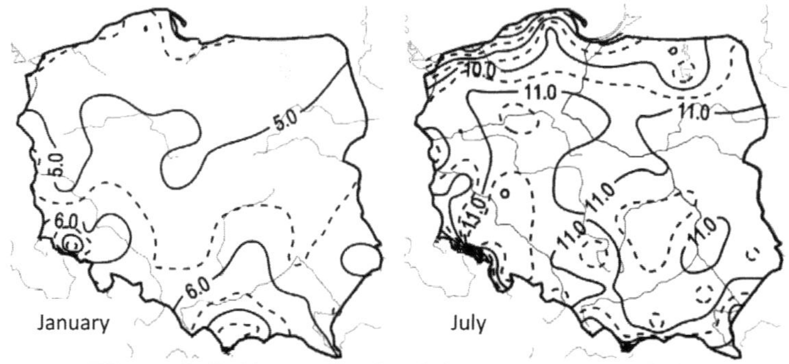

Figure 2. Daily amplitude of air temperature in Poland for January and July [°C], [23].

Analysing different regions of the country, it can be concluded that the greatest variation in temperature circuits occur on the coast (north) and in mountain areas (south). The major part of Poland with a relatively unchanged distribution (2-3°C) is the central area. Therefore, the area of the country as a whole has been treated for the practical design of the pavement without dividing the areas.

However, detailed attention was paid to temperature changes in the different seasons of the year. Fig. 2 shows the distribution of the daily temperature amplitude in Poland for selected months. The listings also come from data from thirty years. January, characterized by small fluctuations and July, in which the largest daily temperature fluctuations occur were presented [23]. Typical daily variations in air temperature for different seasons in Poland have values: spring 10°C, summer 12°C, autumn 8°C, winter 6°C. It should be noted that in Poland, characteristic atmospheric circulations (D2C-south-west and south and E1-north-east and east. Circulations are variable and although they persist in a few days, they will significantly affect the increase in average daily temperature fluctuations (for example in July daily fluctuations reach 24°C). Such even short, incidental, multi-day periods, in which the air temperature will change, will affect the behaviour of the concrete pavement.

4. Results of the temperature distribution tests in concrete pavement

On the basis of long-term own measurements of temperature on the concrete pavement, temperature changes during the day in different places on the thickness of the concrete slab were analysed. Due to the occurrence of the largest daily fluctuations in the summer months, attention was focused on measurements in these months. Fig. 3 shows a typical temperature change in the analysed concrete pavement during the day in the summer. Measurements were recorded continuously on pavement in central Poland (Łódź airport) and in the west on the A4 motorway.

The largest temperature change in the concrete slab occurs in the afternoon and at night. Then the so-called the highest positive and negative temperature difference between the upper and lower surfaces of the concrete slab is observed.

In the morning hours 5.00-8.00 and in the afternoon 16.00-19.00 you can observe a change in the temperature difference in value, the slab does not deform, but there may still be small axial stresses due to the remaining temperature in the centre of the slab.

Due to the fact that higher values have a positive temperature difference (appearing in the midday hours) and cause higher stress in the slab, it was included in further analyses. Using the results of the measurements from July in Figure 4, an example of the dependence of diurnal change in air temperature on the positive temperature difference in the slab is shown.

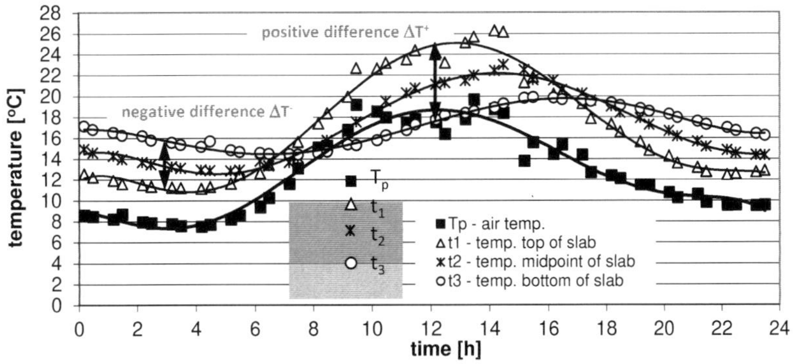

Figure 3. Daily amplitude of air temperature in Poland.

In the further analysis, taking into account the dependence presented in Fig. 5 and the temperature distribution on the slab thickness (Fig. 4), the equation allowing to determine the maximum daily slab temperature differences depending on its thickness and the daily temperature change of the air were determined:

$$\Delta T_h^+ = (0.01412 \cdot \Delta T_p + 0.16037) \cdot h \quad (1)$$

where:
ΔT_h^+ - positive temperature difference for variable slab thickness [°C],
ΔT_p - daily change of air temperature [°C],
h – thickness of a concrete slab [cm].

As a result of the analyses, it was found that the average gradient in our climatic conditions equals 0.4°C /cm.

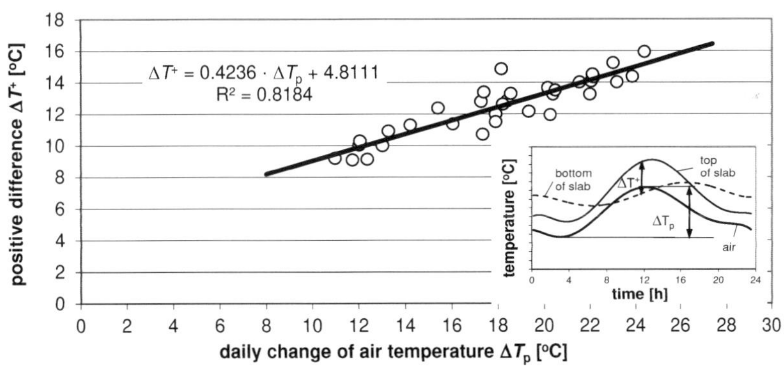

Figure 4. Diurnal change in air temp. depending on the positive temp. difference in the slab.

5. Numerical analysis of stresses in a concrete slab under the influence of temperature

The concrete pavement model was developed in the Cartesian three-dimensional system xyz using the finite element method. In the cross-section, a slab width equal to 4.0 m was adopted. In the longitudinal section, boards with a length of 4.0 m and 5.0 m were analysed. A model scheme of the pavement is shown in Fig. 5.

The adopted model corresponded to the concrete pavement made in whitetopping technology. The concrete slab was located on a layered system, characterized by equivalent elastic modulus, which can be determined using FWD and inverse calculations [24]. For a wider analysis, the variable range of the equivalent module (E_z) of layers from 90 to 1200 MPa was taken into account. The model assumes the modulus of elasticity of the slab from the temperature changes E_t = 26,000 MPa. Numerical calculations were made for slabs with a thickness of 10, 20, 25 and 30 cm, as well as a length of 4.0 m and 5.0 m. The case was also considered with dowels and without dowels.

In dowelled slab dowels with a diameter of 25 cm, length 50 cm, and spacing every 25 cm were included. In the analyses, linear-elastic materials were described with the modulus of elasticity and the Poisson coefficient. In order to perform the relevant calculations with regard to temperature, additional parameters for the concrete slab were adopted: thermal conductivity: 2.5 W/m°C, specific heat: 700 J/kg°C, thermal expansion coefficient: 0.00001 m/(m°C), density: 2400kg/m^3. The slab was loaded with a positive difference of 0.4°C /cm. Loads of gravity were also considered.

In addition, the model includes the appropriate boundary conditions and the cooperation of the concrete slab with the foundation in the form of a slip layer characterized by cohesion 0.2 MPa which corresponds to a coefficient of friction around 0.7. Here, the appropriate GAP-contact elements have been used. Two-node GAP elements are often used in 2D and 3D contact problems, in which the elements are in contact with each other due to the influence of external forces.

In order to determine the share of temperature-induced durability in relation to the total durability of the pavement, which also includes the vehicle's load, the load from the vehicle's wheel located in the centre of the slab was additionally taken into account. A load radius of 0.15 m and a force of 57.5 kN were assumed.

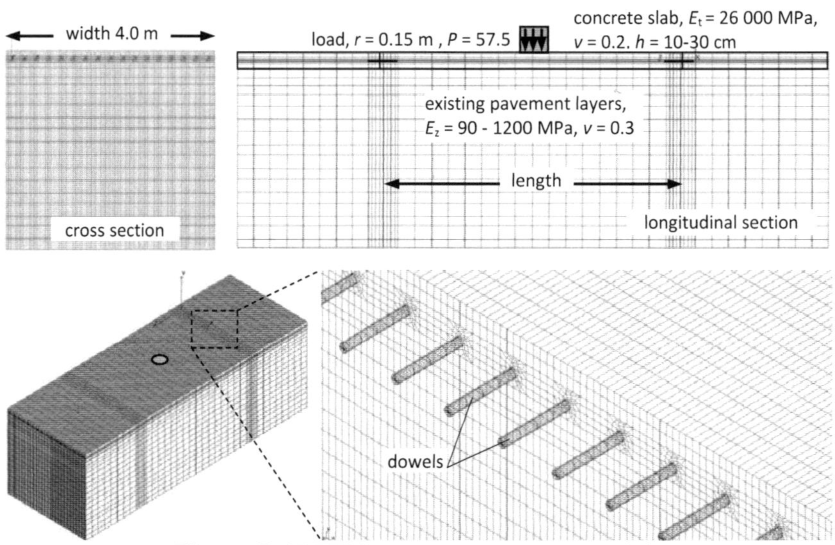

Figure 5. 3DFEM concrete pavement model.

The thermal effect was taken into account by applying appropriate temperature gradients in selected nodes of the slab. COSMOS/M [25] was used in the calculations, which makes it possible to analyse structures in terms of statics as well as linear and non-linear dynamics. Eight-cubic volume SOLID elements were used in the model. The appropriate discretization was also used. The division into 30,000 elements ensured sufficient convergence of results. The iterative procedure in the calculation was based on the modified Newton - Raphson method.

6. The results of FEM calculations

In the first stage of the calculation, the maximum tensile stresses from the temperature occurring on the surface of the concrete slab were assessed. In Fig. 6 for the smallest equivalent module 90 MPa of the existing pavement layers an exemplary distribution of stresses σ_x and deformation for a dowelled slab 20 cm thick and 5.0 m long are shown.

In the case of 5.0 m long dowelled slabs, it was found that as their thickness increases, they increase and then decrease the tensile stress caused by the temperature (Fig. 7). The optimum (maximum) stress has been observed for a thickness of approx. 25 cm. It can also be noticed that for thin slabs (10-20 cm) the stress values, despite the large scope of the equivalent module, do not differ significantly. For a 30 cm slab, the difference is bigger and amounts to about 0.4 MPa. The thicker slab, due to the greater uneven heating of its entire volume, has higher values of the temperature difference between the upper and lower surface and at longer lengths, it is more sensitive to thermal effects. It is also heavier which is of great importance in affecting the various support conditions for the subgrade.

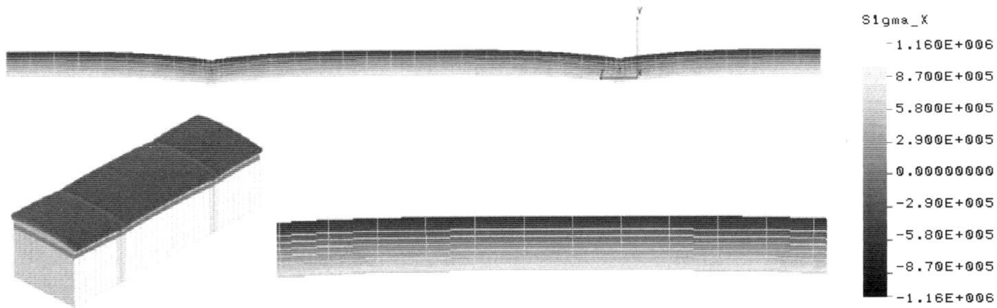

Figure 6. Deformation and stress distribution σ_x for a 20 cm dowelled plate loaded with a temperature gradient (scale factor x 200).

For a slab with a thickness of 25 cm, there is also a large influence of the value of the equivalent module of the subgrade on the values of stresses. It should also be noted that for the 25 cm thick slab, the highest tensile stresses were obtained, about 1.40 MPa, and the smallest for a thin slab 10 cm thick, which is affected by the smallest temperature difference (due to its small thickness).

In the case of a shorter slab with a length of 4.0 m (Fig. 8), smaller stress values were obtained by approximately 80%, however only for thicker slabs (above 20 cm). For a slab with a thickness of 10 cm, the sensitivity to temperature and replacement of the replacement module is smaller. It should be noted that in this case, the highest stress values occur for slabs with a thickness between 20 - 25 cm, depending on the value of the module of the subgrade (visible extreme in the figure). The optimum was obtained similarly to that for slabs with a length of 5.0 m.

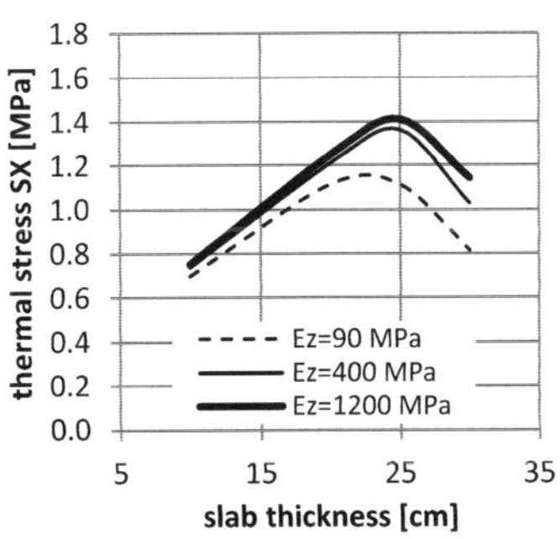

Figure 7. Change in the value of thermal stresses for slab thicknesses, for dowelled slabs with a length of 5.0 m.

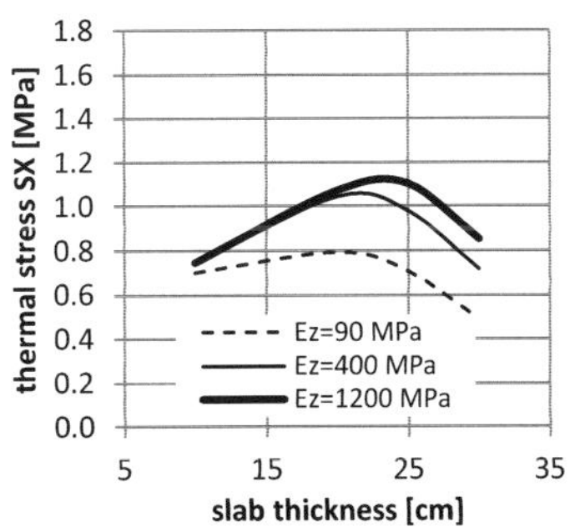

Figure 8. Change in the value of thermal stresses for slab thicknesses, for dowelled slabs with a length of 4.0 m.

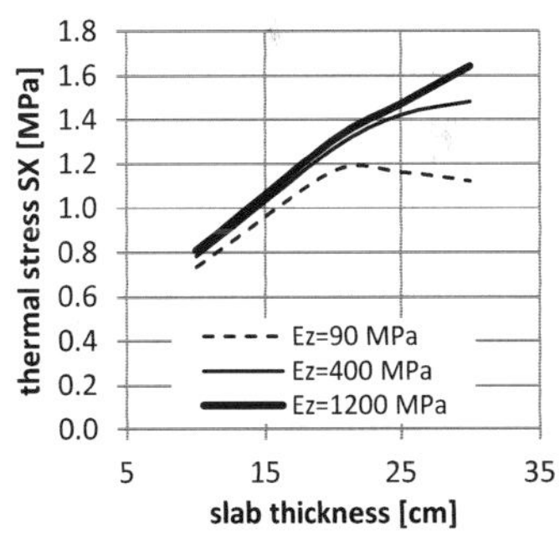

Figure 9. Change in the value of thermal stresses for slab thicknesses, for undoweled slabs with a length of 5.0 m.

Analysis of calculations for undoweled slabs with a length of 5.0 m (Figure 9) allows concluding that dowelling has a significant effect on decreasing stresses from temperature only for thicker slabs (above 25 cm), whose global stiffness is higher and there is more effective cooperation with dowels when the temperature is affected. Smaller thick panels are more flaccid. For such slabs, the dropping of the slab in the central part was observed under the influence of its own weight. Therefore, for thin slabs with only very small lengths, the dowelling effect will have a significant effect on thermal stresses. For undoweled slabs, there is no visible extreme of stresses as in doweled slabs.

7. Analysis of fatigue life

The authors used the fatigue criterion used in the catalogue of rigid pavement [26] to assess the durability of the analysed concrete pavement. This is a stress criterion for determining the durability of a concrete pavement based on its strength.

$$f_f \cdot m_1 \cdot \frac{1}{\gamma_m} \geq (\gamma_p \cdot n_1 \cdot \sigma_p) + (\gamma_t \cdot n_1 \cdot \sigma_t) \qquad (2)$$

where:
f_f – a characteristic tension of concrete for bending [MPa],
m_1 – factor taking into account the repeatability of loads = 1-0,078log$N$,
N – number of loads that concrete pavement at a given level of stress from load and temperature,
σ_p – maximum tensile stresses in the slab from the wheel load [MPa],
σ_t – maximum tensile stresses in the slab from temperature changes [MPa],
γ_m – material factor =1.06, γ_p – safety factor from load =1.1, γ_t – safety factor from thermal =1.1,
n_1 – a coefficient of slabs cooperation (0.9 – undoweled, 0.7 - doweled).

Using the dependence (2), the durability derived from the temperature N_t was determined on the basis of the share of stresses against temperature in relation to total stresses:

$$N_t = N - \left(N \cdot \frac{\sigma_t}{\sigma_p + \sigma_t} \right) \qquad (3)$$

Such dependence allowed to sensitize the sensitivity of different slab thicknesses and subgrade stiffness on the pavement durability associated only with temperature. Fig. 10 shows the results of calculations of durability from the temperature for the undoweled concrete slab, while in Fig. 11 for the doweled plate.

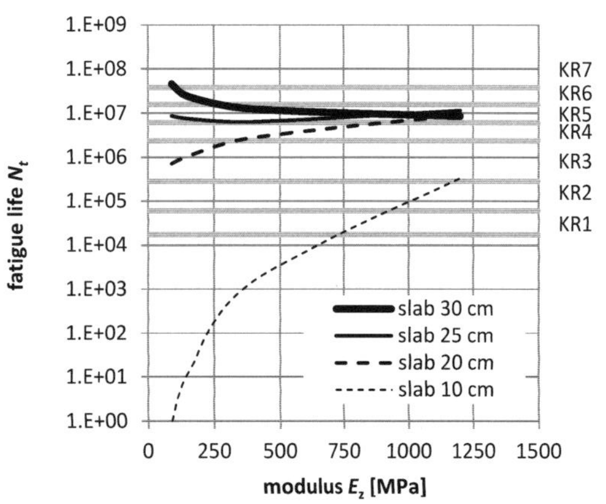

Figure 10. Changing the fatigue life N_t for a undoweled slab depending on the module E_z (slab length 5.0 m, strength 5.5 MPa).

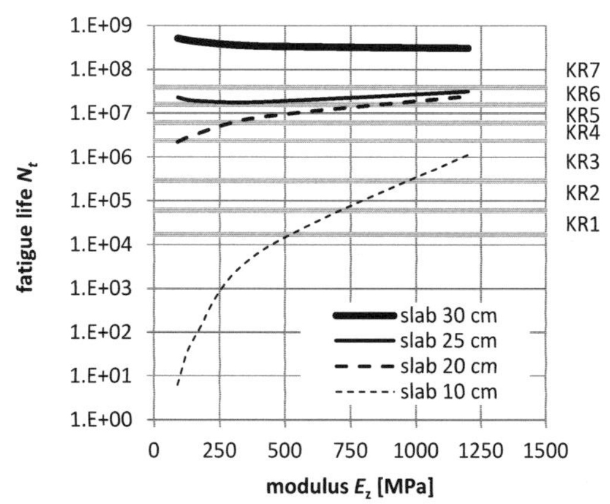

Figure 11. Changing the fatigue life N_t for a doweled slab depending on the module E_z (slab length 5.0 m, strength 5.5 MPa).

The results presented concern slabs with a length of 5.0 m and a concrete strength of 5.5 MPa. In addition, the durability intervals corresponding to the categories of movement contained in the catalogue of rigid pavement for the axes 115 kN are marked on the figures. It was found that the doweled slabs have much higher durability compared to the undoweled slabs. For a 30 cm doweled slab, there was a change of the traffic category from KR5 to KR7. The thinner slabs (20 and 25 cm) "moved" with KR3 / 5 to KR4 / 6. The smallest offset occurred for a thinner 10 cm slab from KR2 to

KR3 and only for a high modulus replacement subgrade. The 10 cm thin slab shows the greatest sensitivity to changing the equivalent module and above the value of 500 MPa it achieves KR1 (undoweled slab). Sensitivity to the change of the subgrade module is also exhibited by a slab with a thickness of 20 cm and in the module compartment from 90 to 1200 MPa changes the durability by 3 categories of movement. The smallest sensitivity is shown by the thickest 30 cm slab. It is worth noting that with low stiffness of the subgrade, its durability decreases slightly.

It should be noted that the calculations performed are simulations and, in engineering practice, not all slab thicknesses are suitable for dowelling. The calculations were made for average daily temperature fluctuations in the year. It should also be noted that the presented calculations refer to thermal interactions also associated with the additional impact of vehicles. The durability cannot be determined separately for each impact, as it depends "very non-linearly" on various factors (e.g. changes in the stiffness of the subgrade, stress values). Further analyses should also consider the total accumulation of stresses as a result of simultaneous temperature impact and unfavourable vehicle position on the concrete slab.

In tables 1 the relation between durability and temperature impact to total durability is presented, expressed in [%]. It should be noted that for thin slabs, the share of durability from the temperature in relation to the total durability, which is additionally durable from the load, is significant and amounts to about 70-80%. This means that the stresses are small, and the stresses from the load determine the loss of total durability. In the case of thick slabs (30 cm), the durability of the load transfer from the wheel rather than the sensitivity to temperature changes is more decisive. In this case, durability from temperature is only 30 - 40% of total durability. The effect of dowelling or reducing the length of the slab causes that for thicker slabs the stresses from temperature decrease with respect to stresses from load, which increases their share in the global durability of the concrete slab, to 50 – 60%.

The authors also conducted an evaluation of the impact of dowelling on the change in durability (Figures 12 and 13).

Table 1. Ratios of durability for temperature to total durability [%].

	undoweled slab – 5.0 m				doweled slab – 5.0 m				doweled slab – 4.0 m			
E_z [MPa]	90	153	400	1200	90	153	400	1200	90	153	400	1200
slab 10 cm	**85.9**	**84.1**	**80.0**	**72.4**	**86.5**	**84.8**	**80.9**	**73.8**	**82.9**	**81.3**	**77.1**	**71.0**
slab 20 cm	**57.8**	**54.7**	**49.2**	**42.1**	**58.8**	**55.7**	**50.2**	**43.1**	**67.8**	**63.3**	**55.2**	**49.0**
slab 25 cm	**49.1**	**45.0**	**38.0**	**31.7**	**50.1**	**45.8**	**38.9**	**32.6**	**62.2**	**56.2**	**48.1**	**39.7**
slab 30 cm	**42.8**	**37.7**	**30.6**	**23.9**	**50.8**	**46.4**	**38.9**	**31.1**	**63.6**	**58.4**	**48.5**	**38.8**

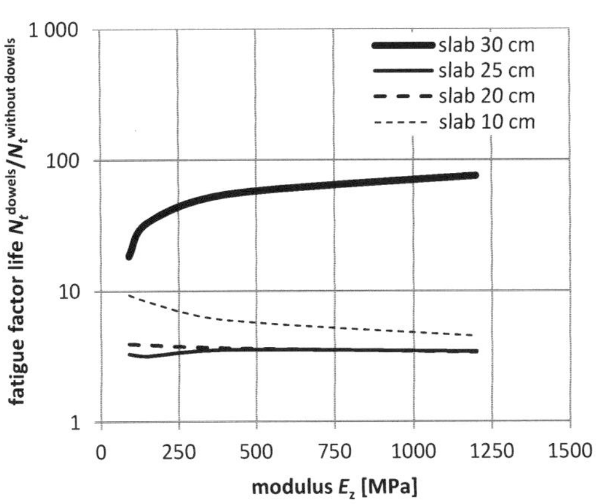

Figure 12. Change in the dowelling factor depending on the equivalent module (slab length 5.0 m, strength 4.5 MPa).

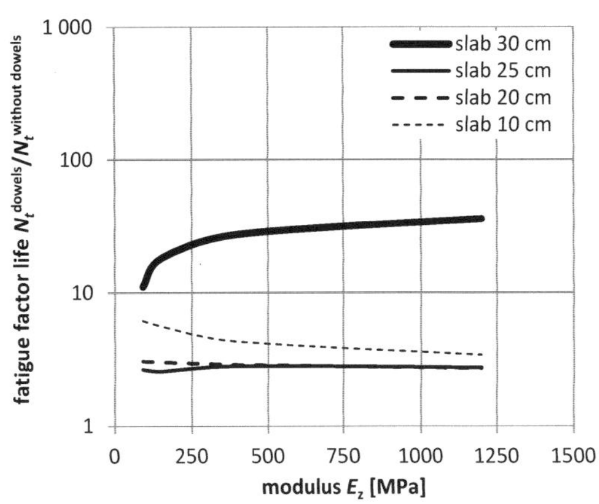

Figure 13. Change in the dowelling factor depending on the equivalent module (slab length 5.0 m, strength 5.5 MPa).

The dowelling factor was determined as the ratio of temperature stability for dowel panels to the durability of panels without dowels. It was found that the dowelling factor reaches higher values for thicker slabs (30 cm) from about 10 to 80. The effect of dowelling is more visible for lower strength concrete. For thinner slabs (20 and 10 cm), the stiffness of the subgrade for the dowelling is not significantly affected. The dowel index varies from just 3 to 8. For the thinnest slabs, 10 cm, a larger index (from 6 to 9) was obtained than for a 20 and 25 cm slabs. It should be noted, however, that in practice, for technological reasons, there is no point in making such thin slabs.

8. Summary

The experience gained so far in the world of road engineering shows that the whitetopping method allows for the pavement characterized by a long service life, low maintenance costs and a high degree of safety for car traffic. This technology allows also repairing typical damage of asphalt pavement, such as: ruts, surface flakes, temperature and fatigue cracks. However, it should be remembered that when reinforcing old pavement by whitetopping method, it is important to correctly identify pavement parameters, because depending on the value of the equivalent module, such values will be stresses from vehicle movement and temperature, and consequently durability.

The biggest impact on the change in durability due to the value of the equivalent module of the subgrade is noticeable for slabs with a thickness of 20 cm (and below). For such slabs, for the module from 90 to 1200 MPa, durability changes by about a dozen million. This means a change of 2 - 3 traffic categories. The smallest influence on the change of the traffic category in relation to the equivalent module value is shown by thicker 25 and 30 cm slabs. Such slabs practically do not change the movement category for both small and large values of the replacement module. For example the 25 cm dowelled slabs is for the movement category KR6 and 30 cm for KR7. Although it should be noted that for high traffic categories, i.e. KR6, the durability range is up to 20 million axles.

Dowelled slabs have significantly higher durability compared to the boards being drilled. For boards with a thickness of 30 cm, the use of dowels increases the durability of more than several tens of millions of axes. Dowelling a thinner slab with a thickness of 20 and 25 cm increases the durability by about several million axes. The smallest effect is obtained for thinner slabs, e.g. 10 cm (several hundred thousand). It was found that the repeatability of thermal stresses significantly reduces the durability of the concrete pavement in the long-term period of impact. Even small temperature fluctuations along with the load from the vehicle can be sufficient to initiate local damage in concrete. Especially high tensile stresses were found for 25 cm thick slabs, for which there is still a relatively large temperature difference. It is therefore recommended to use dowels for slabs above 25 cm, while for the design of thinner slabs (less than 25 cm) for low traffic categories (below KR4) pay attention to the values of the substitute module of the reinforced pavement.

The conducted analyses can be helpful in the design of new concrete pavement, and above all in the assessment of the load-bearing capacity of existing pavement in whitetopping technology. The optimal choice of thickness, slab length or dowelling needs should be taken into account together with the impact of the load and the need to transfer the required traffic at the design stage.

The subject of further work will be an analysis of the impact of the interconnection variable between the slab and the ground and the analysis of stress caused by the load on vehicles and their unfavourable location.

References

[1] Tarr S M Sheehan M J and Ardani A 2000 Mechanistic Design of Thin Whitetopping Pavements in Colorado *Transp. Res. Rec.* No. 1730 pp 64–72

[2] Cole L 1997 Pavement Condition Surveys of Ultrathin Whitetopping Projects *Sixth International Purdue Conference on Concrete Pavement* **2** pp 175–187

[3] Rasmussen R O and Rożycki D K 2004 Thin and Untra-Thin Whitetopping NCHRP Synthesis of Highway Practice 338 *Nat. Cooper. High. Res. Prog.* NRC Washington DC

[4] Roesler J Bordelon A Ioannides A Beyer M and Wang D 2008 Design and Concrete Material Requirements for Ultra-Thin Whitetopping *Publication FHWA-ICT-08-016* ICT June
[5] AASHTO 1998 Design of Pavement Structures. Part II. Rigid Pavement Design & Rigid Pavement Joint Design
[6] SETRA 2000 Chaussees en beton *Guide technique* LCPC
[7] Jasieński A 2002 Drogi z betonu cementowego - przykłady obecnych zastosowań na terenie Belgii *Konferencja Dni Betonu*
[8] Westergaard H M 1926 Stresses in Concrete pavements computed by Theoretical Analysis *Public Roads* **7** (2) pp 25–35
[9] Westergaard H M 1927 Analysis of stresses in concrete due to variations of Temperature *Proceedings of the 6th Annual Meeting Highway Research Board* NRC **6** pp 201–215
[10] Teller L W Sutherland E C 1935 The structure design of concrete pavements, part 2: observed effects of variations in temperature and moisture on the size, shape, and stress resistance of concrete pavement slabs *Public Roads* **16** (9) pp 169–197
[11] Choubane B Mang T 1995 Analysis and Verification of Thermal-Gradient Effects on Concrete Pavement *Journal of Transportation Engineering* **121** (1) pp 75–81
[12] Tabatabai A M Barenberg E J 1978 Finite-element analysis of jointed or cracked concrete pavements *Transportation Research Record* 671 TRB Washington DC
[13] Chou Y T 1981 Structural Analysis Computer Programs for Rigid Multicomponent Pavement Structures with Discontinuities: WESLIQUID and WESLAYER *Technical Report* GL-81-6 Vicksburg MS: US Army Engineer Waterways Experiment Station
[14] Huang Y H 1993 *Pavement Analysis and Design* Prentice Hall Englewood
[15] Khazanovitch L Yu H T Beckemeyer C 2000 Application of ISLAB2000 for Forensic Studies *Proceeding of the 2nd International Symposium of 3D Finite Element for Pavement Analysis, Design, and Research* Charleston West Virginia pp 433–450
[16] Shoukry S N William G W Riad M 2002 Characteristics of Concrete Stresses in Doweled Transverse Joints *International Journal of Pavement Engineering.* **3** (2) pp 117–129
[17] Lytton R L Pufahl D E Michalak C H Liang H S Dempsey B J 1993 An integrated model of the climatic effects on pavements FHWA-RD-90-033 FHA November
[18] Larsen G Dempsey B J 1997 Enhanced integrated climatic model Version 2.0 Minnesota Department of Transportation October
[19] Riad M Y Shoukry S N William G W Fahmy M R 2009 Effect of skewed joints on the performance of jointed concrete pavement through 3D dynamic finite element analysis *Inter. Journal of Pavement Engineering* **10** (4) pp 251–263
[20] Luoke L Yiqiu T Xiangbing G and Yunliang L 2012 Characterization of Contact Stresses Between characterization of Contact Stresses Between Dowels and Surrounding Concrete in Jointed Concrete Pavement *Inter. J. Civil and Envir. Eng.* **12** (5) pp 23–27
[21] Shoukry S N William G W Riad M Y 2007 Effect of thermal stresses on mid-slab cracking in dowel jointed concrete pavements *Structure and Infrastructure Engineering* **3** (1) pp 43–51
[22] Mackiewicz P 2014 Thermal stress analysis of jointed plane in concrete pavements *Applied Thermal Engineering* **73** pp 1167–1174
[23] Baranowski D 2005 Daily amplitudes of air temperature in Poland and their relationship from types of atmospheric circulation (1971-1995) *Słupskie Prace Geograficzne* 2
[24] Szydlo A 1995 Statistical identification of parameters of airport pavements *IIL PWr* Wrocław
[25] COSMOS/M 1993 *Advanced modules user guide* Santa Monica CA
[26] Szydło A 2004 Nawierzchnie drogowe z betonu cementowego. Teoria, wymiarowanie, realizacja *Polski Cement* Kraków

Rehabilitation reliability of the road pavement structure with recycled base course with foamed bitumen

P Buczyński [1]

[1] Kielce University of Technology Department of Transportation Engineering, Faculty of Civil Engineering and Architecture, al. Tysiąclecia Państwa Polskiego 7 25-314 Kielce

E-mail: p.buczynski@tu.kielce.pl

Abstract. This article presents a new approach to reliability assessment of the road structure in which the base layer will be constructed in the process of cold deep recycling with foamed bitumen. In order to properly assess the reliability of the structure with the recycled base, it is necessary to determine the distribution of stress and strain in typical pavement layer systems. The true stress and strain values were established for particular structural layers using the complex modulus (E*) determined based on the master curves. The complex modulus was determined by the direct tension-compression test on cylindrical specimens (DTC-CY) at five temperatures (-7°C, 5°C, 13°C, 25°C, 40°C) and six loading times (0.1 Hz, 0.3 Hz, 1 Hz, 3 Hz, 10 Hz, 20 Hz) in accordance with EN 12697-26 in the linear viscoelasticity (LVE) range for small strains ranging from 25 to 50 $\mu\varepsilon$. The master curves of the complex modulus were constructed using the Richards model for the mixtures typically incorporated in structural layers, i.e., SMA11, AC16W, AC22P and MCAS. The values of the modulus characterizing particular layers were determined with temperature distribution in the structure taken into account, when the surface temperature was 40°C. The stress distribution was established for those calculation models. The stress values were used to evaluate the fatigue life under controlled stress conditions (IT-FT). This evaluation, with the controlled stress corresponding to that in the structure, facilitated the quality assessment of the rehabilitated recycled base course. Results showed that the recycled base mixtures having the indirect tensile strength (ITS_{DRY}) similar to the stress in the structure under analysis needed an additional fatigue life evaluation in the indirect tensile test ITT. This approach to the recycled base quality assessment will allow eliminating the damage induced by overloading.

1. Introduction

Cold deep recycling with foamed bitumen [1, 2] and bitumen emulsion [3] is widely used for road rehabilitation in Poland and elsewhere. This technology helps eliminate the effects of fatigue life losses in the entire pavement layer system and restores the required service parameters of the structure [4]. A comparison of the recycled base to that constructed using conventional technologies, i.e., a base layer with unbound aggregate and cement concrete, shows a considerable advantage of cold recycling [5, 6, 7] in that it provides better cohesion relative to unbound mixture bases and lower stiffness relative to cement concrete bases. Hence, the cold recycled base layer with foamed bitumen (MCAS) contributes to the improved quality and reliability of the rehabilitated road pavement structure. Reliability of the structure is defined as its ability to satisfy the design life requirements.

The selection of rehabilitation technique requires a correct assessment of the existing surface condition [8]. In the case of structural strengthening, it is necessary to calculate fatigue life with regard to material constants (stiffness modulus) for new layers. True values of material constants applied in the calculations will provide more accurate determination of the fatigue life of the rehabilitated road pavement [9]. To determine the stiffness modulus for describing the material used in the structural layers, specialized testing is necessary.

The specialized testing mentioned above was conducted by the authors of [10] for assessing the properties of a mineral-cement mixture with a bitumen emulsion for use in the base layer. The test results for the properties of the MCE mixture in terms of the curing period, 28 days and 18 months, provide information on changes in the stiffness modulus over time and can be taken into account in the design of the structure. The aspect of dosing synthetic fibres to the bituminous mixture was discussed in [11]. Temperature aspects and the effects of various additives used in the bituminous mixture were discussed extensively in [12, 13].

Nevertheless, the design that employs standard parameters alone fails to take advantage of the information about fatigue characteristics that vary between materials intended for the same layer. It is thus advisable to evaluate the quality of structural layers using the fatigue life test under cyclic load (controlled stress or strain). These issues have been described and discussed by many researchers [14-16]. As shown in [17], the addition of 35% reclaimed asphalt pavement (RAP) material to the hot mix asphalt (HMA) mixtures required the assessment of fatigue life to confirm that 35% RAP content would not cause the fatigue life loss. The addition of 30% RAP in combination with a bio-agent improved the fatigue life [18]. Brzeziński et al. [19] studied the effect of cyclic loads on the static strength of the hydraulically bound mixture for use in the pavement structural layers. Results showed the improvement in the strength of the material. It follows from the above that the fatigue life assessment allows the detailed evaluation of the quality of materials to be incorporated in the structural layers. Ensuring the proper values of standard parameters is insufficient to guarantee reliable service of the pavement structure.

The article highlights the importance of assessing the fatigue life of a recycled base with foamed bitumen in the aspect of the stress levels that will probably occur in the rehabilitated pavement structure. The materials used in the structural layers were described by complex modulus master curves. The analysis was performed for the unfavourable temperature distribution at the wearing layer surface temperature of 40°C. This approach made it possible to accurately estimate the complex modulus in the aspect of temperature distribution and depth of the layers. The information obtained was used to build a computational model for the analysis of the tensile stress distribution in the structures. Due to the brittle behaviour of the recycled base layer, it was necessary to determine its durability for the true/modelled stress values. This approach to fatigue life assessment of rehabilitated pavements will secure the recycled base layer with foamed bitumen against cracking due to the cyclic load induced overload.

2. Object of study

The fatigue life analysis of the pavement structure was performed by checking the fatigue life of the recycled mineral-cement mixture with foamed bitumen in the laboratory-based indirect tensile test (ITT). Typical pavement structures (table 1) [20] were selected for the analysis in terms of the 100 kN standard axle load for ranges $5x10^5<ESAL_{100kN}<2.5x10^6$ (in Poland assigned to the KR3 traffic category) and $2.5x10^6<ESAL_{100kN}<7.3x10^6$ (in Poland assigned to the KR4 traffic category).

Table 1. The pavement layer system for the analysed road rehabilitation.

Layer - material	Layer thickness in [cm] for $5x10^5<ESAL_{100kN}<2.5x10^6$	Layer thickness in [cm] for $2.5x10^6<ESAL_{100kN}<7.3x10^6$
Wearing course - SMA11 PMB 45/80-55	4.0	4.0
Binder course - AC16W 35/50	8.0	5.0
Base course - AC22P 35/50	-	7.0
Base course - MCAS 50/70	20.0	20.0
Subgrade $E_0\geq100$ MPa	-	-

3. Methods

The procedure applied involved constructing a computational model for which the stiffness modulus E* (MPa) was determined in terms of temperature distribution; the value of stiffness modulus was determined using master curves models for the temperature calculated in the mid-thickness of the layer; the horizontal stress (σ_{x-x}) distribution at the underside of the structural layers was calculated for the adopted models; and finally, the stress values were used in the recycled base fatigue testing in stress-controlled mode ITT.

3.1. Temperature distribution in the pavement layer system

To analyse the temperature distribution across the thickness of the asphalt layers, typical pavement structures were considered [20]. The history of temperature changes in the layers was determined for the road surface temperature of 40°C (at about 10:30). The temperature distribution was determined using the relationship shown in formula (1) and described in detail in [21, 22, 23]:

$$T(t,z) = T_M + T_A \cdot exp(-\lambda z) \cdot \sin\left(2\pi \cdot \frac{10{,}4}{24} - \lambda z + \tau\right) \tag{1}$$

where: T_M- average road surface temperature equal to T_A = 33.9°C; T_A- daily road surface temperature change amplitude equal to T_A = 16.7°C; λ – a factor dependent on thermal properties of the material; z – the depth of the layer; τ – correction resulting from the shift of the first measurement time relative to the sine function describing daily temperature changes, equal to $\tau = 1.25\pi$.

*3.2. Master curves of complex modulus E**

The master curve models were developed using the Richards model [24] which is a modification of the model presented in the NCHRP 9-29 report: PP 02 [25]. This model can be classified as a non-symmetrical sigmoid mathematical model owing to the addition of lambda parameter (λ) as an asymmetry coefficient. The form of the sigmoid non-symmetrical function is described by equation (2). Master curves are constructed using the principle of time-temperature superposition. For this purpose, it is necessary to introduce a temperature shift coefficient (α_T) implemented in formula (2) [26]:

$$\log|E^*| = \delta + \frac{\alpha}{\left[1+\lambda e^{\beta+\gamma \log(a+b\cdot T+c\cdot T^2)}\right]^{1/\lambda}} \tag{2}$$

where: $|E^*|$ – complex modulus, ω –angular frequency, δ –the lower asymptote, (master curve fitting parameter), α – the difference between the upper and lower asymptotes, (master curve fitting parameter), λ, β, γ – master curve fitting parameters, T – test temperature, T_{ref} – reference temperature, a, b, c – model parameters.

The complex modulus was determined in the direct compression and tension test (DTC-CY) where the sample is subjected to a cyclic sinusoidal load with a low strain value in the range from 25 to 50 με [27-29]. Testing carried out beyond this strain range may cause irreversible deformations in the sample, leading to errors in the complex modulus. The tests were carried out at four temperatures (5°C, 13°C, 25°C, 40°C) and six loading times (0.1 Hz, 0.3 Hz, 1 Hz, 3 Hz, 10 Hz, 20 Hz).

3.3. Distribution of the horizontal strains (σ_{x-x}) in the pavement layer system

To calculate the cooperation between the asphalt layers, the base layer and the underlying native subgrade, a model of statically loaded elastic layers with the assumed full interlaying bonding was used [30]. The load transmitted to the surface was adopted according to [20, 31] in the form of a vertical, uniformly distributed contact pressure of 850 kPa induced by a wheel load of 50 kN (100 kN/axle). The loading time 10 Hz. The distribution of stresses (ε) and strains (σ) in this system was determined.

3.4. Recycled base fatigue in the indirect tensile test ITT

The fatigue of the recycled base course was investigated in accordance with the requirements set forth in PN-EN 12697-24 Annex E [29]. The test samples were subjected to cyclic, compressive loading with a semi-sinusoidal signal. The length of the pulse was 1.5 s and the loading time was 0.124 s. The load was applied through the diametral vertical plane, with a specimen failing by fracturing along the vertical diameter. The strength assessment was performed under stress-controlled conditions. The value of the normal stress used in the test is the result of the stress distribution in the structure calculated at the underside of the recycled base layer as the tensile stress (σ_{x-x}; σ_{y-y}). The test was carried out up to 30 load cycles or to the failure of the specimen. Prior to testing, the specimens were preconditioned for four hours at 25°C.

4. Materials

4.1. Basic parameters of the bituminous mixtures and of the recycled mixture MCAS

Materials used in the structural layers intended for rehabilitation in the roadway designed for traffic category KR3 ($5x10^5<ESAL_{100kN}<2.5x10^6$) and KR4 ($2.5x10^6<ESAL_{100kN}<7.3x10^6$) included the SMA11 bituminous mixtures in the wearing layer and asphalt concrete AC bituminous mixtures in the binder and base layers. In the base layer for the KR3 and KR4 traffic, a cold foamed bitumen mineral-cement mix was also used. The gradation curves of these mixtures are shown in Fig. 1.

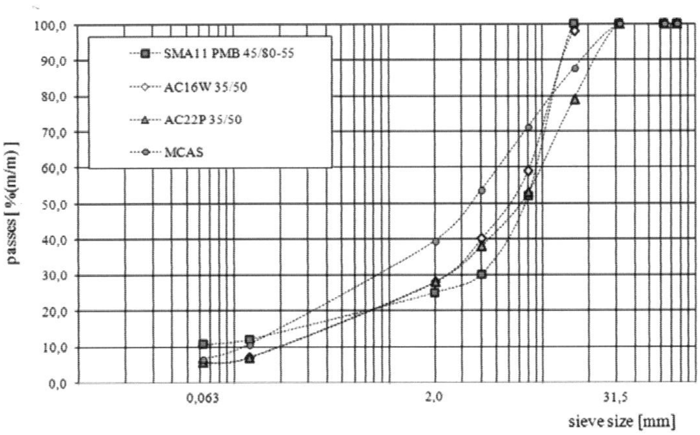

Figure 1. Grading curves of the mineral mixtures.

The bituminous mixtures SMA11, AC16W and AC22P were designed to the requirements set forth in WT-2 2014 [32], and the mineral-cement mix with foamed bitumen (MCAS) was designed to the specifications contained in [33, 34]. The results of the basic properties of the mixtures under analysis are summarized in Table 2.

Table 2. Basic parameters of the mixtures.

Property	Symbol	u.m.	SMA11	AC16W	AC22P	MCAS
Bitumen/cement content	B / C	[%]	B=6.3	B=4.6	B=3.9	B=2.4 C=2.0
Water sensitivity, PN-EN 12697-12, + [32]	ITSR	[%]	95.0	89.9	84.2	-
Moisture resistance, Wirtgen 2006 [33]	TSR		-	-	-	99.9
Void characteristics, PN-EN 12697-8	V_m	[%]	3.0	4.3	5.6	10.6
Indirect tensile strength PN-EN 12697-23 (T=25°C)	ITS	[kPa]	822	1133	994	546

Proportional rut depth (small device, 60°C, 10000 cycles) PN-EN 12697-22	PRD_{AIR}	[%]	5.4	3.9	4.8	-
Wheel-tracking slope (small device, conditioning in air, 60°C, 10000 cycles), PN-EN 12697-22	WTS_{AIR}	[mm/ 1000cycle]	0.06	0.05	0.08	-
Binder drainage, % m/m, (at production temperature, Schellenberg's method)	D	[% (m/m)]	0.20	-	-	-

4.2. Complex modulus master curves for the mixtures

The principles of time-temperature superposition [25, 35] were used to build the complex modulus master curves of the bituminous and MCAS mixtures [25, 35]. The sigmoid function was optimized by minimizing the sum of squared deviations for the complex moduli determined. The master curve coefficients (α, β, γ, δ, λ, a, b, c) had to be determined from the observed data in the optimization process. The criterion of the normalized root-mean-square error RMSE, applied by the authors to the evaluation of the SMA behavioral changes [36], was used to assess the goodness-of-fit of the sigmoid function. The parameter assessment results are shown in Table 3.

Table 3. Master curves fitting parameters.

Mixture	Master curve fitting parameters							Goodness-of-fit
	α	β	γ	δ	λ	a	b	R^2
SMA11	2.278	0.362	-0.443	2.278	0.143	2.01	-0.048	0.97
AC16W	2.177	0.299	-0.492	2.177	0.21	2.24	-0.047	0.95
AC22P	2.083	0.35	-0.400	2.083	0.3	2.32	-0.034	0.97
MCAS	2.149	1.035	-0.224	2.149	0.1	2.572	-0.014	0.98

The very high value of R^2 for the non-symmetrical sigmoid function model fitting parameters confirms the good fit of the function to the determined values of the complex modulus shifted by a value of α_T. Analysis of the master curve "λ" parameter showed variations in the behaviour of the mixtures. The highest complex modulus variation at low and high temperatures was observed in the SMA11 mix. It should also be noted that SMA11 in the case of a long loading time (at high temperatures) showed the lowest value of the complex modulus in comparison to the other analysed mixtures. This is a disadvantageous phenomenon from the point of view of plastic deformation.

The master curves in the reduced frequency domain at a reference temperature of 25°C are shown in Fig. 2.

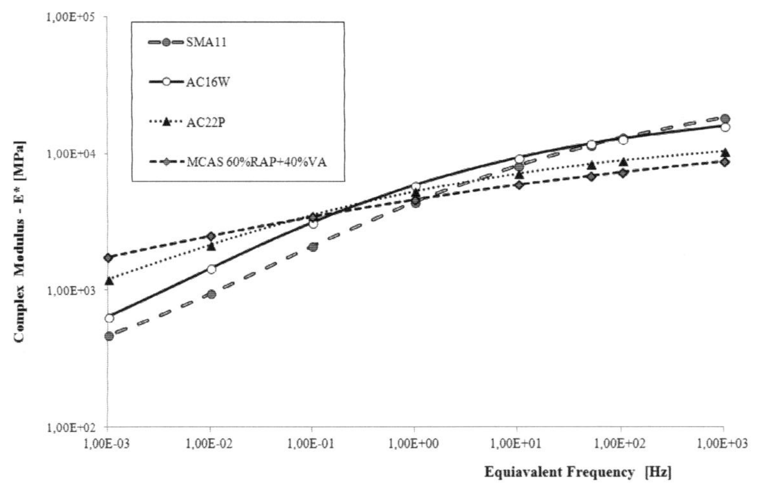

Figure 2. Master curves (T_{ref}=25°C) for the tested mixtures.

It is evident (Fig. 2) that at very low frequencies (below 0.1 Hz) the complex modulus correlates with the quantity of bitumen in the bituminous mix. The SMA11 mix, containing the highest bitumen amount, shows the lowest module. This behaviour adversely affects the bituminous mix layer designed for the wearing course, where long loading time will quickly change its properties. A similar relationship in terms of loading time in SMA mixes was reported in [31]. Also, the asphalt concrete mixture for use in the binder layer, AC16W 35/50, containing 4.6% bitumen, behaves in a similar manner at the early stage of loading. However, below 10 Hz, the AC16W mixture has a higher stiffness modulus than the SMA11 mixture. This phenomenon may be related to the higher quantity and softer type of the bitumen in the SMA11 mixture. Mixtures designed for base layers, such as AC22P and MCAS, demonstrate a similar rate of complex modulus variation as a function of loading time. Nevertheless, the recycled base with foamed bitumen MCAS will have better parameters at long and short loading times. Under very slow traffic (vehicle speed of less than 0.5 km/h), the complex modulus of MCAS is higher than the modulus of the AC22P mixture by over 1000 MPa. This behaviour will reduce the vertical strain in the subgrade and thus increase the fatigue life.

5. Results of the calculations and analyses

5.1. Computational model

The computational model for the typical structures being analysed was constructed using the dependence (1) described in detail in [21, 22, 23]. The temperature in the rehabilitated roadway structures was defined at the mid-thickness of the layer. The value of the complex modulus at a given temperature in the layer was determined using the characteristics obtained from the master curve models. As a result, comprehensive information was collected for the determination of the stress distribution in the pavement structures. The results of the analysis are presented in Table 4.

Table 4. Computational model for the structures being analysed.

Layer - material	Layer thickness for $5x10^5<ESAL_{100kN}<2.5x10^6$				Layer thickness for $2.5x10^6<ESAL_{100kN}<7.3x10^6$			
	h [m]	t [°C]	v [-]	E * (10Hz) [MPa]	h [m]	t [°C]	v [-]	E * (10Hz) [MPa]
Wearing - SMA11 PMB 45/80-55	0.04	37.0	0.40	4453	0.04	37	0.40	4453
Binder - AC16W 35/50	0.08	32.0	0.40	6483	0.05	32	0.40	6483
Base - AC22P 35/50	-	-	-	-	0.07	30	0.40	5855
Subbase - MCAS 50/70	0.20	31.0	0.40	4925	0.20	31	0.40	4925
Subgrade $E_0≥100$ MPa	-	-	0.35	100	-	-	0.35	100

The values of complex modulus obtained from the master curves are lower than those recommended for the calculation of fatigue life in Poland [20]. The differences between measurement temperatures affect those values.

5.2. Determination of the stress and strain states using the FEM

The layer system was modeled in ABAQUS. The mesh of elements was constructed with the assumption that the minimum dimension of the element is 4 cm and there is a minimum of four elements in one layer of the pavement structure. A basic eight-node finite element with quadratic shape functions (CAX8) [37] was used for analysis. The computational region in the analysis was 5 m (height) and 3.5 radius (rotationally symmetric model). The side edges and the base had pinned joint (connection). The distribution of stresses for the analyzed typical structures was determined. The

IOP Conf. Series: Materials Science and Engineering **356** (2018) 012016 doi:10.1088/1757-899X/356/1/012016

results of the analysis in the form of stress distribution maps for the structure are shown in Fig. 3. The values of horizontal stresses (σ_{x-x}; σ_{y-y}) for individual points in the structure are provided in Table 5.

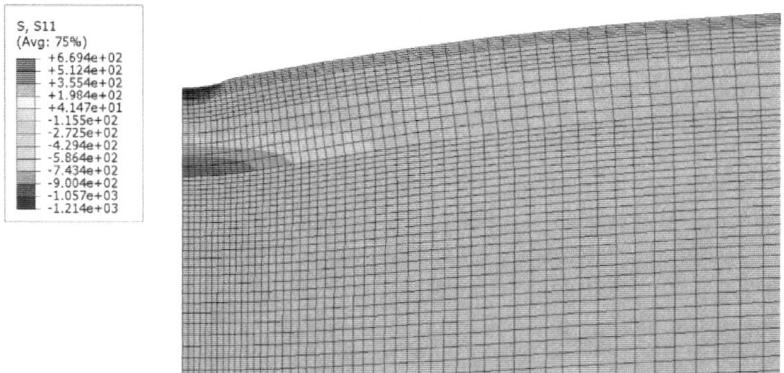

Figure 3. Distribution of stresses in the KR3 ($5x10^5$<$ESAL_{100kN}$<$2.5x10^6$) pavement structure.

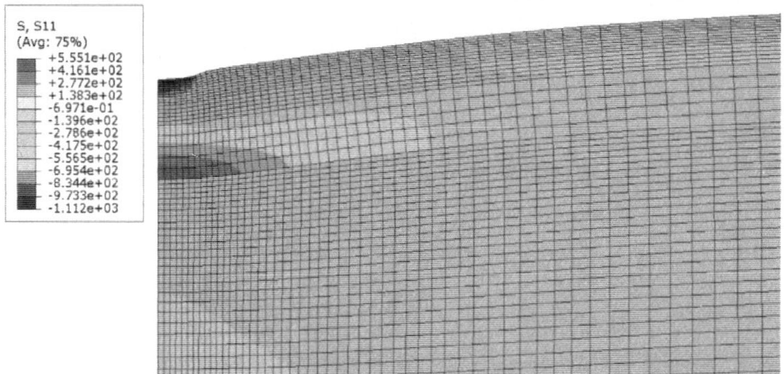

Figure 4. Distribution of stresses in the KR4 ($2.5x10^6$<$ESAL_{100kN}$<$7.3x10^6$) pavement structure.

Table 5. Computational model of the structures being analysed.

Layer - material	KR3 $5x10^5$<$ESAL_{100kN}$<$2.5x10^6$		KR4 $2.5x10^6$<$ESAL_{100kN}$<$7.3x10^6$ [kPa]	
	Depth pkt. [m]	[kPa]	Depth pkt. [m]	[kPa]
Wearing - SMA11 PMB 45/80-55	0.0000	- 1214.3	0.0000	- 1112.3
	0.0399	- 924.9	0.0399	- 860.7
Binder - AC16W 35/50	0.0401	- 919.3	0.0401	- 873.1
	0.1199	- 218.9	0.0899	- 472.7
Base - AC22P 35/50	----------------------------		0.0901	- 384.9
			0.1599	- 53.2
Subbase - MCAS 50/70	0.1201	- 103.7	0.1601	1.9
	0.3199	669.3	0.3599	555.1
Subgrade $E_0 \geq 100$ MPa	0.3201	- 2.4	0.3601	- 1.8

The tensile stress values (σ_{x-x}; σ_{y-y}) at the underside of the recycled base layer for the KR3 structure ($5x10^5$<$ESAL_{100kN}$<$2.5x10^6$) is 669.3 kPa and 555.1 kPa for the KR4 structure ($2.5x10^6$<$ESAL_{100kN}$<$7.3x10^6$) Increased thickness of the package of bituminous mix-based layers decreases tensile stresses in the base layer produced with the mineral-cement mixture with foamed bitumen, MCAS, thus contributing to reduced vertical strain in the subgrade and improved fatigue life of the system as a whole. The values of the tensile stress are similar to the values of indirect tensile strength (ITS_{DRY}=546 kPa) characteristic of the recycled base with foamed bitumen. As this may confirm a probability of overloading/cracking of the base layer due to unfavourable weather conditions (high temperature), it seems necessary to evaluate the fatigue life of the mineral-cement mixture with foamed bitumen for use in the base layer. Determining the fatigue life under dynamic loading in the indirect tensile test will contribute to a more accurate quality evaluation of the mineral-cement mix with foamed bitumen designed for use in the base layer during road rehabilitation works.

The fatigue test was performed on cylindrical samples in an indirect tensile test in accordance with the requirements of PN-EN 12697-24, Annex E [29], for two stress levels, 375 kPa and 500 kPa. These values are close to the values of stresses occurring in the structures under analysis. The fatigue characteristics of the recycled base with foamed bitumen are shown in Figure 5.

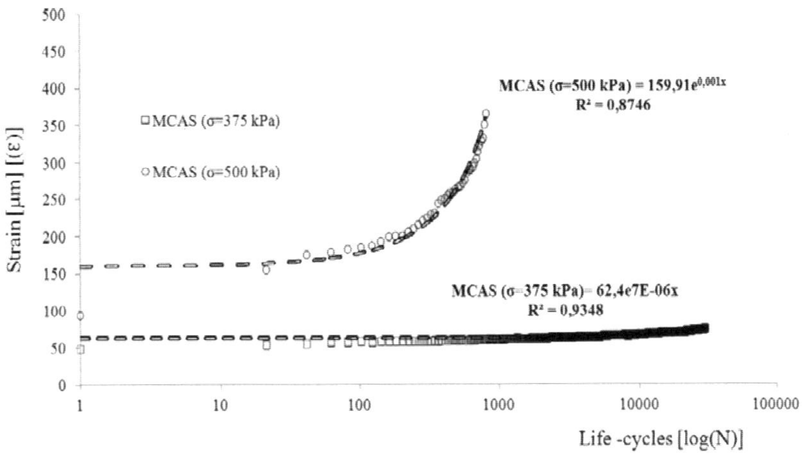

Figure 5. Fatigue curves of the MCAS recycled base.

Note that the increase in the stress level in the tested recycled base samples increased the initial strain value. At a stress of 375 kPa, the strain after the 100th load cycle was 58 μm, while at 500 kPa the initial strain was 185 μm. Furthermore, throughout the testing period, i.e., 30,000 load cycles at a stress of 375 kPa, only a 29% increase in strain relative to the initial value was found. Different fatigue characteristics were obtained in the study of the recycled base with foamed bitumen at a stress of 500 kPa, where after 800 load cycles the sample failed.

The results above support the observation that the increase in the stress level has a destructive effect on the quality of the recycled base layer and that attaining the required basic quality parameters does not guarantee road rehabilitation reliability. The indirect tensile strength values (ITS_{DRY}) exceeding the value of the stress level used in this study does not ensure the required life. Dynamic testing is a better tool for discriminating the quality of the recycled base.

6. Conclusions

Based on the results of this study, the following conclusions are offered:

- As for the asphalt mixes evaluated, at very low frequencies below 0.1 Hz, the complex modulus correlates with the bitumen content. The lowest value of the complex modulus E * was obtained for the SMA11 mixture (with B = 6.4%) and the highest value E* for the asphalt concrete AC22P (B = 3.9%) and MCAS (B = 2.0%). This is due to the sensitivity of the bitumen to the deformation in a long loading time,
- An increase in the thickness of asphalt layer package contributes to the reduction of tensile stress ($\sigma_{x\text{-}x}$; $\sigma_{y\text{-}y}$) in the recycled base layer made with mineral-cement mixture and foamed bitumen, MCAS,
- Increased stress level in the fatigue life test of the recycled base with foamed bitumen affected the initial strain value and showed a deleterious effect on the durability/strength of the mineral-cement mixture with foamed bitumen,

- application of the mineral-cement mixture with foamed bitumen to the KR3 traffic category ($5x10^5$<$ESAL_{100kN}$<$2.5x10^6$) roads, where the horizontal stress derived from the computational model is 669.1 kPa, does not guarantee the reliability of the road rehabilitation works. The stress level of 500 kPa led to the premature failure of the MCAS sample after 800 cycles of dynamic loading,
- quality assessment through dynamic testing should be employed when the value of indirect tensile strength (ITS_{DRY}) is close to the value of horizontal stress in the structure in which the recycled base layer is planned to be used.

Acknowledgment
The study results were developed within the framework of the project entitled "The use of recycled materials" under "RID" co-funded by the National Centre for Research and Development and the General Directorate for National Roads and Motorways in Poland.

References

[1] Li Z, Hao P, Liu H, Xu J, Chen Z 2016 Investigation of early-stage strength for cold recycled asphalt mixture using foamed asphalt, *Const. Build. Mater.* vol 127, pp 410-417. https://doi.org/10.1016/j.conbuildmat.2016.09.126

[2] Iwański M, Buczyński P, Mazurek G 2014 Statistical evaluation of water and frost resistance of deep cold recycled bases with foamed bitumen, hydrophobic activators and fines, *3rd International Conference on Transportation Infrastructure. Pisa, ITALY, Apr. 22-25, 2014,* pp 203-212. doi: 10.1201/b16730-30

[3] Dołżycki B 2017 Polish experience with cold in-place recycling, *IOP Conf. Ser.: Mater. Sci. Eng. 236 012089.* https://doi:10.1088/1757-899X/236/1/012089

[4] Iwański M, Chomicz-Kowalska A 2015 Evaluation of the pavement performance, Bulletin of the Polish Academy of Sciences Technical Sciences, vol. 63(1), pp. 97-105. doi:10.1515/bpasts-2015-0011

[5] Buczyński P 2016 The Frost Resistance of Recycled Cold Mixes with Foamed Bitumen and Different Types of Road Binders, *Proced. Eng.* vol. 161 pp 54–59. http://dx.doi.org/10.1016/j.proeng.2016.08.497

[6] Iwański M, Chomicz-Kowalska A 2016 Application of the foamed bitumen and bitumen emulsion to the road base mixes in the deep cold recycling technology, *The Baltic Journal of Road and Bridge Engineering*, vol. 11(4), pp 291-301. http://dx.doi.org/10.3846/bjrbe.2016.34

[7] Kukiełka J 2007 Durability of mineral-cement-emulsion mixtures bases (MCEM), *Budownictwo i Architektura,* vol. 1, pp 45-56. (In Polish)

[8] Sybilski D, et. all 2001 Katalog wzmocnień i romontów nawierzchni podatnych i półsztywnych, Warszawa, Instytut Badań Dróg i Mostów, Generalna Dyrekcja Dróg Publicznych, p. 224. (In Polish)

[9] Ryś D, Judycki J, Jaskuła P 2016 Determination of vehicles load equivalency factors for Polish catalogue of typical flexible and semi-rigid pavement structures, *Transportation Research Procedia*, vol. 14, pp 2382–2391. doi: 10.1016/j.trpro.2016.05.272

[10] Dołżycki B, Jaczewski M, Szydłowski C 2017 The long-term properties of mineral-cement-emulsion mixtures, *Constr. Build. Mater.*, vol. 156, pp 799–808. doi.org/10.1016/j.conbuildmat.2017.09.032

[11] Chomicz-Kowalska A, Iwański M.M, Mrugała J 2017 *Basic Performance of Fibre Reinforced Asphalt Concrete with Reclaimed Asphalt Pavement Produced in Low Temperatures with Foamed Bitumen*, IOP Conf. Series: Mater. S. Eng. 245, 032092 doi:10.1088/1757-899X/245/3/032092

[12] Chomicz-Kowalska A, Gardziejczyk W, Iwański M.M, 2017 Analysis of IT-CY stiffness modulus of foamed bitumen asphalt concrete compacted at 95°C, *Procedia Eng.*, vol. 172, pp 550 – 559. doi: 10.1016/j.proeng.2017.02.065

[13] Pokorski P, Radziszewski P, Sarnowski M 2015 Rheological Properties of Asphalt Mixtures for Bridge Pavements, *Procedia Eng.*, vol. 111, pp 637-644. doi.org/10.1016/j.proeng.2015.07.060

[14] Yan J, Ni F, Yang M, Li J, 2010 An experimental study on fatigue properties of emulsion and foam cold recycled mixes, *Constr. Build. Mater.*, vol. 24, pp. 2151–2156. doi.org/10.1016/j.conbuildmat.2010.04.044

[15] Buczyński P, Iwański M 2016 Fatigue life comparison of cold recycled bases with foamed bitumen and with bitumen emulsion, *Procedia Engineering*, vol. 172, pp. 135-142. http://dx.doi.org/10.1016/j.proeng.2017.02.035

[16] Bańkowski W, Gajewski M, Pronk A.C 2017 Processing of four point bending test results for visco-

elasticity and fatigue models, *International Journal of Pavement Engineering,* pp 1-5. https://doi.org/10.1080/10298436.2017.1398549
[17] Mannan U. A, Islam M. R, Tarefder R. A 2015 Effects of recycled asphalt pavements on the fatigue life of asphalt under different strain levels and loading frequencies, *International Journal of Fatigue*, vol. 78, pp. 72-80. https://doi.org/10.1016/j.ijfatigue.2015.04.004
[18] Kowalski K, Król J, Bańkowski W, Radziszewski P, Sarnowski M 2017 Thermal and Fatigue Evaluation of Asphalt Mixtures Containing RAP Treated with a Bio-Agent, *Applied Sciences*, vol. 7(3):216, pp. 1-11. doi:10.3390/app7030216
[19] Brzeziński K, Adamczewski G, Tutka P 2017 The influence of cyclic loading on the static strength of cement-bound sand – qualitative analysis, *Roads And Bridges,* vol. 16(1), pp 37-46. doi:http://dx.doi.org/10.7409/rabdim.017.003
[20] Judycki J, Jaskuła P, Pszczoła M, Alenowicz J, Dołżycki B, Jaczewski M, Ryś D, Stienss M 2014 Catalogue of typical flexible and semi-rigid pavements. Gdańsk University of Technology, Poland (in Polish) (General Directorate for National Roads and Motorways)
[21] Pszczoła M, Ryś D, Jaskuła P 2017 Analysis of climatic zones in Poland with regard to asphalt performance grading. *Roads And Bridges,* vol. *16*(4), pp 245-264. doi.org/10.7409/rabdim.017.016
[22] Judycki J, Pszczoła M, Jaczewski M, Mejłun Ł, Ryś D 2013 Badanie wpływu zastosowania warstw betonu asfaltowego o wysokim module sztywności (AC-WMS) w konstrukcjach nawierzchni na spękania niskotemperaturowe i na zmniejszenie powstawania deformacji trwałych, *Politechnika Gdańska, Raport z trzeciego etapu,* Gdańsk 2013, pp. 111. (in Polish).
[23] Judycki J 1991 Drogowe asfalty i mieszanki mineralno-asfaltowe modyfikowane elastomerem, *Zeszyty Naukowe Politechniki Gdańskiej,* vol. 452,
[24] Rowe G.M, Sharrock M.J 2011 Alternate Shift Factor Relationship for Describing the Temperature Dependency of the Visco-Elastic Behaviour of Asphalt Materials. *Transportation Research Record,* vol. 2207, pp 125-135.
[25] Witczak M. W and Root R. E 1976 Summary of Complex Modulus Laboratory Test Procedures and Results. *STP 561, American Society for Testing and Materials*, pp. 67-94.
[26] Yusoff N. I. Md, Airey G.D, Hainin M.R 2010 Predictability of Complex Modulus Using Rheological Models. *Asian Journal of Scientific Research*, vol. 3, pp 18-30.
[27] Gandi A, Carter A, Singh D 2017 Rheological behavior of cold recycled asphalt materials with different contents of recycled asphalt pavements. *Innov. Infrastruct. Solut.* vol. 2:45. https://doi.org/10.1007/s41062-017-0094-3
[28] Godenzoni C, Graziani A, Perraton D 2017 Complex modulus characterisation of coldrecycled mixtures with foamed bitumen and different contents of reclaimed asphalt, *Road Materials and Pavement Design*, vol. 18 (1), pp. 130–150. http://dx.doi.org/10.1080/14680629.2016.1142467
[29] EN 12697-26 Bituminous mixtures. Test methods for hot mix asphalt. Stiffness
[30] Jaskuła P, Ryś D 2017 Effect of interlayer bonding quality of asphalt layers on pavement performance, *Bestinfra, IOP Conference Series: Materials Science and Engineering*, vol. 236, 012005. doi:10.1088/1757-899X/236/1/012005
[31] Nagórski R, Błażejowski K, Nagórska M 2015 Comparative analysis of deflections and strains of two road pavements with high traffic load. *Roads And Bridges*, vol. 14(1), pp. 31-46. http://dx.doi.org/10.7409/rabdim.015.003
[32] WT-2 z 2014 - cześć I. Mieszanki mineralno-asfaltowe. Wymagania techniczne. GDDKiA. Warszawa 2014. p 52. (in Polish) (General Directorate for National Roads and Motorways)
[33] Wirtgen Cold Recycling Technology (3re ed.), Windhagen: Wirtgen GmbH, 2006
[34] OST D-04.10.01a, "The detailed technical specifications: The road base of the mineral-cement with foamed bitumen mixture made in cold recycling technology". Warsaw, GDDKiA, 2013. (in Polish)
[35] Mazurek G, Iwański M 2017 Modelling of Asphalt Concrete Stiffness in the Linear Viscoelastic Region, *IOP Publishing. IOP Conf. Series: Materials Science and Engineering,* vol. 245:032029, pp 1-10. doi:10.1088/1757-899X/245/3/032029.
[36] Mazurek G, Iwański M 2017 Relaxation Modulus of SMA with Polymer Modified and Highly Polymer Modified Bitumen, *Procedia Eng.*, Vol 172, pp 731-738. doi.org/10.1016/j.proeng.2017.02.093
[37] Górnaś P, Pożarycki A, 2014 Selected properties of FEM numerical models for inverse analysis of road pavement structures, *Roads and Bridges,* vol. 13(3), pp 203-222.

IOP Conf. Series: Materials Science and Engineering **356** (2018) 012017 doi:10.1088/1757-899X/356/1/012017

Effectiveness of the steel mesh track in repairing asphalt pavements in Małopolska region

P Zieliński

Cracow University of Technology, Poland

E-mail: pzielin@pk.edu.pl

Keywords: Steel mesh track, FWD deflections basin, Bearing capacity of pavement, fatigue life of pavement, asphalt pavement strengthening.

Abstract. The aim of this publication is to present and evaluate the effectiveness of the steel mesh track during reconstruction of the pavement on national roads in Małopolska. The paper presents the condition of the pavement before reconstruction, applied design solutions and the current state after 6-10 years of operation. To assess the effectiveness of pavement reinforcement, the results of central deflection tests using the FWD apparatus before and a few years after the reconstruction were compared, it was found that the reinforcement effect was achieved, what has been demonstrated by means of significance analysis of differences in Statgraphics program. Additionally the analyses were extended with parameters characterizing the FWD deflection basin. For selected parameters the values of tensile strains at the bottom of asphalt layers were determined on the basis of correlations given in literature and then the fatigue life was calculated using the criteria of the USA Asphalt Institute and compared with the results of design calculations. The pavement fatigue life estimated on the basis of FWD measurements is generally greater than the one calculated for the design solutions. The assessment of the influence of the steel mesh track on the bearing capacity of the pavement was carried out indirectly, by comparing the central deflections of the structures measured after the reconstruction, with theoretical deflections calculated using the pavement model in the BISAR program, without taking into account the presence of the steel mesh. In some cases the deflections measured are significantly smaller than the deflections calculated for the model without mesh, which can be explained by the reinforced effect of the steel mesh track, especially for sections with the lowest bearing capacity before reconstruction, and where the steel mesh track is placed in the tension zone of the asphalt layers.

1. Introduction

Geosynthetics and related products, such as steel mesh track, have been used in asphalt repairs for over 30 years. The basic functions of these products are delaying reflected cracks, waterproofing the surface and strengthening the asphalt layers. Studies on the effectiveness of using these materials in repairs of asphalt pavements were the subject of many publications, however, most of these works relating to laboratory tests, i.e. works conducted in Belgian Road Research Centre in Brussels [1] show that steel nets are the most effective materials used as stress relieving layer, delaying the reflective cracking propagation in the semirigid asphalt pavements. Computer finite element modelling of asphalt pavement with steel wire grid was performed by Hohamady at al [2], the results show that the performance of steel mesh reinforced sections is better than that of geosynthetics grid reinforced sections and almost close to rigid section. In addition, the above studies confirmed the results of earlier works [3], [4], that the best location of reinforcement is at the bottom of asphalt layers. Testing of the different interlayers on real road sections [5] confirmed the best performance in delaying reflective cracking by glass fibre grid and steel mesh track, other types of intermediate layers have proved less effective. Similar results were achieved in Portuguese research [6], where steel mesh with slurry seal had the best performance, and the bitumen impregnated geotextile sections were the second best.

Initial tests of the efficiency of steel mesh tracks in the reinforcements of asphalt pavements in Małopolska have already been studied on the example of 5 sections, the results were published at the CETRA 2016 Conference Proceedings [7]. Condition of pavements of all analysed road sections after several years of exploitation is very good. No damages were observed, what confirms the effectiveness of applied solution. Bearing capacity of the tested sections evaluated according to the pavement condition evaluation system, i.e. DSN [8], classifies all road tested sections in class A, what means that the remained fatigue life is equal to minimum 20 years. Increase of the bearing capacity of reinforced pavements evaluated with the FWD method for all sections is very substantial. Central deflection values measured on the pavements reinforced with the steel mesh track and adjusted to static load conditions are lower than deflections calculated with BISAR program for the pavement structure without the steel mesh. The differences are substantial for 2-3 out of 5 tested sections, depending on the assumed significance level 95% or 90 %. The best effectiveness of the steel mesh applying is observed for the sections where the bearing capacity before rebuilding was the lowest, and where the steel mesh is placed in the tension zone.

In the present paper, the scope of the analyzed sections was enlarged by 3 additional ones, and the analyses were extended with additional parameters characterizing the FWD deflection basin, such as SCI, BDI, BCI, AUPP and AREA. For selected parameters, i.e. BDI and AUPP, the values of tensile strains at the bottom of asphalt layers were determined on the basis of correlations given in [9] and [10] and then the fatigue life was calculated using the criteria of the USA Asphalt Institute [11] and compared with the results of design calculations.

2. Test program

For the assessment, several road sections were selected, which before the reconstruction were characterized by poor technical condition of the pavement. Depending on the section, the following failures were found: ruts, corrugations, single and alligator cracks, reflected transverse cracks, patches and potholes.

The test program includes:

- Characterization of surface condition before reconstruction, i.e. testing the bowl of dynamic deflections using Falling Weight Deflectometer - FWD (sections 1, 2 and 3 for DK94 and section 5 for DK4) or static deflections using the Benkelman Beam (section 6 for DK28 and sections 4 and 7 for DK44) as well as visual assessment of the surface condition;
- Description of the applied design solutions and calculations of the pavement structure fatigue life;
- Assessment of the pavements condition a few years after the reconstruction, including the measurement of the dynamic deflection basin with the use of FWD as well as visual assessment, evaluation of the pavement bearing capacity according to DSN system [8], comparison of the central deflections of pavement measured before and after the reconstruction with theoretical values calculated acc. to design solutions, application of statistical tests to access the significance of differences;
- Evaluation of the deflection basin parameters before and after the reconstruction, e.g. SCI, BDI, BCI, AUPP and AREA with statistical evaluation of the significance of differences, then calculations of the tensile strains at the bottom of asphalt layers acc. to literature correlations and finally determining the fatigue life of the pavement based on the calculated values along with the comparison with the results obtained at the design stage;
- Discussion of the results and formulation of the conclusions.

3. Results

3.1. Characterization of road sections

The road sections selected for the tests were typical flexible structures with asphalt layers on granular base. All sections before rehabilitation were submitted to the diagnostic tests of their conditions then the existing pavement structures as well as subgrade were identified.

The list of the analyzed road sections with layers thicknesses (H) measured on pavement cores and layers stiffness modulus (E) determined from the FWD back analysis and converted to static load condition and the equivalent temperature of +20°C acc. to Catalogue [12] are given in Table 1. Sections No. 1÷4 were already partially analyzed before [7], while sections 5-7 have not been analyzed so far.

Table 1. Summary of the analyzed sections of roads.

Road number	Section	Chainage		Asphalt layers	Aggregate subbase	Subgrade
94 - roadway right	1A	285+488 286+300	÷	H= 28 cm E=4111 MPa	H=17 cm E=165 MPa	H=infinity E=92 MPa
	1B	286+300 287+450	÷	H= 28 cm E=3855 MPa	H=31 cm E=370 MPa	H=infinity E=71 MPa
	1C	287+450 288+320	÷	H= 26 cm E=3349 MPa	H=51 cm E=159 MPa	H=infinity E=73 MPa
94 - roadway left	2A	285+488 286+650	÷	H= 22 cm E=3580 MPa	H=40 cm E=308 MPa	H=infinity E=84 MPa
	2B	286+650 287+870	÷	H= 22 cm E=2726 MPa	H=47 cm E=249 MPa	H=infinity E=76 MPa
	2C	287+870 288+320	÷	H= 25 cm E=6594 MPa	H=47 cm E=240 MPa	H=infinity E=132 MPa
94	3	305+100 307+100	÷	H= 21 cm E=3000 MPa	H=29 cm E=400 MPa	H=infinity E=108 MPa
44	4	101+900 102+900	÷	H= 17 cm E= 2000 MPa	H=40 cm E= 200 MPa	H=infinity E= 58 MPa
4	5	482+800 483+900	÷	H= 31 cm E=2000 MPa	H=70 cm E=220 MPa	H=infinity E=75 MPa
28	6	142+070 146+530	÷	H= 20 cm E=2000 MPa	H=100 cm E=100 MPa	H=infinity E=40 MPa
44	7	58+500 61+100	÷	H= 15 cm E= 1500 MPa	H=45 cm E= 182 MPa	H=infinity E= 50 MPa

3.2. Designed solutions of pavement structures

Design activities of the rehabilitation included milling of the asphalt layers to a specific depth (between 2 and 17 cm), possibly laying the profiling layer (AC 8 with a thickness of 3 cm), laying the steel mesh track (with a tensile strength of 40/50 kN/m) and fitting it to the lower layer with Slurry Seal mixture (with a thickness of 1 cm), next applying the new asphalt layers of total thickness 11 ÷ 18 cm. Details of the designed structures on each road section are given in Table 2.

Table 2. Designed pavement structures for the analyzed sections of roads.

Road number	Chainage	Milling depth	Profiling layer	Interlayer	New asphalt layers
94 - roadway right (section 1)	285+488 ÷ 288+320	2 ÷ 5 cm	-	Steel mesh track with 1cm of Slurry Seal	7 ÷ 9 cm AC 16 + 4cm SMA 11
94 - roadway left (section 2)	285+488 ÷ 288+320	3 ÷ 5 cm	-	Steel mesh track with 1cm of Slurry Seal	7cm AC 16 + 4cm SMA 11
94 (Section 3)	305+100 ÷ 307+100	4 cm	-	Steel mesh track with 1cm of Slurry Seal	8 cm AC 16 + 4 cm SMA 11
44 (Section 4)	101+900 ÷ 102+900	5 cm	3 cm AC 8	Steel mesh track with 1cm of Slurry Seal	8 cm AC 16 + 4 cm SMA 11
4 (Section 5)	482+800 ÷ 483+900	17 cm	3 cm AC 8	Steel mesh track with 1cm of Slurry Seal	9 cm AC 16 + 4 cm SMA 11
28 (section 6)	142+070 ÷ 146+530	5 cm	-	Steel mesh track with 1cm of Slurry Seal	8 cm AC 20 + 4 cm SMA 11
44 (Section 7)	58+500 ÷ 61+100	4 cm	3 cm AC 8	Steel mesh track with 1cm of Slurry Seal	2 x 7 cm AC 16 + 4 cm SMA 11

Designed structures were verified by mechanical-empirical method, using the USA Asphalt Institute fatigue criteria [11] at the temperature of 10°C, which is reliable temperature for pavement design in Poland, acc. to [12]. Material parameters for old pavement layers were assumed acc. to Table 1, while for new asphalt layers they were adopted acc. to design documentation. Stress and strain states in the pavement structures were calculated with the computer program BISAR 3.0, then fatigue durability was estimated, results are given in Table 3. Obtained results have satisfied the requirements for the design traffic category, calculated for a period of pavement exploitation equal to 20 years.

Table 3. Calculated results of the strains and the fatigue life of pavement for the analyzed sections of roads.

Road number	Section	Year of reconstruction	Chainage [km]	Horizontal strain in asphalt layers [$*10^{-6}$]	Vertical strain on subgrade [$*10^{-6}$]	Fatigue durability of pavement [Millions of 100 kN/axle] calculated	required
94 - roadway right	1A	2007	285+488 ÷ 286+300	52.8	-132	52.8	27.5
	1B		286+300 ÷ 287+450	47.8	-124	78.8	
	1C		287+450 ÷ 288+320	64.3	-127	33.5	
94 - roadway left	2A	2007	285+488 ÷ 286+650	59.4	-122	46.0	
	2B		286+650 ÷ 287+870	74.3	-131	28.1	
	2C		287+870 ÷ 288+320	39.2	-118	95.8	
94	3	2007	305+100 ÷ 307+100	65.5	-149	29.0	17.3
44	4	2012	101+900 ÷ 102+900	98.0	-202	10.9	5.0
4	5	2011	482+800 ÷ 483+900	76.5	-99	24.6	24.1
28	6	2009	142+070 ÷ 146+530	127.0	-128	7.4	2.7
44	7	2012	58+500 ÷ 61+100	73.1	-158	36.6	6.5

3.3. Evaluation of the pavement condition after rebuilding

During a few years after rebuilding, condition of pavements evaluated visually were good, no damage was seen. Additionally, in 2013 the tests of FWD deflection basins were carried out on all road sections, what allowed to compare the bearing capacity of pavements before and after rebuilding. All calculations were made at the equivalent temperature +20°C acc. to [12]. Example of the comparison of the corrected central deflections measured before and after the reconstruction for section No. 2 is shown in Fig.1.

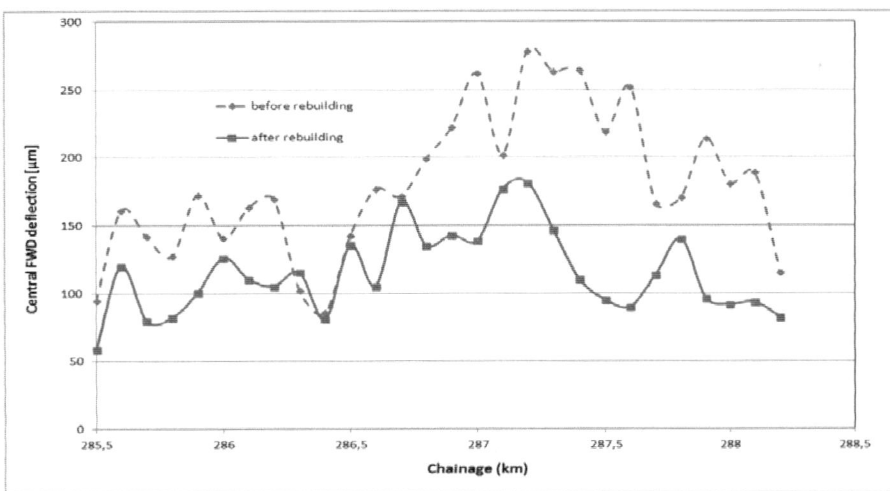

Figure 1. Comparison of the central FWD deflections before and after rebuilding for the road section DK 94, km 285+488-288+320 – roadway left (section 2).

Example of the average FWD deflection basin results, measured before and after rehabilitation are presented in Figure 2 (the highest improvement – section with asphalt overlay) and Figure 3 (the lowest improvement – section with replacement of asphalt layers, without pavement thickening). It was noticed that the differences in deflections measured before and after the rebuilding decrease with the distance from the load axis.

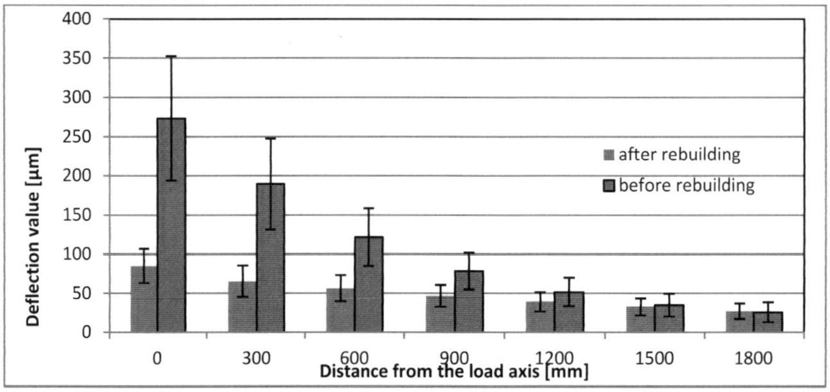

Figure 2. Comparison of the average and standard deviation values of pavement deflection basin before and after rebuilding for the road section DK 94, km 305+100 - 307+100 (section 3).

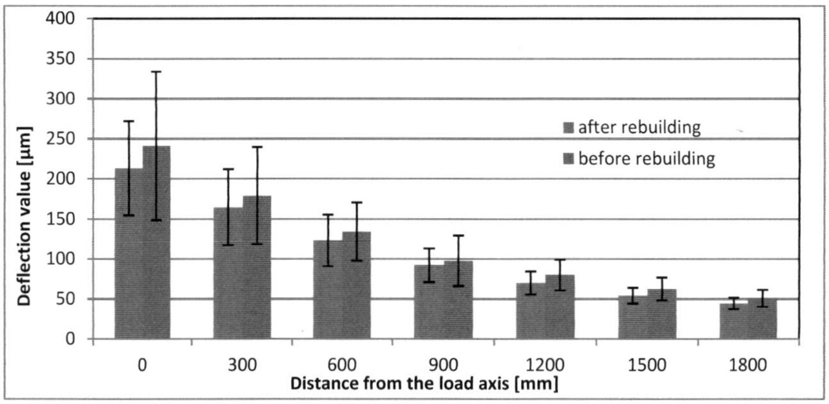

Figure 3. Comparison of the average and standard deviation values of pavement deflection basin before and after rebuilding for the road section DK 4, km 482+800 - 483+900 (section 5).

To evaluate the statistical significance level of the dynamic central deflection (d_0) changes (before and after reconstruction), the tests of multiple comparisons with LSD procedures in the Statgraphics program were carried out. For that test 95% confidence level was used, the results are presented in Table 4.

The results given in Table 4 were used to determine reliable FWD deflections, calculated as the sum of the average value and standard deviation, which are the basis for assessing the bearing capacity of the pavements in Poland, according to the pavement condition evaluation system, i.e. DSN [8]. The results given in Figure 4 show that pavement condition is on the required level, for six sections it is in class A (good condition), only for section 5 (DK4) results are in class B (satisfactory condition).

To evaluate the influence of the steel mesh track on the rebuilt structures bearing capacity, the comparison of the central FWD deflection value with the results for the pavement structure without the steel mesh, which were calculated using analytical model in program BISAR, was done. For this purpose, the deflection values measured using FWD have been converted to static load conditions by coefficient f, acc. to equation (1) from work [13]. The results of deflections measured before and after reconstruction with the values calculated for design solutions are summarized in Table 5.

$$f = -0{,}002 \times r + 1{,}3313 \qquad (1)$$

where: r - distance from load axis [cm]

Table 4. Analysis of significance of FWD central deflection [μm] differences before and after reconstruction, (T=20°C).

Time of measuring	average	standard deviation	coefficient of variation [%]	difference	+/- limits (95%)
DK 94, km 285+488 – 288+320, roadway right (section 1)					
Before reconstruction	181	73	40.6		
After reconstruction	122	39	31.6	59	38*
DK 94, km 285+488 – 288+320, roadway left (section 2)					
Before reconstruction	180	53	29.4		
After reconstruction	115	31	26.7	65	38*
DK 94, km 305+100 – 307+100 (section 3)					
Before reconstruction	273	80	29.1		
After reconstruction	85	22	25.6	188	31*
DK 44, km 101+900 – 102+900 (section 4)					
Before reconstruction	539	157	29.1		
After reconstruction	230	65	28.1	309	45*
DK 4, km 482+800 – 283+900 (section 5)					
Before reconstruction	241	93	38.6		
After reconstruction	213	59	27.6	28	61
DK 28, km 142+070 – 146+530 (section 6)					
Before reconstruction	570	173	30.4		
After reconstruction	224	49	21.9	346	30*
DK 44, km 58+500 – 61+100 (section 7)					
Before reconstruction	630	150	23.8		
After reconstruction	228	56	24.8	402	39*

*denotes a statistically significant difference

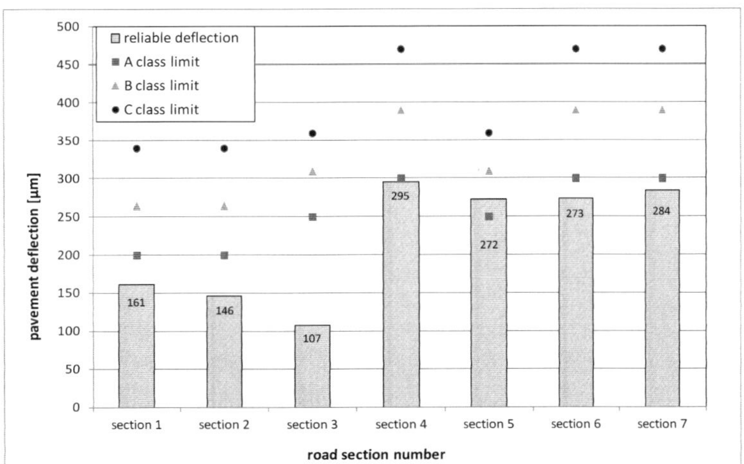

Figure 4. Evaluation of pavement FWD central deflection acc. to DSN [8].

Table 5. Results of the central static deflections in [μm] for the analyzed sections of roads after reconstruction (20°C).

Road number	Chainage [km]	Calculated	Measured	Difference	+/- limits (95%)
94 - roadway right (section 1)	285+488 ÷ 288+320	190	163	37	32*
94 - roadway left (section 2)	285+488 ÷ 288+320	183	153	30	27*
94 (Section 3)	305+100 ÷ 307+100	251	113	138	22*
44 (Section 4)	101+900 ÷ 102+900	397	306	93	54*
4 (Section 5)	482+800 ÷ 483+900	289	284	5	51
28 (section 6)	142+070 ÷ 146+530	484	299	185	Not tested
44 (Section 7)	58+500 ÷ 61+100	390	303	87	Not tested

*denotes a statistically significant difference

3.4. Analyses of the deflection basin parameters

In the next step, analyses of the deflection basins were made using the parameters given in Table 6 acc. to work [14]. The list of calculated average dynamic deflection basin parameters and their standard deviations for chosen individual sections before and after pavement rehabilitation, along with statistical analysis of differences, is shown in Table 7 (calculations at 10°C). In the case of the section 4, 6 and 7 due to the lack of FWD measurements before reconstruction only the results of FWD basin parameters after are given. Figure 5 shows the mean values and standard deviations for the SCI parameter for sections before and after rebuilding.

Table 6. FWD deflection basin parameters.

Deflection parameter	Formula	Parameter's objective
Surface Curvature Index (SCI)	$SCI = d_0 - d_{300}$	Condition of bound layer
Base Damage Index (BDI)	$BDI = d_{300} - d_{600}$	Condition of bound layer
Base Curvature Index (BCI)	$BCI = d_{600} - d_{900}$	Condition of subbase layer
Area Under Pavement Profile (AUPP)	$AUPP = \frac{5d_0 - 2d_{300} - 2d_{600} - d_{900}}{2}$	Condition of the pavement upper layers
Area (AREA)	$AREA = \frac{150(d_0 + 2d_{300} + 2d_{600} + d_{900})}{d_0}$	Shape of the deflection basin close to the load by the normalized area on the top of the deflection basin

Note: d_i – deflection at *i* mm from the center of loading plate in μm

Table 7. Average (standard deviation) deflection basin parameters for sections before and after pavement rehabilitation.

Time of measuring	d_0 [μm]	SCI [μm]	BDI [μm]	BCI [μm]	AUPP [μm]	AREA [mm]
DK 94, km 285+488 – 288+320, roadway right (section 1)						
Before rehabilitation	145 (59)	39.3 (26.1)	31.6 (16.7)	21.8 (8.0)	156 (92)	586 (54)
After rehabilitation	98 (31)	28.3 (10.7)	19.1 (8.7)	14.6 (5.2)	108 (41)	569 (51)
difference	47	10.6	12.5	7.2	48	17
+/- limits (95%)	43*	7.4*	8.3*	4.8*	37*	28
DK 94, km 285+488 – 288+320, roadway left (section 2)						
Before rehabilitation	144 (42)	39.7 (15.9)	31.1 (12.2)	21.9 (7.0)	157 (58)	574 (58)
After rehabilitation	92 (25)	25.0 (7.6)	17.0 (6.2)	12.7 (4.1)	94 (29)	590 (59)
difference	52	14.7	14.1	9.2	63	16
+/- limits (95%)	43*	7.4*	8.3*	4.8*	37*	28
DK 94, km 305+100 – 307+100 (section 3)						
Before rehabilitation	219 (64)	64.0 (28.3)	57 (23.1)	34.5 (16.4)	263(102)	544 (60)
After rehabilitation	68 (17)	15.8 (4.4)	7.4 (3.8)	7.4 (3.4)	54 (16)	656 (60)
difference	151	48.2	50	27.1	209	112
+/- limits (95%)	36*	5.0*	6*	3.9*	30*	23*
DK 44, km 101+900 – 102+900 (section 4)						
After rehabilitation	184 (52)	47.0 (20.2)	39.2 (17.9)	26.2 (8.9)	189 (81)	601 (45)
DK 4, km 482+800 – 283+900 (section 5)						
Before rehabilitation	193 (74)	49.7 (32.9)	36.2 (24.8)	29.0 (23.6)	193 (126)	615 (71)
After rehabilitation	170 (47)	38.9 (10.9)	33.3 (13.1)	24.9 (9.5)	160 (49)	619 (30)
difference	23	10.8	2.9	4.1	33	4
+/- limits (95%)	48	11.9	7.4	5.3	41	31
+/- limits (85%)	36	8.7*	6.8	3.9*	30*	26
DK 28, km 142+070 – 146+530						
After rehabilitation	179 (39)	36.4 (14.7)	35.0 (12.7)	25.7 (8.0)	156 (57)	644 (48)
DK 44, km 58+500 – 61+100 (section 7)						
After rehabilitation	182 (45)	43.6 (20.5)	38.7 (15.5)	27.1 (6.7)	180 (76)	612 (51)

*denotes a statistically significant difference

Area Under the Pavement Profile (AUPP) is used as a parameter for the determination of the tensile strain at the bottom of asphalt layer (ε_{ac}), which allows to calculate the pavement fatigue life. Since the AUPP is a geometric property of the deflection basin, the use of the AUPP for the prediction of ε_{ac} is not affected by the type of subgrade and pavement. For aggregate base pavements, the relationship between ε_{ac} and AUPP developed by Thompson [9] is as follows (AUPP in mils):

$$\log(\varepsilon_{ac}) = 0.821 \times \log(AUPP) + 1.210 \quad (2)$$

Another method to predict the ε_{ac} is presented by Kim and Park [10], where two approaches were used. The first approach uses a statistical regression method to relate ε_{ac} with BDI (in mils) and asphalt layers thickness (H_{ac} in inches), which is expressed in equation (3). Another method to predict ε_{ac} values is plotted in equation (4) with utilization of AUPP.

$$\log(\varepsilon_{ac}) = 1.082 \times \log(BDI) + 0.259 \times \log(H_{ac}) + 1.409 \quad (3)$$

$$\log(\varepsilon_{ac}) = 1.034 \times \log(AUPP) + 0.932 \quad (4)$$

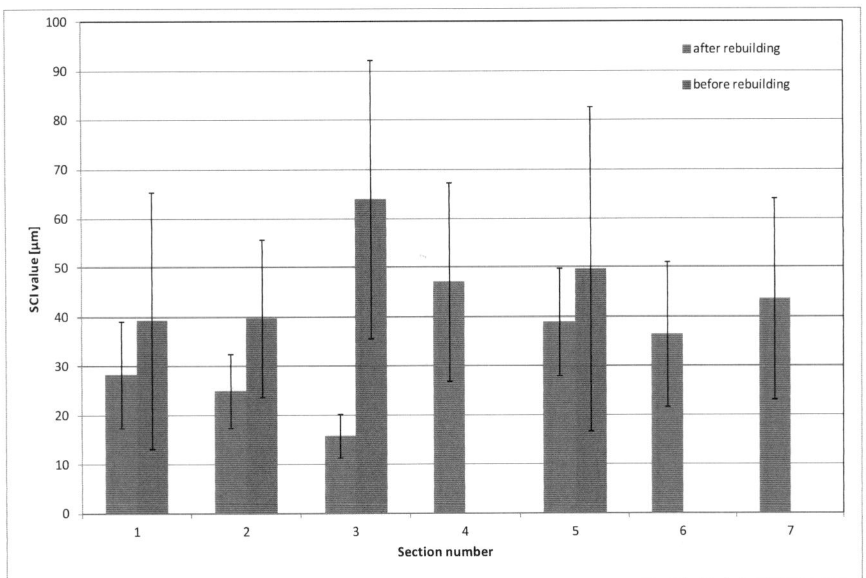

Figure 5. The mean values and standard deviations for the SCI parameter for sections before and after rebuilding.

The results of the determination of tensile strains at the bottom of the asphalt layers ε_{ac}, calculated according to the formulas (2÷4) and according to the design as well as the fatigue life determined on their basis for specific 11 sections of road (at temperature 10°C) are presented in Table 8.

Table 8. Calculated results of ε_{ac} and the fatigue durability of pavement for the analyzed sections of roads

Road number (section)	ε_{ac} [$*10^{-6}$], calculated acc. to				Fatigue life of pavement [Millions of 100 kN/axle], acc. to			
	equation *(2)*	equation *(3)*	equation *(4)*	Design calculations	equation *(2)*	equation *(3)*	equation *(4)*	Design calculations
94 - roadway right (section 1)	61	46	47	53	38.0*	96.1	89.5	52.8
	46	32	31	48	101.5	335.3	372.2	78.8
	54	39	40	64	67.5	197.2	181.4	33.5
94 - roadway left (section 2)	46	32	29	59	108.2	357.1	493.7	46.0
	52	38	39	74	91.2	256.0	235.0	28.1
	36	24	22	39	143.8	546.3	727.5	95.8
94 - (section 3)	30	19	13	65	379.2	1704.8	5943.8	29.0
44 (Section 4)	83	68	79	98	18.8	36.2	22.1	10.9
4 (Section 5)	73	57	66	76	28.7	64.8	40.0	24.6
28 (section 6)	71	56	69	127	31.5	68.7	34.6	7.4
44 (Section 7)	80	65	77	73	27.2*	53.8	30.8*	36.6

* sections with lower calculated fatigue life according to correlation equations than design value

4. Results discussion

For all analyzed road sections, where overlays were made, central deflection values after reinforcement are suitably lower than before that treatment, the highest differences were observed for the section of previously the lowest bearing capacity and the thickest of the asphalt overlay (DK44, km 58+500 – 61+100), where steel mesh track is situated in the tension zone of the asphalt layers. In the case of a section where only the replacement of asphalt layers was carried out (DK4), without their thickening, there was no statistically significant reduction of central deflections, but the homogeneity of the results measured after rehabilitation was improved, which was also confirmed on other rebuilt sections as diminishing of the values of deflection variability. This improvement of the uniformity of deflections, even with a similar level of average values, results in a favorable reduction in the value of reliable deflections, thus obtaining a higher bearing capacity rating, acc. to diagnostic system [8].

With the exception of section 5, significantly lower values of deflections calculated in relation to the measured ones were observed, which may indicate the influence of the applied steel mesh track on the behavior of the pavement structure bearing capacity.

The results contained in Table 7 indicate a statistically significant improvement of all parameters describing the condition and the bearing capacity of the upper layers of pavement (SCI, BDI and AUPP) for sections after reinforcement with a steel mesh track, except for the section No. 5, where only the exchange of asphalt layers has been done, without thickening the pavement structure. Additionally, in the case of all sections with a steel mesh track, a statistically significant increase in the BCI parameter was found, which indicates improved condition of subbase layer. This improvement was much more visible for section 3, where an asphalt overlay was applied as compared to section 5, where only the asphalt layers were replaced. In the case of the AREA parameter, which takes into account a change in the stiffness of individual pavement layers, except the section No. 3, no significant differences after the reconstruction were found. After reconstruction, all parameters of the deflection basin with the exception of AREA have much smaller dispersion, which is a desirable effect.

The results of the tensile strain at the bottom of the asphalt layers, obtained on the basis of the correlation from the measured parameters of the deflection basin are in most cases much smaller than those specified in the design documentation. This means, that pavement durability estimated on the basis of FWD measurements is greater than originally calculated for the planned pavement solutions. This effect of the expected extension of durability of the pavement structure can be attributed to the use of steel mesh track under the new asphalt layers, which improves the parameters of the deflection basin, describing the state of the upper asphalt layers (SCI, BDI and AUPP). At the same time, it was found that equation *(2)* gives the largest expected tensile strains of asphalt layers and thus their lowest durability, while the durability determined by equation *(3)* is by far the largest.

5. Conclusions

Presented in this paper tests of the road sections and analyses of the results allowed to draw the following conclusions:

- Evaluation of pavements of all analysed road sections reconstructed with the steel mesh track after several years of exploitation shows good condition, without any damages and imperfections. Bearing capacity of the tested sections determined acc. to the diagnostic system, classifies all road tested sections in class A, except section No. 5, where class B was established.
- Average central FWD deflections of pavements after reconstruction with steel mesh track and asphalt overlay decreased significantly. Additionally an improvement of the uniformity of deflection basins was observed, which results in a more homogenous bearing capacity and then gives better condition of pavement maintenance.
- The positive effect of the use of steel mesh track has been demonstrated by comparing the deflections measured and calculated for the pavement model without a mesh, for most cases the reduction of central deflection turned out to be statistically significant.

- In 10 out of 11 analyzed road sections, fatigue life of rebuilt pavements with the application of steel mesh track, determined on the basis of measured deflection basins, turned out to be higher than those calculated at the design stage.
- In general, the best effectiveness of strengthening the pavement was observed on the sections with the initially lowest bearing capacity, where the mesh was located in the tensile zone of asphalt layers.

References

[1] Vanelstraete A, Francken L 1996 *Laboratory testing and numerical modelling of overlay systems on cement concrete slabs* Proceedings of the 3rd RILEM Conference on Reflective Cracking in Pavements, Maastricht, pp 170-181.

[2] Mohamady A, ElhadyA-B M, Eisa M S 2013 *Improving of Asphalt Pavement Performance using Steel wire Grid Reinforcement* International Journal of Engineering Inventions, Vol. 2, Issue 10, pp 12-22.

[3] Grzybowska W, Górszczyk J, Zieliński P 2007 *Modeling of the asphalt pavement structure behavior in different conditions of interlayer bonding, with using of geosynthetics reinforcement*, Research Project financed by Polish Ministry of Education and Computerization, Cracow Univercity of Technology, p 243 (Edition in Polish).

[4] Górszczyk J 2010 *The influence of reinforcing geosynthetic interlayer on the fatigue durability of asphalt road pavement*, PhD thesis, Cracow Univ. of Technology, p 222 (Edition in Polish).

[5] Vervaecke F, Maeck J, Vanelstraete A 2008 *On site validation and long term performance of anti-cracking interfaces* Proceedings of the 6th RILEM International Conference on Cracking in Pavements, Chicago, pp. 761-768.

[6] Antunes M L, Fontul S, Pinelo A M 2008 *Anti-reflective cracking solution for asphalt overlays: 8 years performance monitoring*, Proceedings of the 6th RILEM International conference on cracking in pavements, Chicago, pp. 791-798.

[7] Zieliński P, Grzybowska W 2016 *Effectiveness of the steel mesh track in strengthening cracked asphalt pavements* Proceedings of the 4rd International Conference on Road and Rail Infrastructure - CETRA, Šibenik, Croatia, pp. 195-202.

[8] GDDKiA 2015 The pavement condition evaluation system – usage guidelines, Warsaw (Edition in Polish).

[9] Thompson M R 1989 *Area Under the Pavement Profile to Predict Strain* Informal Presentation at FWD Users Group Annual Meeting, Indianapolis, IN.

[10] Kim Y R, Park H 2002 *Use of Falling Weight Deflectometer Multi-Load Data for Pavement Strength Estimation* Final Report to North Carolina Department of Transportation, Raleigth, NC, p 180

[11] Research and development of the Asphalt Institute's 1982 Thickness Design Manual (MS-1) Ninth edition, The Asphalt Institute, Research Report No. 82-2, RR-82-2.

[12] IBDiM 2013 *Catalogue for designing of flexible and semiridig pavement structures rebuilding and renovation,* Warsaw, p 240, (Edition in Polish).

[13] Krawczyk B 2012 *Identyfication of parameters for road pavement models derived from dynamic impulse test*, PhD Thesis, Wrocław Univ. of Technology, p 149 (Edition in Polish).

[14] Kim Y R, Lee Y C, Ranjithan S R 2000 *Flexible pavement condition evaluation using deflection basin parameters and dynamic finite element analysis implemented by artificial neutral networks* in ASTM STP 1375 Nondestructive testing of Pavements and Backcalculation of Moduli: Third Volume, pp 514-530.

ACKNOWLEDGEMENTS

The FWD tests have been carried out by courtesy of General Directorate for National Roads and Motorways, Krakow Branch.

IOP Conf. Series: Materials Science and Engineering **356** (2018) 012018 doi:10.1088/1757-899X/356/1/012018

Environmental aspects of the implementation of geogrids for pavement optimisation

Jacek Kawalec[1,3], Michał Gołos[2], Piotr Mazurowski[2]

[1]Tensar International s.r.o., Lipova 1965, 737 01 Cesky Tesin, Czech Republic
[2]Tensar Polska sp. z o.o., ul. Azymutalna 9, 80-298 Gdańsk, Poland
[3]Silesian University of Technology, Faculty of Civil Engineering, Department of Geotechnics and Roads, ul. Akademicka 5, 44-100 Gliwice, Poland

kawalec@tensar.pl

Abstract. Technological developments in highway construction should not only result in durable, safe and cost-effective solutions for roads and pavements but also, and perhaps above all, lead to solutions that minimise the negative impact of construction on the environment. One of the ways to ensure these requirements are met is to apply technology using geosynthetics. This paper discusses the stabilisation of aggregate with hexagonal geogrids and the benefits - from the point of view of reducing the emission of harmful gases to the atmosphere - which can be realised from this approach, compared with traditional approaches. Solutions for the improvement of weak subgrades and optimisation of the entire pavement structure are discussed, along with the presentation of sample calculations of greenhouse gas emissions, carried out with the use of specialized software related to the construction of the structures in various technologies.

1. Introduction

Geosynthetics have been used widely in pavement engineering for several decades, to improve drainage efficiency and as asphalt interlayers, providing three different functions: reinforcement, stress relief (as a stress absorbing membrane interlayer - SAMI) or to act as a water barrier. The stabilising function has been recently identified by ISO as a new, separate function, different to reinforcement. As discussed in this paper, geogrids can also be used for stabilising aggregate placed on firm foundations, which is quite an innovative approach. Furthermore, the positive effect of the stabilisation of unbound aggregate layers within the pavement structure can result in reductions in layer thickness (including asphalt layers), without reducing the design life of the pavement. Verification of theoretical assumptions, made through trafficking tests, provides opportunities for pavement optimisation. This is one of possible ways for sustainable development of roads. When combined with the environmental benefits, this can be seen as a modern approach for designing asphalt pavements.

2. Mechanical stabilisation of unbound aggregate with geogrid

When unbound aggregate is placed and compacted on a layer of stiff geogrid, a geogrid/aggregate composite is created, with aggregate particles interlocked and confined within the geogrid's stiff apertures. This composite is often referred to as "Mechanically Stabilised Layer" (MSL). In an MSL, lateral restraint is provided by the geogrid, reducing strain and thereby increasing the stiffness of the layer, compared with a non-stabilised aggregate layer.

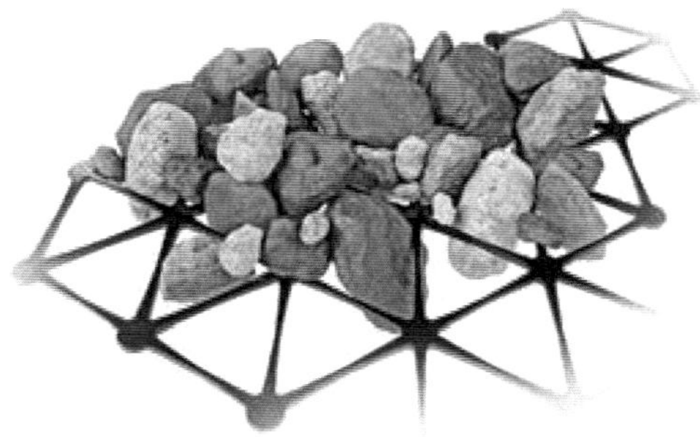

Figure 1. Aggregate particles interlocked within a geogrid's apertures.

Stiff geogrid ribs resist particle movement, also under cyclic loading, preventing layer deformation. One of the effects of stabilisation of an aggregate with a geogrid is an increase in stiffness or modulus of the layer. The modulus of unbound aggregate is a function of the stress state (Θ). Several relationships allowing for the calculation of aggregate modulus exist in literature, one of them is presented in AASHTO Guide for Design of Pavement Structures [1]:

$$E = k_1 \Theta^{k_2} \quad (1)$$

where:
Θ - the sum of principal stresses ($\sigma_1 + \sigma_2 + \sigma_3$)
k_1, k_2 – constants that depend upon aggregate type.

When a trafficking load is applied to a layer, the stiff ribs of the geogrid react, preventing aggregate particles from moving laterally. This increases the horizontal principal stresses, σ_2 and σ_3, and Θ increases as a result. When Θ increases, the modulus of a layer increases and so does its bearing capacity.

Many laborato/ry and in-situ tests have confirmed the stiffness increase of aggregate stabilised with hexagonal geogrid. Kwon et al. [5] used a test combination of AASHTO T307 (Determining the Resilient Modulus of Soils and Aggregates) and NCHRP 598 (Repeated Load Permanent Deformation) to test both stabilised and non-stabilised, poorly graded silty gravel samples. Stabilised samples had a modulus 5% to 20% higher than non-stabilised samples. In-situ resilient modulus tests conducted with Automated Plate Load Tests (APLT) demonstrated a 5% to 30% increase in the modulus of geogrid-stabilised sections [8], [9].

3. Pavement optimisation with geogrid stabilised aggregate base

Increasing the pavement design life or reducing its thickness for a given design life, through the use of geosynthetics in an aggregate base layer is often referred to as a "Pavement optimisation". Pavement optimisation can be described as obtaining the pavement design objectives at the most economic cost. The design objectives in most cases will be to reach the minimum traffic life requirements but the objective could also be to meet the construction programme or to meet environmental requirements, such as reducing greenhouse gas emissions during construction.

Pavement optimisation has been accepted and used to varying degrees around the world. In the US, the use of geosynthetics for pavement optimisation has been acknowledged and has been in common use for many years, being covered by the AASHTO R50-09 Standard [2]. This states: "Geosynthetics are used in the pavement structure for structural support of traffic loads over the design life of the pavement. The geosynthetic is expected to provide one or both of these benefits: (1) improved or extended service life of the pavement, or (2) reduced thickness of the structural section". The document does not provide detailed design requirements or recommendations instead, it outlines the procedure that should be undertaken by the geosynthetic manufacturer to incorporate it in a pavement design.

One of the important requirements stated in [2] is the necessity to perform full-scale pavement testing for any geosynthetic being considered for pavement optimisation. AASHTO R50-09 states: "Because the benefits of geosynthetic reinforced pavement structures may not be derived theoretically, test sections are necessary to obtain benefit quantification." Different types of geosynthetics, and even different types and grades of geogrids, can perform differently. Physical characteristics, such as tensile strength, are not a good indicator of geosynthetic performance in pavements and soft soil improvement. Strong evidence exists [3], including full scale trafficking tests, that confirms geogrids with similar tensile strength characteristics can have quite different levels of performance. In some cases, one type of stiff, monolithic geogrid even outperformed flexible, welded geogrid, despite the former having twice the tensile strength.

AASHTO R50-09 recognises that design methods based upon performance testing of one type of geosynthetic may only be applied to the use of that specific geosynthetic: "Design procedures use experimentally derived input parameters that are often geosynthetic specific. Thus, computed engineering designs and economic benefits are not easily translated to other geosynthetics."

4. Trafficking tests

Between 2012 and 2016, a series of trafficking tests were performed to quantify the benefits of using one type of geogrid in flexible pavements, in line with AASHTO R50-09 requirements. Testing of the hexagonal triaxial stabilisation geogrid was conducted at the US Army Engineer Research and Development Centre (ERDC) and consisted of three stages, in which multiple full-scale pavement sections were constructed and trafficked. Details of the test sections' arrangement, instrumentation, trafficking and results are available in reports [4],[6],[7]; the key points are summarised below.

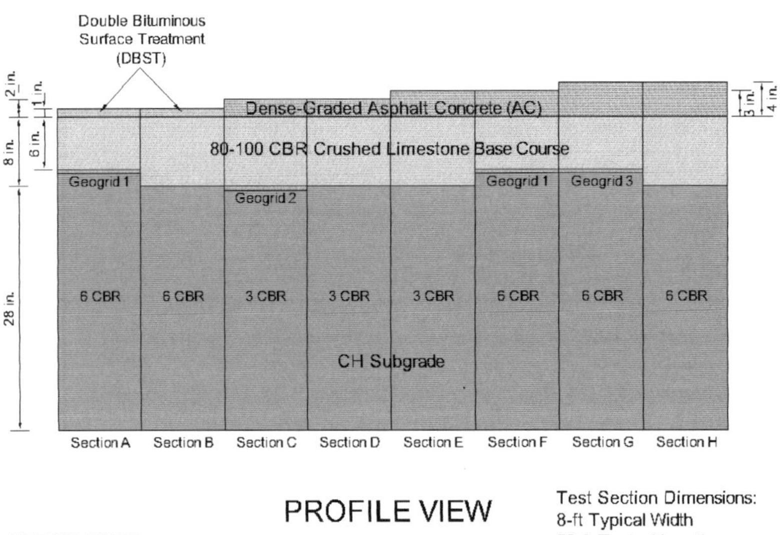

Figure 2. ERDC Test sections build-up.

Test sections were constructed using typical construction equipment and technology and consisted of 2.44m by 15.2m areas. They were built inside a hangar, to eliminate the potential influence of variable environmental factors. The subgrade consisted of clay of either 3% or 6% CBR. The granular base layer consisted of either 15cm or 20cm of crushed limestone. Four of the sections had a base layer stabilised with hexagonal geogrids. Two sections had a double bituminous surface treatment (these sections are not discussed in this paper) and six had surfacing of 5cm, 7.5cm or 10cm of asphalt concrete (AC). The section build-up is shown in Figure 2.

The test sections were trafficked with the Heavy Vehicle Simulator (HVS) (Figure 3), which simulates typical heavy highway traffic for accelerated pavement testing (APT). The set-up consisted of a tandem-axle dual wheel gear loaded 90kN (Figure 4). An environmental chamber surrounded the trafficking gear, maintaining a constant temperature of $25^{O}C$ +/-$5.5^{O}C$, to minimise the temperature change influence on the asphalt modulus. Throughout the tests, rut depth measurements were collected. Rutting was the decisive failure mechanism of all the sections tested. Figures 5 and 6 present the results from the trafficking of test sections built on 3% and 6% CBR subgrade, respectively.

Figure 3. HVS used in tests in hangar at ERDC facility.

Figure 4. Tandem-axle dual wheel gear used in ERDC tests.

All the sections with an aggregate base layer stabilised with hexagonal geogrids outperformed the non-stabilised control section significantly. Stabilised sections with thinner asphalt and/or base layers, also performed better than the non-stabilised sections.

A good reference for analysis of APT results is to compare traffic required to produce a rut depth of 12.5mm (0.5"). 12-12.5mm rut depth is considered a failure point in many Transfer Functions (TF) used in M-EPDM, for example in the Asphalt Institute TF, Nottingham University TF and Shell TF. For sections constructed on 3% CBR subgrade, 100,000 passes of Standard Axle was needed to produce a 12.5mm rut for a section stabilised with hexagonal geogrid, while for non-stabilised sections with the same layer thickness, this rut depth was reached after 6,400 passes or 12,000 passes for a non-stabilised section with a 2.5cm thicker asphalt layer. For sections constructed on 6% CBR subgrade, 12.5mm rut depth was reached after 104,000 Standard Axle passes in the non-stabilised section, while for sections with geogrid-stabilised base layers this rut depth was never reached, even after 800,000 Standard Axle passes.

Until now, geogrids have been used typically for construction on weak and very weak soils. The ERDC test results show that this is not the only application for geogrids. A subgrade of 6% CBR (~ Ev2 = 60MPa) is a relatively firm subgrade, for which typically, the use of geogrid would not be considered. However, this research demonstrates that the use of hexagonal geogrids to stabilise aggregate base layers can bring benefits to pavements constructed on firm subgrade – a novel application for geogrids. In practice, geogrids can either increase the life of a pavement, without changing its thickness; reduce the thickness of pavement layers, while maintaining required life, or a combination of the two.

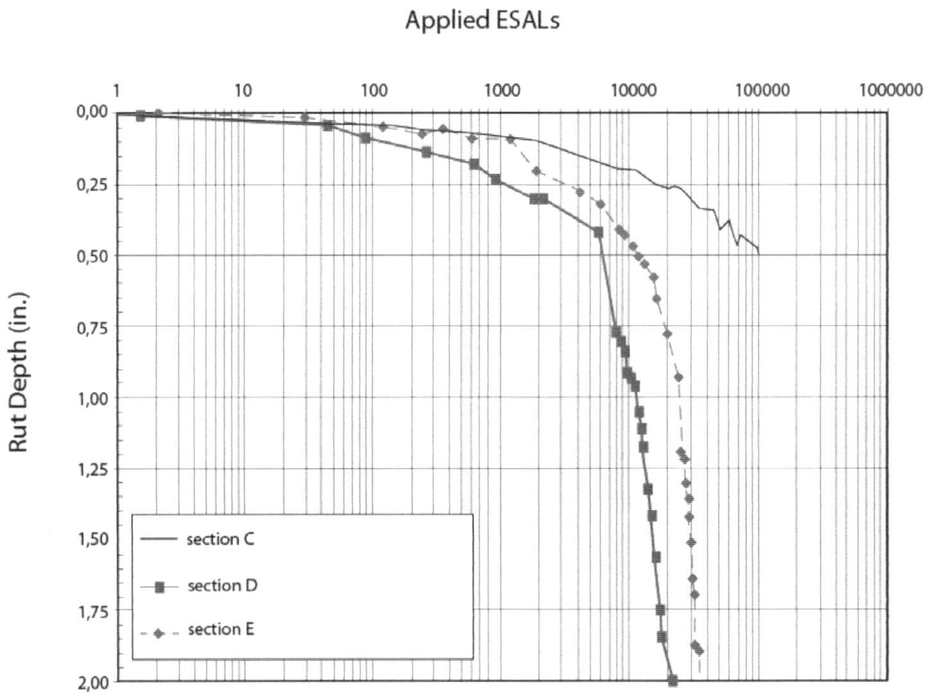

Figure 5. Rut depth vs applied ESALs for sections on 3% CBR subgrade.

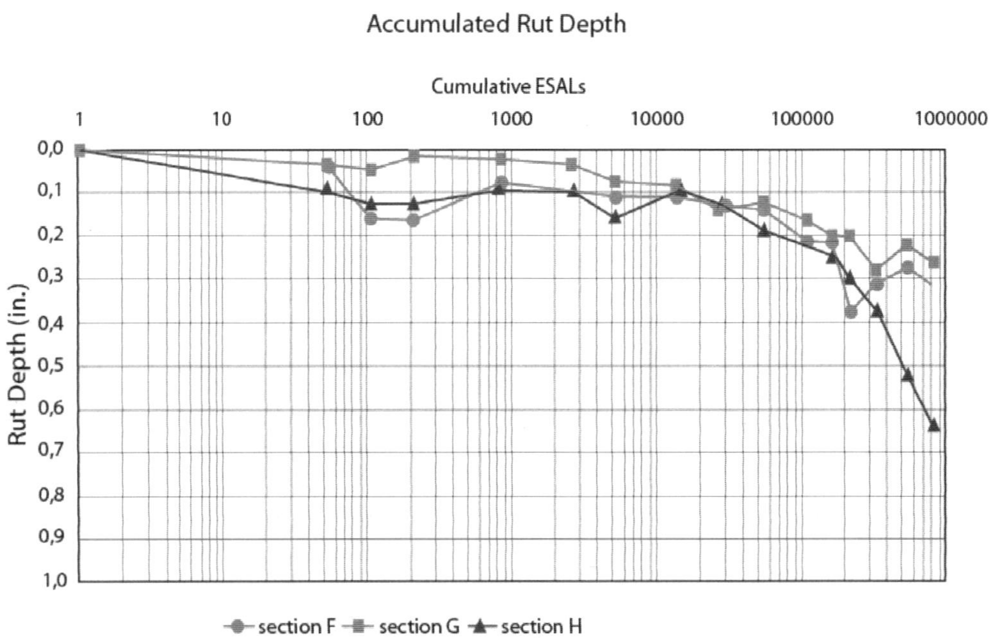

Figure 6. Rut depth vs applied ESALs for sections on 6% CBR subgrade.

Data gathered during APT tests, as well as data from other tests not described in this paper, has been the basis for modifying pavement design methods to allow for the influence of hexagonal geogrid on pavement life. A series of back calculations and calibrations have been performed to set design parameters in a way that calculated results match results obtained in the tests. Two design methods have been modified: the AASHTO'93 empirical design method and the Mechanistic-Empirical Design Method.

The influence of geogrid in the AASHTO'93 pavement design method has been taken into account by modifying material factor, a, of an aggregate layer stabilised with hexagonal geogrid. Factor a is increased, however the magnitude of increase is not constant; it depends upon the bearing capacity of the subgrade, the thickness of stabilised layer and the type of hexagonal geogrid. A typical value of a is 0.14 for a crushed aggregate base, which can be increased to between 0.18 and 0.27 by incorporating a geogrid.

The Mechanistic-Empirical Design Method was modified by the addition of two separate factors reflecting the beneficial effects of a geogrid, which are taken into account simultaneously in the design.

First, the modulus of the aggregate layer stabilised with geogrid is increased by multiplying the initial value by a modulus enhancement factor. This is done in the pre-linear elastic analysis stage. Strains in the asphalt layers and the subgrade at critical points and life are then calculated, as in standard M-EPDM procedures. Finally, so-called Life Shift Factors are applied to the calculated life, both for asphalt and subgrade, to determine the final life of pavement being designed.

Special algorithms have been developed for calculating both the aggregate modulus enhancement factor and the Life Shift Factors. These algorithms take into account the thickness of asphalt layer, the stabilised layer thickness, the distance from the geogrid to the bottom of lowest asphalt layer, the distance from the geogrid to the top of subgrade, the subgrade modulus and the type of hexagonal geogrid. Sample stabilised pavement structures described later in this paper have been designed with this modified M-EPDM method.

5. Evaluating the carbon footprint in highway applications of hexagonal geogrids. The Carbon calculator tool

Incorporating geogrid into a pavement can result in the optimisation of its design. Extension of pavement life or reduction of layer thickness, while keeping the same pavement life, can bring significant financial benefits, through the reduced cost of materials, reduced cost of waste and generally more efficient use of resources, compared with traditional solutions using aggregates, concrete or steel. Important environmental benefits resulting from the use of geosynthetics can also be realised when the entire life cycle of a civil engineering project is considered[10]. This paper focuses on the environmental benefits of incorporating geogrid in a pavement structure.

The carbon footprint is the amount of carbon dioxide released into the atmosphere as a result of the activities of a particular individual, organisation or community. In terms of civil engineering, it is a assessed value relating to the amount of CO_2 released into the atmosphere as a result of all building processes. Carbon footprint is measured in kilograms kCO_2e or tonnes tCO_2e of carbon dioxide equivalent [11]. To assess the total carbon footprint for any civil engineering process, e.g. highway construction, it is necessary to take into account the CO_2 emissions for every activity relating to the process, i.e:

- Quarrying of natural aggregate fill
- Manufacturing of all materials and products e.g. concrete, steel, chemical binders, asphalt, asphalt mixes and geosynthetics
- Transportation and delivery of above materials from quarries, manufacturing facilities and plants to the project site
- Installing, placing, compacting and assembling all above materials into construction.

One geogrid manufacturer has developed software for calculating carbon emissions, the "Carbon calculator", which uses methodology provided in PAS 2050:2011, Specification for the assessment of the life cycle greenhouse gas emissions of goods and services [12]. An independent consultant [13] was commissioned by the hexagonal geogrid manufacturer to develop this tool, which evaluates the carbon emissions savings to quantify the environmental benefits of using a MSL.

The carbon calculator estimates the carbon savings gained by using a MSL incorporating hexagonal geogrids for ground stabilisation, compared with the construction of a non-stabilised granular layer of equivalent performance. Savings are the result of a reduction in aggregate layer thickness, through the use of an MSL.

The carbon saving calculation considers an MSL and an associated non-stabilised layer for a given set of input parameters. The evaluation process can be split into three sections, as follows:

1. Determine the embodied carbon within the materials used in the construction of both highway structures at their respective "factory gates".
   - In case of the geogrid, the polypropylene is delivered to geogrid manufacturing facility by a combination of sea and road transport (depending on the supplier's location). The geogrid manufacturing process uses electricity and natural gas to power machinery. Embodied carbon and energy are assessed and calculated in accordance with [18] and [19] and expressed in equivalent $kgCO_2e/km$ (polypropylene transportation) and in accordance with [18] and expressed in a value of $kgCO_2/m2$ (energy use).
   - For the aggregate component of the MSL, the calculation is based on emissions related to quarrying of the aggregate. The embodied carbon of natural aggregate is calculated based on [16] and is expressed in $kgCO_2/kg$.
2. Assess the carbon emissions from delivering materials to site, based on the distance from the manufacturing facility (geogrid) and from the quarry (aggregate) to the project site.
3. Assess the carbon emissions for construction of the highway structures (stabilised and non-stabilised):
   - Installation of the geogrid is a manual operation and so no embodied carbon can be measured
   - The aggregate component of an MSL is typically placed and compacted using mechanical plant and these operations generate their own carbon footprint, which are taken into account in the final assessment. This is calculated in accordance with [19] and expressed in $kgCO_2/m2$.

There are also some assumptions which the carbon calculator is based, the most important of which are:

1. The user has to be given a stabilisation scheme, providing the MSL thickness, the number of layers of geogrid required and the strength of the subgrade
2. Carbon emissions data is based on available grades of hexagonal geogrids
3. The equivalent non-stabilised thickness is calculated based on the input data
4. The calculator assumes that deformation due to construction traffic is limited to 40mm in accordance with [14].
5. Carbon emissions for the construction of both stabilised and non-stabilised layers are calculated separately and the percentage difference is presented to the user
6. Geogrid is delivered from the factory in the UK or China to the site, as chosen by the user – outward distance only
7. Aggregate is delivered to the site from quarry at a distance specified by the user – outward and return distance
8. Aggregate is compacted in accordance with [15]
9. Embodied carbon in aggregate is calculated in accordance with [16]
10. Embodied carbon in the geogrid is calculated in accordance with [12], including embodied carbon of polypropylene in accordance with [17]
11. Embodied carbon and conversion factors for fuel consumption and efficiencies are in accordance with [18]
12. Embodied carbon for container ship transportation and aggregate properties are in accordance with [19].

Table 1. Inventory of Carbon and Energy main summary [16].

Lp.	Material	*EE - Embodied Energy* [MJ/kg]	*EC - Embodied Carbon* [kgCO_2/kg]
1	Aggregate (gravel or crushed stone)	0.083	0.0048
2	Bitumen, general	51.0	0.38-0.43
3	Asphalt, 4% (bitumen) binder content (by mass)	2.86	0.059
4	Asphalt, 8% (bitumen) binder content (by mass)	5.00	0.076
5	Polypropylene	95.40	4.98
6	Cement stabilised soil layer 5%	0.68	0.06
7	Concrete 25/30MPa	0.78	0.106
8	Concrete blocks – 8MPa compressive strength	0.59	0.059
9	Glass fibre	100.00	8.10
10	Sand	0.081	0.0048
11	Granite aggregate	11.00	0.64
12	Lime aggregate	1.50	0.087

6. The carbon footprint of an entire pavement. An example of using the carbon calculator for an optimised pavement

The incorporating an MSL in a pavement design can provide environmental and economic benefits. the information gathered to determine embodied energy for materials, delivery, placement and compaction can be used to estimate the carbon footprint of an MSL when used as a part of whole pavement assessment. The data related to embodied carbon for the aggregate and bitumen components of asphalt is presented in [16] (some of the example building materials are presented in Table 1). Information included in [18] allows compaction operations to be included in the resulting evaluation of carbon footprint for asphalt layers of manufactured and installed asphalt layers (in kgCO2/kg).

The results of sample analysis performed with the carbon calculator are presented below. The analysed pavement was designed for heavy traffic (KR7- according to [20]). The following assumptions were made:

- Heavy traffic: KR7, pavement life > 52M 100kN axles/lane
- Subgrade strength: E2=25 MPa (G4 - according to [11])
- Length of road section: 1km
- Width at road level: 20m
- Maximum excavation for the non-stabilised and non-optimised pavement structure: 1.25m
- Maximum excavation due to stabilised (MSL) and optimised (PO) pavement structure: 0.94m
- Width at deepest applicable formation level: 25m
- Width at deepest applicable formation level due to both MSL and PO reduction: 23.7m
- The pavement is designed below the existing ground level (in the cut)
- Site: Bialystok, Podlaskie voivodeship, Poland
- Distances: between quarry and site: 25km; between the asphalt plant and site: 10km; between the site and disposal site of unacceptable soils: 10km.

The primary pavement structure with a non-stabilised granular layer (according to [20]) and a structure with an (1) MSL incorporating hexagonal geogrid (for ground stabilisation) and with (2) asphalt and aggregate base course layers incorporating hexagonal geogrid (for PO) are presented in Table 2. Both pavements were designed with an equivalent performance, to reach the same level of required life.

Table 2. Comparison of both structures - primary according to [20] and optimised with hexagonal geogrids.

Structure layers	Thickness of primary pavement structure [11] (cm)	Thickness of optimised pavement structure (cm)
Wearing course, asphaltic concrete	4	4
Binder course, asphaltic concrete	8	7
Asphalt base course, asphaltic concrete	18	16
Aggregate base course, C90/3	20	17
Hexagonal geogrid to stabilise aggregate base course layer, Type 1 (PO)	-	X
	E2=120MPa	E2=120MPa
Sub-base aggregate course CBR>=60%	35	25
	E2=50MPa	E2=50MPa
Aggregate capping layer for existing soil improvement, CBR>=20%	40	25
Hexagonal geogrid to stabilise aggregate capping layer, Type 2 (MSL)	-	X
Non-woven geotextile for separation and filtration	X	-
Existing sub-soil, (G4)	E2=25MPa	E2=25MPa
Total thickness	125	94
Thickness reduction	-	31

Analysis using the carbon calculator obtained the following costs, plus the amount of embodied carbon, for both construction approaches (Table 3).

Table 3. Calculation results – costs and carbon emissions.

Parameter	Primary pavement structure [11]	Optimised pavement structure
Cost of executing construction (million PLN)	6,383	5,500
Carbon emissions during construction (million kgCO_2e)	4,977	3,822

The use of an MSL and pavement optimisation resulted in construction cost savings of 13.8% and carbon emission savings of 23.3%.

7. Conclusion

The use of hexagonal geogrid for pavement optimisation can bring significant economic savings and environmental benefits to projects. In the opinion of the authors, the environmental aspects, particularly the calculation of the reduction of CO_2, should become a standard check procedure at the design stage for any kind of road. The example calculations shown in the paper confirm the potential benefits of pavement optimisation in terms of protecting the environment.

References:

[1] American Association of State Highway and Transportation Officials 1993 *AASHTO Guide for Design of Pavement Structures*, Washington, USA
[2] American Association of State Highway and Transportation Officials *2009 Standard Practice for Geosynthetic Reinforcement of the Aggregate Base Course of Flexible Pavement*

Structures AASHTO R 50-09, Washington, USA
[3] Jenner C G, Watts G R A and Blackman D I 2002 *Trafficking of Reinforced, Unpaved Subbases Over a Controlled Subgrade*, Proceedings of 7th International Conference on Geosynthetics, Nice, France
[4] Jersey S R, Tingle J S, Norwood G J, Kwon J and Wayne M 2012 *Full-Scale Evaluation of Geogrid-Reinforced Thin Flexible Pavements*. Transportation Research Record, Journal of the Transportation Research Board, No. 2310, TRB of the National Academies, Washington, USA
[5] Kwon J, Wayne M, Norwood G J and Tingle J S 2012 *The Implementation of Findings From Accelerated Pavement Testing in Pavement Design And Construction Practice*, Advances in Pavement Design through Full-scale Accelerated Pavement Testing – Jones, Harvey, Mateos & Al-Quadi (Eds.), Taylor&Francis Group, London, UK
[6] Norwood G J and Tingle J S 2014 *Performance of Geogrid-Stabilized Flexible Pavements. Final Report*. EDRC/GSL TR-14-28. U.S. Army ERDC, USA
[7] Robinson W J, Tingle J S and Norwood G J 2017 *Full-Scale Accelerated Testing of Multi-Axial Geogrid Stabilized Flexible Pavements*. Draft Final Report. EDRC/GSL TR-17-X. U.S. Army ERDC, USA
[8] White J D 2014 *In Situ Performance Verifiaction of Geogrid-Stabilized Aggregate Layers*: Dewitt County, Cuero, Texas – Schlinke Rd., Geomatters Technologies Inc., Ames, USA
[9] White J. D 2014 *Preliminary APLT Comparision of Ev Strain Modulus and In-situ Resilient Modulus/Permanent Deformation*, Geomatters Technologies Inc., Ames, USA
[10] Jones R and Dixon N 2011 *Sustainable development using geosynthetics: European perspectives*. Geosynthetics (Vol. 29, No. 3)
[11] The Cut Costs and Carbon Calculator, https://www.carbontrust.com/resources/tools/cut-costs-and-carbon-calculator-catering/, accessed 25 February 2018
[12] Publically Available Specification PAS 2050:2011 *Specification for the assessment of the life cycle greenhouse gas emissions of goods and services* http://shop.bsigroup.com/upload/shop/download/pas/pas2050.pdf accessed 25 February 2018
[13] Coffey Geotechnics Ltd. *Tensar Carbon Calculator Assumptions – based on International Carbon Calculator V4.0*, 701.2AB_M_001B_Assumptions for Carbon CalculatorV4.0
[14] Powell W D, Potter J F, Mayhew H C and Nunn M E 1984 *The structural design of bituminous roads* (No. LR 1132 Monograph)
[15] Manual of Contract Documents for Highway Works 2016 Vol. Specification for Highway Works Series 600 Earthworks, Amendment – February 2016; Table 6/1 and Table 6/4. http://www.standardsforhighways.co.uk/ha/standards/mchw/vol1/pdfs/600.pdf
[16] Hammond G and Jones C 2011 *Inventory of Carbon & Energy (ICE)* Version 2.0 Sustainable Energy Research Team, Dept. of Mechanical Engineering, University of Bath, www.bath.ac.uk/mech-eng/sert/embodied
[17] Plastics Europe Eco-profiles http://www.plasticseurope.org/application/files/7415/1747/5136/eco-profile_methodology_version2-0_April2011.pdf accessed 25 February 2018
[18] DEFRA Conversion Factors (UK Government Department for Environment, Food and Rural Affairs https://www.gov.uk/government/publications/greenhouse-gas-reporting-conversion-factors-2017
[19] Waste and Resources Action Programme (WRAP) CO2 Emission Estimator Tool http://www.wrap.org.uk/search/gss/co2%20emission%20estimator%20%20 accessed 25 February 2018
[20] Pavement Design catalogue *Katalog Typowych Konstrukcji Nawierzchni Podatnych i Półsztywnych* 2014 Załącznik do zarządzenia nr 31 GDDKiA z dn. 16.06.2014r. https://www.gddkia.gov.pl/userfiles/articles/z/zarzadzenia-generalnego-dyrektor_13901/zarzadzenie%2031%20zalacznik.pdf

IOP Conf. Series: Materials Science and Engineering **356** (2018) 012019 doi:10.1088/1757-899X/356/1/012019

Incorporating the transverse profile of the wearing course into the control of the hot in-place recycling of asphalt concrete

Michalina Makowska[1], Eeva Huuskonen-Snicker[2], Pauli Alanaatu[2], Kalle Aromaa[3], Abhishek Savarnya[4], Terhi Pellinen[1], Animesh Das[4]
[1] Department of Civil Engineering, Aalto University, Espoo, Finland
[2] Destia Oy, Turuntie 207, 02740 Espoo, Finland
[3] Anton Paar Nordic AB, Tekniikantie 14, FI- 02150 Espoo, Finland
[4] Department of Civil Engineering, IIT Kanpur, Kanpur, India 208016

michalina.makowska@aalto.fi

Abstract. The hot in-place recycling (HIR) of asphalt concrete (AC) is one of the least CO_2 emissive reuse techniques. It allows for 100% reuse of material in-situ in the same application, at a reduced need for the material transport to and back from the construction site, as well as the reduced price in comparison with the fresh wearing course overlay. Finland uses the technique predominantly to fill wheel path ruts caused by the studded tire abrasion, to retain structural capacity and prevent hydroplaning. During the HIR process, the aged AC material is heated up in-situ, milled to the approximate depth of 40 mm, blended with fresh AC admixture and rejuvenator. However, the amount of the aged material and the amount of the aged bitumen that undergoes rejuvenation depends on the pavement transverse profile. The rut depth, width and shape determine the minimum volume of admixture necessary for refill during the process in order to retain the structural capacity, as well as the amount of the aged binder requiring rejuvenation. In favor of achieving homogenous rheological properties in the final product, the proportion between the aged binder and the fresh binder should be controlled, as it influences the required amount of rejuvenator. Therefore, the rut cross-sectional area and furthermore, the rut volume is one of the previously unrecognized or ignored major variables of the hot in-place recycling process in Finland that should be incorporated to the HIR process control. This article demonstrates the methodology of incorporating the transverse road profile measurements by 17 vehicle-mounted laser sensors into the calculation of required rejuvenator amounts. This can be done during the procurement preparation phase or during the paving work as a continuous in-situ process control. In the rheological optimization the apparent Newtonian viscosity concept and the rotational viscosity are utilized in the viscosity based blending equation, which then allows the use of oily rejuvenators. The method reduces the need for aged pavement sampling compared with the determination of the calibration curve between rejuvenator concentration and the rheological response. Additionally, the apparent Newtonian viscosity corrects the complex viscosity by the phase angle derived correction factor, opening a previously unexplored opportunity of targeting desired viscoelastic characteristics. The approach is less sensitive to the frequencies and temperatures at which the shear measurements are conducted. This makes proposed calculative method of the desired proportioning of the aged binder, the fresh binder and the rejuvenator a promising tool for the industry. The combined algorithm presented allows for: the discrimination of sites where HIR type maintenance of pavement in question would result in a substandard product; the choice of the most promising material combination of the admixture and rejuvenator, as well as for the adjustment of the admixture and rejuvenator amount in-place.

IOP Conf. Series: Materials Science and Engineering **356** (2018) 012019 doi:10.1088/1757-899X/356/1/012019

1. Introduction

The hot in-place recycling (HIR) of asphalt concrete (AC) is one of the least CO_2 emissive reuse techniques [1] [2] [3]. It allows for 100% reuse of AC in-situ in the same application, at a reduced need for the material transport to and back from the construction site, as well as the reduced price in comparison with the fresh wearing course overlay [3]. Finland uses the technique predominantly to fill wheel path ruts caused by the studded tire abrasion.

During the HIR process, the aged AC material is heated up in-situ, hot milled to the depth of ca. 40 mm, blended with fresh AC admixture (typically with bitumen penetration grade 70/100, about 25 kg/m^2) and rejuvenator (paving grade soft bitumen 650/900, up to 250 g/m^2). As a result it is estimated that the final product contains approximately 70-80% aged AC, which makes it one of the highest AC reuse techniques available. Nevertheless, the past research reports suggest that pavements deterioration increases in the consecutive cycles of reuse [4] [5], and that especially for AC with stiff aged binder achieving good performance in the final product is troublesome. In addition to the previously discussed shift towards decreased viscous response due to the multiple recycling of bitumen [6], one of the simpler aspects that has been recognized now is the lack of optimization of rheological properties during HIR. The progressive increase of stiffness of binder in the final product due to that is preventable.

In favor of achieving homogenous rheological properties in the final product, the proportion between the aged binder and the fresh binder should be controlled, as it influences the required amount of rejuvenator. The rut depth, width and shape determine the minimum volume of admixture necessary for refill during the process in order to retain the structural capacity, as well as the amount of the aged binder requiring rejuvenation. A necessity to develop continuous in-situ control system for binder consistency optimization, based on the existing pavement transverse profile (TRP) is proposed.

The purpose of this study, funded by the Finnish Transport Agency, is to demonstrate how along the length of the road section the true variation in TRP can affect the ratio between the aged and fresh binder in the HIR, and as a result influence the required rejuvenator amount.

At this point the correction of the aggregate gradation during the HIR process is not considered. The HIR conducted correctly is protecting the gradation in-place [7] as the separation of aggregates is happening through the softening of the bitumen rather than the mechanical force (e.g. during cold planning [2]). Due to that and brevity, the focus of present paper is in linking the TRP to the optimization of the rheological characteristic of the binder after HIR.

2. Binder rheology

The data used to evaluate the blending rules presented below comprised characteristics of bitumen samples extracted and recovered from cores, straight run bitumen of various grades (viscosity and penetration) before and after aging, as well as oil based rejuvenators. More specific information about the bitumen samples and their rheological characteristics are provided elsewhere [8] [9] [10] [7].

Bitumen samples were measured with dynamic shear rheometer to establish complex modulus ($|G^*|$) and phase angle (δ) (MCR 302 device with a Peltier temperature control system, Anton Paar, Austria) in frequency sweeps 0,01-10 Hz (FS) for temperatures 0-90°C at 10°C increments, with additional FS at 25°C. The parallel plate geometry of diameter 8 and 25 mm, using 2 mm and 1 mm gap, was used respectively. The applied shear strains were 0.05 % with 8 mm plate and 1% with 25 mm plate and within the Linear Viscoelastic range assured by amplitude sweep measurements.

For rotational viscosity (η_{rot}) measurements, the same equipment, but using the cone-plate (CP) geometry of 50 mm diameter (cone truncation 0.1 mm, CP angle 1° [ISO 3219:1995]), was used for oil based rejuvenators and soft bitumen samples at shear rates of 0.01-100 [1/s] between temperatures of 20-70°C at 10°C intervals, with additional measurement at 25°C.

Penetration (Pen) for all the bitumen samples reported below was measured at 25°C according to standard SFS-EN 1426 typically on the 100 g size sample. The Pen of 25 1/10 mm is noted as 25Pen.

For the below calculations, rheological characteristics of three extracted and recovered bitumen (BEX) samples from field collected material were used to represent field aged material. One BEX of

35Pen comes from a core collected from the road before the process. The two additional samples, 3R and 4R [8] [10], both are BEX of 25Pen from the reheated and scarified material in the HIR process, before addition of the admixture or rejuvenator. They differ by history [10] and |G*| to δ [8] relationship.

As reported previously, for BEX of Pen>32Pen the lower raveling and thermal cracking were observed [9]. Thus, two rheological targets for final product properties were considered, 40Pen (BEX from core from good performing area after the HIR process, 3FC in [8]) and 47Pen (rolling thin film oven aged 70/100 bitumen, similar to those found in freshly recovered AC). To represent the properties of fresh bitumen in admixture the characteristic of straight run bitumen 70/100 (79Pen) was used [7].

The currently used rejuvenator in the HIR process is soft bitumen penetration grade 650/900 (Penetration between 180-360 1/10 mm measured at 15°C according to EN 12591:2009 and [11]), but the motivation of this study is to switch towards a softer viscosity graded product of viscosity at 60°C between 1000-2000 mm^2/s (V1500). The previous reports of using V1500 in the place of 650/900, without the amount adjustment due to the difference in the viscosity of those products, reported problems with bleeding and over softening [12].

The purpose of hereby work was to develop the rules of blending which could allow substitution of 650/900 with V1500, while preventing the over softening. Coincidentally, the rules apply as well to other low viscosity products which could be potentially used as rejuvenators. The bio-based low viscosity oil was used to demonstrate the applicability of the approach to this types of products as well.

2.1. Blending equations

Because of the complex structure of the asphalt binder it is very difficult to model the interaction. Several attempts have been made by different researchers to come up with the mixing rules and their review is provided elsewhere [13]. Nevertheless, the Finnish Asphalt Specifications (FAS) [11] allow the use of four blending equations:

$$\log(Pen25°C_{blend}) = a * \log(Pen25°C_{aged}) + b * \log(Pen25°C_{fresh}) \quad (1)$$

$$SP_{blend} = a * SP_{aged} + b * SP_{fresh} \quad (2)$$

$$log|G^*_{blend}| = a * log|G^*_{aged}| + b * log|G^*_{fresh}| \quad (3)$$

$$loglog(visc_{blend}) = a * loglog(visc_{aged}) + b * loglog(visc_{fresh}) \quad (4)$$

where a+b =1, Pen25°C stands for penetration measured at 25°C [1/10 mm], SP stands for softening point [°C], $|G^*_x|$ stands for complex modulus of blend component, measured at 1,78 Hz and 30°C, or 1,59 Hz and 25°C [Pa], "visc" stands for viscosity (typically kinematic or dynamic) [units unspecified], "blend" superscript indicates the properties in the obtained mixture/blend, "aged" superscript indicates the properties of the aged binder, e.g. RAP, "fresh" superscript indicates the properties of the freshly added material, e.g. admixture.

However, not every equation is applicable for the use in blending of components such as penetration graded bitumen, viscosity graded bitumen and oils (see Table 1).

Table 1. The measurable rheological properties for type of subcomponent in bituminous binders in the context of equations 1-4. The "x" means the property is applicable for the material, while "-" is not.

Measurable property	Aged binder	Fresh binder	Soft bitumen as rejuvenator	Oil based rejuvenator
Penetration at 25°C	x	x	-	-
Softening Point	x	x	x	-
Complex shear modulus	x	x	x	-
Viscosity	x	x	x	x

Moreover, when the binary system of aged and fresh bitumen is considered, the ratios "a" and "b" are different when derived from viscosity (e.g. measured at 60 °C) and when derived from eq. 1. That issue is demonstrated in Figure 1. Additionally, it was pointed out that equations suggested for use in

FAS [11] are not having a specified unit for viscosity. Both mPa•s and Pa•s are accepted viscosity units elsewhere in the specifications. When using mPa•s and Pa•s, the calculated ratio is different, due to the lower linearity of the temperature susceptibility upon the use of Pascal as a main unit.

Nevertheless, there is still diminished linearity in the temperature susceptibility curves for viscosity expressed in mPa•s, which is related to deviation from Newtonian behavior at lower temperatures and higher frequencies, when the $\delta < 90°$ [14] (Figure 2). For that reason, the concept of apparent Newtonian viscosity (η_{VTS}) [15], which takes account the effects of non-linearity in the viscoelastic binders, was investigated as a plausible input into equation 4 (see section 2.2.).

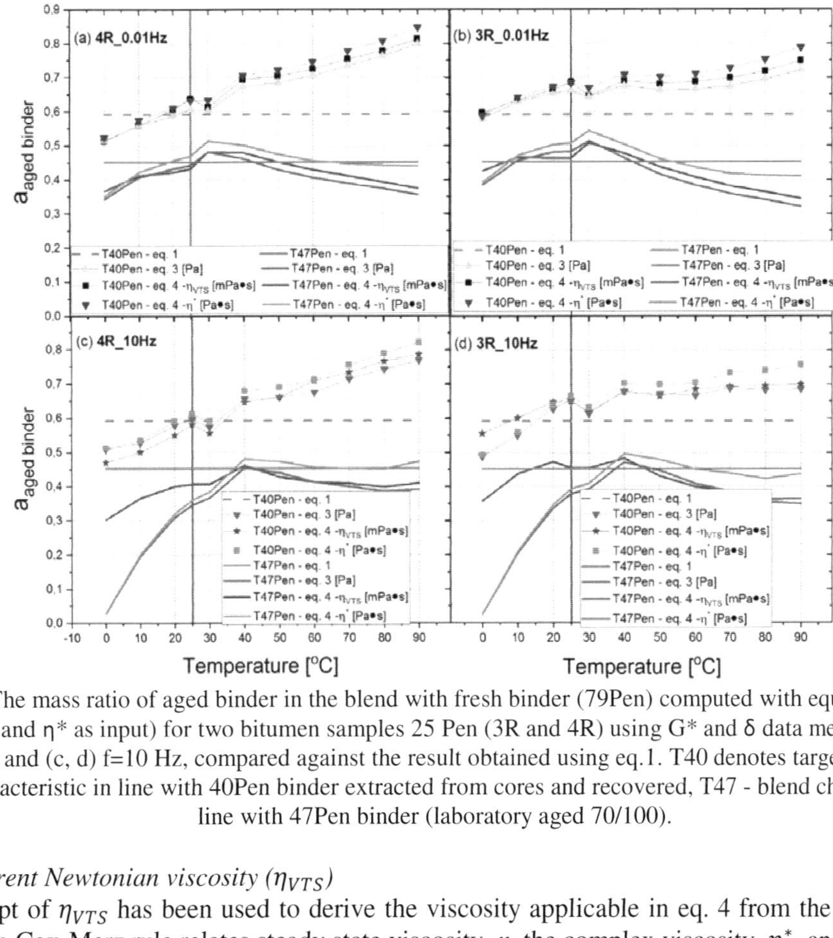

Figure 1. The mass ratio of aged binder in the blend with fresh binder (79Pen) computed with equations 3 and 4 (using η_{VTS} and η^* as input) for two bitumen samples 25 Pen (3R and 4R) using G* and δ data measured at (a, b) f=0,01Hz and (c, d) f=10 Hz, compared against the result obtained using eq.1. T40 denotes target rheological blend characteristic in line with 40Pen binder extracted from cores and recovered, T47 - blend characteristic in line with 47Pen binder (laboratory aged 70/100).

2.2. *Apparent Newtonian viscosity (η_{VTS})*

The concept of η_{VTS} has been used to derive the viscosity applicable in eq. 4 from the |G*| and δ of binder. The Cox-Merz rule relates steady state viscosity, η, the complex viscosity, η^*, and the complex modulus G* of a fluid [16]. It's given as:

$$\eta^* = \frac{G^*}{\omega} \tag{5}$$

where η^* is the complex viscosity, G* is the complex modulus and ω is the angular frequency in rad/sec. At higher temperatures, when δ approaches 90°, the η^* and the steady state viscosity are equal:

$$\eta = |\eta^*| = \frac{|G^*|}{\omega} \tag{6}$$

The asphalt binders do not exhibit Newtonian behavior over the complete range of interest in pavement applications. Hence at low temperatures and high frequencies, correction factor is used to convert |G*| to steady state viscosity. Researchers compared standard rheological measurements to |G*| and proposed the form of correction factor obtaining the apparent Newtonian viscosity (η_{VTS}) or the viscosity from ASTM viscosity-temperature equation [mPa•s] [15] as:

$$\eta_{VTS} = \frac{|G^*|}{\omega}\left(\frac{1}{sin\delta}\right)^{a_0+a_1\omega+a_2\omega^2} \quad (7)$$

where $|G^*|$ is the measured shear modulus, δ is the measured phase angle, ω is the angular frequency used to measure $|G^*|$ and δ and a_0, a_1, a_2 are the fitting parameters, hereby used for non-modified bitumen 3.639216, 0.131373 and -0.000901, respectively [15] [7].

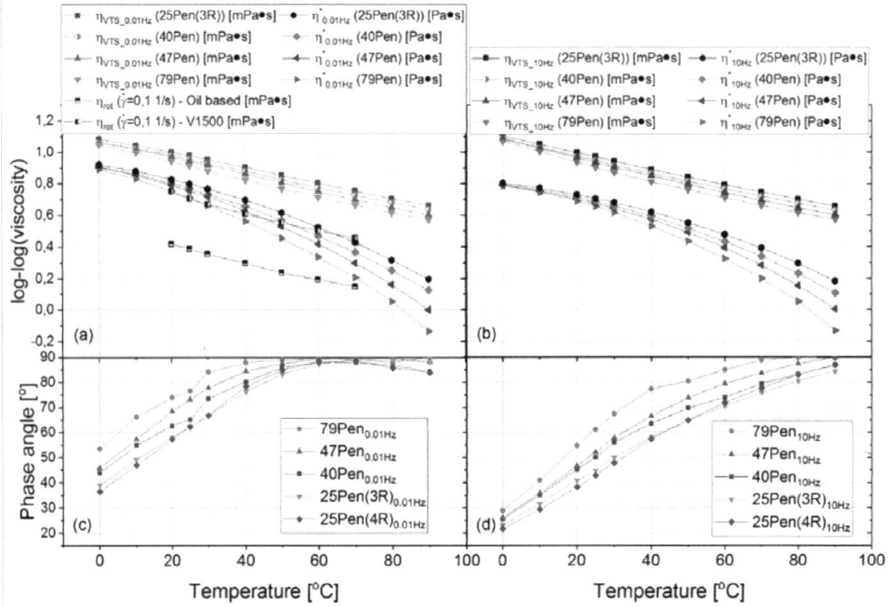

Figure 2. The temperature susceptibility curves of considered materials (25Pen, 40Pen, 47Pen and 79Pen bitumen) are presented by means of apparent Newtonian viscosity [mPa•s] and complex viscosity [Pa•s] at (a) 0.01 Hz ($\dot{\gamma} \rightarrow 0$) and (b) 10 Hz. The rotational viscosity at ($\dot{\gamma} \rightarrow 0$) in this case 0.1 1/s is plotted in (a) for V1500 bitumen and oil based rejuvenator. To demonstrate the deviation from the Cox-Merz rule at lower temperatures and higher frequencies in bitumen samples, the δ at loading frequencies (c) 0.01Hz and (d) 10 Hz is given.

2.3. Input into the calculations

The goal is to recover the viscous properties at the lowest loading frequencies (e.g. 0,005 rad/s [17]) and the lowest shear rates, due to the interesting aspect of recovering the ductile behavior for better low temperature performance. Based on the computations presented in Figure 1, and the above assumption the choice of optimization frequency of bitumen loading from the collected data was 0,01 Hz (0.06283 rad/s) as the lowest measured (not obtained through time-temperature superposition principle).

Both η^* [Pa•s or mPa•s] and η_{VTS} [mPa•s] work in an acceptable manner in eq. 4 when bituminous binders are considered. The problems emerge when η_{rot} needs to be considered as a third component in the blending equation. If the Pa•s is used as unit log-logη_{rot} becomes negative. Additionally, the rejuvenators used in this study are non-Newtonian liquids. Therefore the assumption was made that using viscosity measured at the lowest shear rates measurable ($\dot{\gamma} \rightarrow 0$) should be used in calculations.

As presented in Figure 2, the η^* of bitumen samples at low frequency and low shear rates has a tendency to be highest (e.g. 0.01 Hz vs. 10 Hz). At increased frequency of loading and increased shear rate the η^* is lower, resulting in the higher "a" as a result of the calculations in equation 4 (Figure 1). Using η^*_{10Hz} can lead to a risky assumption that the final product could contain bigger percent of aged AC, leading to decreased viscous response at low temperatures. However, by using frequency as one of the variables, the eq. (7) corrects the temperature susceptibility of η^*, producing overlapping lines.

Whether the shear rate is corrected for due to this transform, remains to be studied. Nevertheless, the "a" calculated using the η_{VTS} at 10 Hz gives similar result to those at 0.01 Hz, which is untrue if η^* is used as input. Additionally, in case of using measurements at higher frequencies as input in calculations, we still need to consider how to choose the viscosity of the rejuvenator measured by η_{rot} if we use η^* as an input. The use of η_{VTS} eliminates that issue.

Nevertheless, it has been experimentally confirmed, using five various rejuvenators, that using this approach of combining equation 4 and 7 for three components, i.e. aged bitumen, fresh bitumen and rejuvenator, we can obtain a blend of rheological characteristic within 10% by |G*| and under 3° by δ difference to the target binder across temperature range 0-90°C and frequencies 0.01-10 Hz, or within +/-2 1/10 mm as measured by Penetration ([7]).

The applicability of different frequencies is still under investigation, but at the moment, suggested frequency and temperature for optimization or rheological properties of blend is at 25 °C and 0.01 Hz, and rotational viscosity of the rejuvenator at $\dot{\gamma} \rightarrow 0$.

3. Transverse profile on the rutted pavements

The trigger for HIR maintenance is defined by the maximum rut depth (h_{max}) on the level of 16-21 mm. The rut depth profile in Finnish pavements is a result of combined effect of initial compaction after construction, permanent deformation in traffic and studded tire wear. It may be additionally contributed to by the consolidation of the subgrade during freeze-thaw cycles [18].

Because of that the road network is systematically monitored. The survey vehicle based on Greenwood Engineering Profiler technology, mounted with 17 lasers in cross profile in the front of the survey vehicle is used. Used Profiler (width 320 cm, $W_{profile}$) has approximately 18 cm average distances between the laser sensors. Although, the lasers are located more frequently in the bottom of the wheel paths (see Figure 3 for position of the lasers) [19].

3.1. The maximum rut depth (h_{max}) calculations

Different methods and algorithms are being applied to calculate the maximum rut depth (see e.g. [20]). The maximum rut depth is calculated by using wire line method (see Figure 4) with equation:

$$h_{max} = \max(w_i - s_i) \tag{8}$$

where i = 1–17 or as a width of the profile i belongs to [0, 320] cm; w_i = wire height at point i; s_i = road heigth at point i; i = location of the transversal point.

3.2. Defining of rut area calculation

It was proposed that by using the rut profile measured at 10 cm interval, using the rut profile area trapezoidal integration, a volume of rut (V_{rut}) could be calculated for each 1 m of the pavement length and 320 cm of measured profile by averaging the result from each 10 cm stretch. Area of the rut is also calculated by using the so-called wire line method [20]. Area of the rut (A_{rut}) is the area which is located between the road surface and virtual wire over the road profile as convex hull over the profile points, and is calculated with the formula:

$$A_{Rut} = \sum_{i=1}^{N-1} (w_i^{width} - w_{i+1}^{width}) \frac{(w_i^{height} - s_i^{height}) + (w_{i+1}^{height} - s_{i+1}^{height})}{2} \tag{9}$$

where N = profile point count, w_i^{width} = transversal length of the wire at point i, w_i^{height} = height of the wire at point i, s_i^{height} = height of the road at point i.

3.3. Test road sections (TRS)

During the summer 2017 the two Finnish roads were surveyed for TRP and rut volume was calculated for them in addition to the typically computed maximum rut depth (Figure 5). The considered TRS were: TRS1 (8561 cars/day, direction 1), TRS2 (8561 cars/day, direction 2), TRS3 (4462 cars/day, direction 1) and TRS4 (4462 cars/day, direction 2).

Unfortunately the information regarding the true bitumen content, bulk density and other relative parameters for those TRS were not available. Therefore the results are used to demonstrate hypothetically what would be the effect of the true differences in TRP to the rejuvenator amount in order to evaluate the usefulness of this type of measurement for the on-line process control activities.

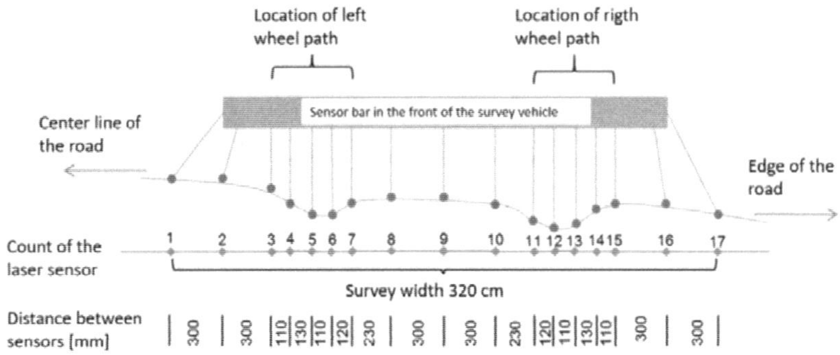

Figure 3. Schematic explanation of the Profiler and the distribution of used 17 lasers along its length of 320 cm [19].

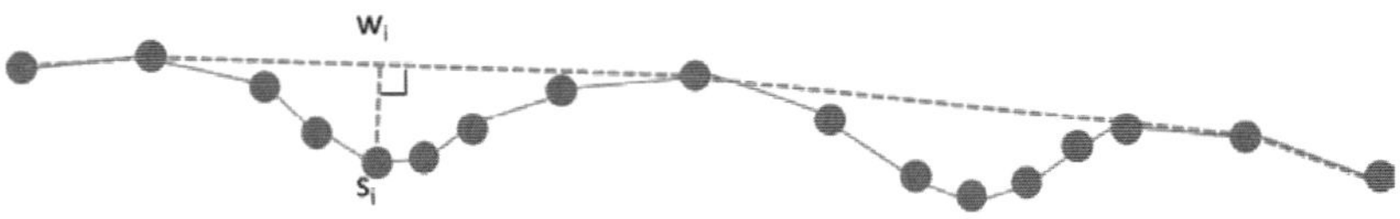

Figure 4. Sample profile how Area of the Rut is defined. Blue points are defining the road profile and the amount of the profile points is N. Red line defines how wire is adjusted over the profile as convex hull over the profile points [19].

4. The effect of the rut volume on the admixture control during recycling

Typically the consumption Consumption [kg/m^2] of admixture in HIR is expressed in mass (m_{admix}) per surface of the road (S_{road}). This value can be established using formula

$$Consumption = \frac{m_{admix}}{S_{road}} = \frac{G_{mb\_admix} * V_{rut}}{S_{road}} = \frac{G_{mb} * A_{rut} * L}{W_{profile} * L} \qquad (10)$$

where G_{mb_admix} is bulk density (G_{mb}) of admixture (calculated using design air voids for considered specifications [11]), V_{rut} is the volume of rut, $W_{profile}$ is the width of the surveyed profile (in this study 3.2 m), L is the length of the road along the driving distance [m], A_{rut} [m^2] is the area of the rut (calculated using eq. 9).

The difference between h_{max} (average per 1 m) and calculated rut volume per 1 m length of the road is presented in Figure 5. Due to differing profile types, the same V_{rut} can be calculated for equal h_{max}. Previously the Consumption was linked on the basis of the h_{max}. It is hereby demonstrated that the value should be more related to the actual V_{rut} than to the h_{max}. Typical practice of purchase is to order e.g. 25 kg/m^2 or 35 kg/m^2 of an average Consumption. How that value is understood or altered during construction depends on the operator on the side of the contractor. The resulting over or under estimation of Consumption may lead to increasing waviness of the longitudinal profile (see Figure 5 i-l). The interest would be to study in the future if calculating the Consumption more adequately would allow for correcting of the road profile during HIR type maintenance.

Survey vehicle can be also mounted with an additional Destia Scanner System [DSS], which is located in the rear of the survey vehicle. DSS has a sample distance of 1 cm in the transversal profile and maximum survey width 400 cm. Starting from 2018, calculating the V_{rut} using this upgraded system, should provide more confidence in the transverse profile measurements. The increased width of the DSS, in comparison to the currently used system, will allow for more realistic estimates of the admixture amount, i.e. in line with the actually HIR processed width of the pavement.

5. The effect of the rut volume on the optimal rejuvenator amount to reach rheological targets

Assuming that a milling operation occurs to the 40 mm beneath the surface on the edges, using the G_{mb} (e.g. 2500 kg/m^3) of the aged pavement, we can also calculate the mass of aged material per S_{road} [kg/m^2]. Knowing the bitumen content (P_b) in both aged material (e.g. P_{b_aged}=5,8%) and fresh admixture (e.g. P_{b_adm}=5,8%), we can easily calculate the ratio between aged and fresh bitumen at each point along the treated distance using formula:

$$a = \frac{V_{aged} * G_{mb_{aged}} * P_{b\_aged}}{V_{aged} * G_{mb_{aged}} * P_{b_{aged}} + V_{rut} * G_{mb_{fresh}} * P_{b_{adm}}} \quad (11)$$

From that and eq. 4 the $\eta_{VTS_{blend}}$ can be calculated. Then the eq. 4 can be expanded by additional term, connected with rejuvenator's viscosity

$$loglog(\eta_{VTS\_target}) = a' * loglog(\eta_{VTS_{blend}}) + c * loglog(\eta_{rejuv}) \quad (12)$$

where a' is a cumulative ratio of aged and admixture bitumen, c is the ratio of rejuvenator in the final blend, η_{VTS_target} represents η_{VTS} that should, or should not, be achieved in the product.

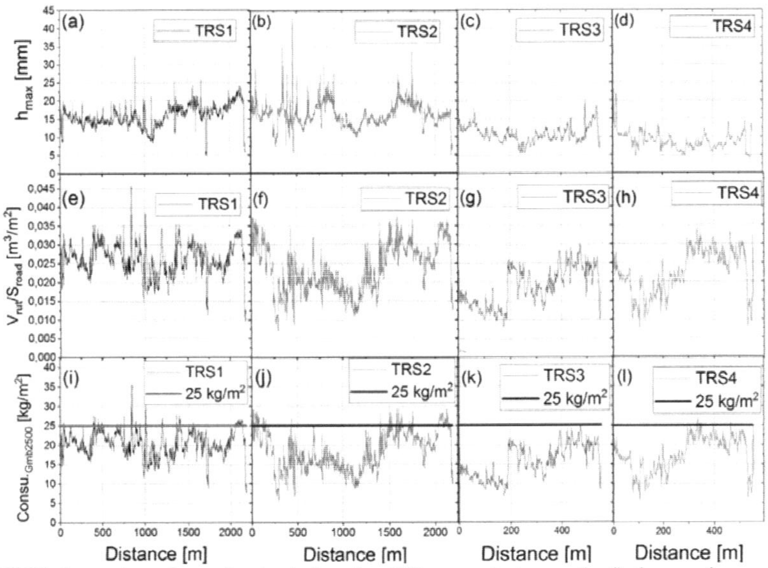

Figure 5. For the TRS1-4 a series of graphs depicting the difference between (a-d) the maximum rut depth (h_{max}) and (e-h) the average volume of rut per area (V_{rut}/S_{road}), in relation to the (i-l) calculated required amount of admixture to refil rut volume under the assumption of bulk density of the admixture at 2500 kg/m^3 ($Consu._{Gmb2500}$).

Hereby, the demonstration was performed by calculating "c" that would allow reaching 40Pen or 47Pen characterized binders (see Figure 6), when starting material is 25Pen or 35Pen, and when (η_{rejuv}) is equal to that of 650/900 or V1500. The "c" is presented for typical operation (Consumption = 25 kg/m^2) or explored Consumption in line with V_{rut} refill (RV).

The bulk density of both aged pavement and desired bulk density of admixture, have actually very high impact on the results of rejuvenator ratio calculations. For example at 0,03 m^3/m^2 of the treated aged pavement at equal P_{b_aged}=5,8%, the difference is between 4,87 kg (G_{mb}=2780 kg/m^3) versus 4,26 kg (G_{mb}=2450 kg/m^3) of aged bitumen per 1 m^2. In the first case, more rejuvenating power is required, which can be realized by a softer rejuvenator/bitumen in admixture or increased admixture amount.

6. Discussion and field plans

Typically, due to volumetric restrains of the AC, shifting P_b by 0.1-0,25% is considered of low risk in terms of mix stability and bleeding. Once the bleeding is observed during HIR process the rejuvenator amount is decreased. Of course the limiting rejuvenator amount depends on the voids filled with bitumen, but for simplification assumption is that rejuvenator addition should remain under 0.1-0.25 kg/m^2. The adjustment within the line segment of 20 m is feasible in that mass range, but the increase is rarely practiced after decrease (e.g. [8]). The possibility of control will be studied during field studies.

Looking at the results presented in Figure 6 it is visually simple to evaluate which maintenance types could be prone to bleeding or over softening risk, e.g. the viable maintenance methods for aged

AC characterized by 25Pen are those using V1500. The purchasing party could make the decision on the choice of materials for the HIR process. The operator of the HIR process on the side of the contractor can also easily distinguish in which regions to increase or decrease the rejuvenator amount, depending on the approach to admixture use (rut volume refill or fixed Consumption) and rejuvenator type.

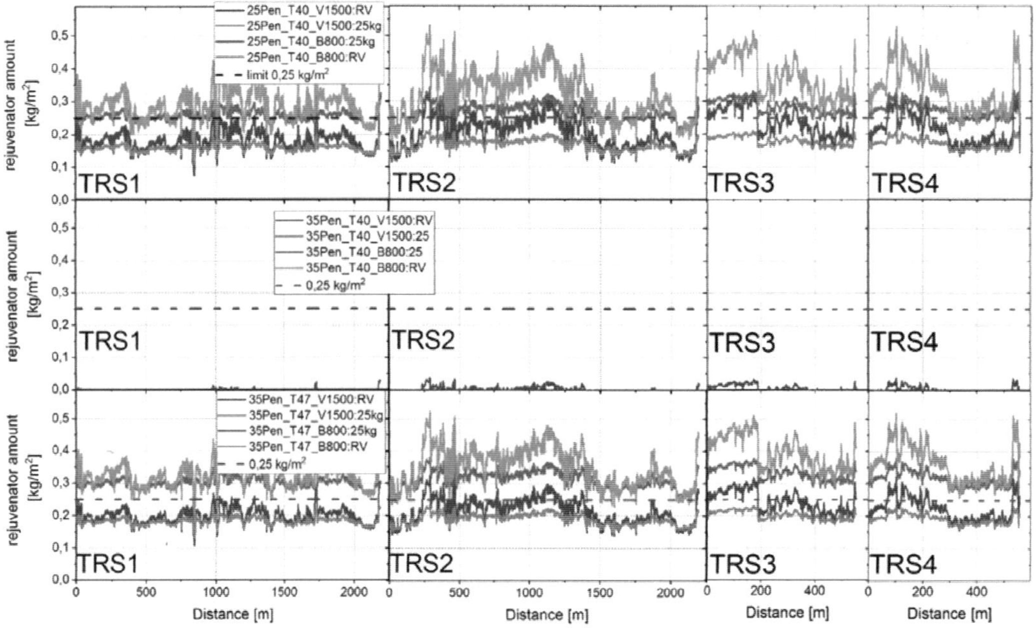

Figure 6. The results of the calculations of required rejuvenator amount (both V1500 and 650/900, a.k.a. B800) at two admixture consumption levels - fixed 25 kg or equal to the rut volume refil (RV) for considered TRS1-4, when both aged and fresh AC have G_{mb}=2500 kg/m^3 and Pb 5,8%. The upper row of frames is calculated for 25Pen aged material targeting to the rheological characteristics consistant with 40Pen type bitumen (T40), the middle row of frames for 35Pen aged material targeted towards T40, while the lower row of frames for 35Pen aged material targeted towards 47Pen type bitumen (T47).

However, currently the calculations provide only the result of the theoretical estimate of required admixture to refill the rut depth, considering the limited width of the measured profile. Currently the rut volume calculation was incorporated into the survey and allows us to gather information and compare with results of the maintenance activities. The use of DSS in future is promising. Despite the fact, that the calculations are hypothetical, the actual TRP was used. It allows for better understanding of the true deviation between sections, or along the distance within the section, in terms of necessity to refill the rut and its effect on the rejuvenator requirement. Moreover, the above is still in need of coupling with the actual milling depth monitoring to assure the correct binder ratios.

7. Conclusions

This article presented promising theoretical background behind the concept of incorporation of transverse road profile, rheological and volumetric properties of the aged pavement to the on-line control of the hot in-place recycling process. The use of apparent Newtonian viscosity concept was proposed for the field applications. The method is least susceptible to changes of frequency and temperature during the shear measurements, and thus at high possibility for coupling with measurement performed already routinely on bitumen. The transverse road profile, parameter previously ignored, was demonstrated as an important variable.

The combined algorithm presented allows for: the discrimination of sites where HIR type maintenance of pavement in question would result in a substandard product; the choice of the most promising material combination of the admixture and rejuvenator, as well as for the adjustment of the admixture and rejuvenator amount in-place.

The above presented research provides an initial framework for further field investigations.

References

[1] Miliutenko S, Björklund A and Carlsson A, 2013, "Opportunities for environmentally improved asphalt recycling: the example of Sweden," *Journal of Cleaner Production*, **43**, pp. 156-165
[2] Asphalt Recycling and Reclaiming Association, Basic Asphalt Recycling Manual, U.S.A.: U.S. Department of Transportation - Federal Highway Administration, 2001
[3] Santos J, Bryce J, Flintsch G and Ferreira A, 2017, "A comprehensive life cycle costs analysis of in-place recycling and conventional pavement construction and maintenance practices", *International Journal of Pavement Engineering*, **18**, no. 7-8, pp. 727-743
[4] Rantanen T and Suikki L, "Uusiopäällysteiden käyttö päällysteiden ylläpidossa (The use of hot in place recycling in the maintenance of the pavements)," Tiehallinto (Finnish Transport Agency), Helsinki, 2009
[5] Suikki L and Spoof H, "Uusiopintausmenetelmien kestoikäanalyysit," Liikennevirasto, Helsinki (In preparation), 2017
[6] Blomberg T, Makowska M and Pellinen T, 2016, "Laboratory simulation of bitumen aging and rejuvenation to mimic multiple cycles of reuse," Transportation Research Procedia, **14**, no. Transport Research Arena TRA2016, p. 694–703
[7] Pellinen T and Makowska M, "Asfaltin uusiokäytön kehittäminen ja REMIX-pintauksen tutkimus in preparation," 2018
[8] Aromaa K, The effects of ageing and rejuvenation on bitumen rheology, Espoo: Aalto University, 2016
[9] Makowska M, Aromaa K and Pellinen T, 2017, "The rheological transformation of bitumen during the recycling of repetitively aged asphalt pavement", *Road Materials and Pavement Design,* **18:sup2**, pp. *50-65*
[10] Makowska M, The physicochemical influence of the inorganic phase on the aging and performance of asphalt pavements in Finland, Aalto University publication series DOCTORAL DISSERTATIONS, 189/2017 ed., Espoo: Unigrafia, 2017
[11] Finnish Pavement Technology Advisory Council, Finnish Asphalt Specifications 2017, Helsinki: Premedia Helsinki Oy, 2017
[12] Patosalmi S, Vanhan asfalttipäällysteen sideaineen elvyttimet, Helsinki: Helsinki University of Technology, 1996
[13] Savarnya A, Makowska M, Pellinen T and Das A, "Determination of blending rules for bituminous binder composed of three components," N/A, in preparation, 2018.
[14] Airey G, Rheological Characteristics of Polymer Modified and Aged Bitumens, Nottingham: University of Nottingham, 1997
[15] Bonaquist R, Pellinen T and Witczak M, "Development of relationship between binder viscosity and stiffness," Department of Civil Engineering, University of Maryland, College Park, 1998
[16] Cox W and Merz E, 1958 "Correlation of dynamic and steady flow viscosities," *Journal of Polymer Engineering and Science*, **28**, pp. 619-621
[17] Rowe G, Prepared Discussion following the paper by Anderson et al. (2011), 2011.
[18] Dawson A and Kolisoja P, "ROADEX II, Northern Periphery. Permanent Deformation, Report on Task 2.1.," EU Council, e-publication: http://www.roadex.org/wp-content/uploads/2014/01/2_1-Permanent-Deformation_l.pdf, 2004
[19] Tiehallinnon selvityksiä, Käsikirja päällysteiden pinnan kunnon mittaamiseen APVM 2006-2007 T&K., Helsinki: Edita Prima Oy, 2007
[20] Wang D, Falchetto A, Goeke M, Wang W, Li T and Wistuba M, 2017, "Influence of computation algorithm on the accuracy of rut depth measurement," *Journal of Traffic and Transportation Engineering (English Edition)*, pp. 156-164

IOP Conf. Series: Materials Science and Engineering **356** (2018) 012020 doi:10.1088/1757-899X/356/1/012020

Geogrid reinforced road subgrade influence on the pavement evenness

A Šiukščius[1], V Vorobjovas[2] and A Vaitkus[3]

[1] JSC ViaCon Baltic, Metalo str. 9a, LT-02190 Vilnius, Lithuania
[2,3] Road Research Institute, Vilnius Gediminas Technical University, Linkmenų str. 28, LT-08217 Vilnius

aurimas.siukscius@viacon.lt

Abstract. As a result of increasing geogrid reinforcement applications in the road subgrade, there are number of projects where geogrid reinforcement is used to control road pavement evenness when there are small layers of peat or mud deeper under road construction. For this task geogrid reinforcement application is not documented but widely used in Lithuania for over a decade. This paper evaluates the long term influence of the geogrid reinforced soil influence on the road surface evenness, when the organic soils stratify in the deeper layers of the subgrade. The geological conditions of the investigated sections are reviewed. The experiment methodology and test results are described, which leads to the conclusions and insights how the pavement evenness depend on the geological conditions and its enhancement. The question is raised about the need for including this geogrid application to the normative documentation. Explanation of the problems that are encountered and the need for further research is given.
Keywords: subgrade, pavement evenness, enhancement, geosynthetics, geogrids, soil reinforcement.

1. Introduction

Due to the deterioration of the condition of Lithuanian roads, the number of road sections, that do not meet the required bearing capacity for the pavement construction and do not provide the sufficient traffic quality, are being reconstructed. During the reconstruction, geometrical parameters of the road are changed and the pavement construction is strengthened, therefore it is important to evaluate the existing geological conditions. After geological investigations, it is often found that a poor state road sections with cracks, ruts and potholes or bumps are located above the organic soils, i.e. peat or mud.

In the roads that were built over organic soils many years ago, the soil loading technology was often used, where the existing organic soil was not replaced, but a certain size of soil mas was filled over it, which had to increase the organic soil consolidation time. The structural works took place after the biggest settlements occurred. Due to such construction practices, carried out several decades ago, reconstructing the existing Lithuanian roads nowadays, the geological situation is often found as the following: on the top, there are filled sandy soils which lay from 1 to 6 m deep and below them there are organic soils (peat or mud) which lay approximately till 10–14 m deep. Problems of this type are usually solved by constructing a pile system and a geogrid reinforced soil platform above them which holds the road construction in place. This method has been globally used for about 25–30 years, and has been well documented in the normative documents. When the peat layer is small and it lays not deeper than 5 m, the pile system cost around the same amount as the excavation and replacement of the soil. However, in any case, this is a costly solution, especially when there is a need to increase the quality of the roads when having the same budget for road maintenance (A. Vaitkus et al. 2016). For this reason, after the geosynthetic materials were being started to be used, a number of Lithuanian road reconstruction projects have been implemented over the last 10 years, where instead of excavation and replacement of soil or pile systems various strength geogrids were used directly under the road construction or a bit lower, making up to 0,6 m soil replacement. With this method of strengthening the subgrade, it was expected that the reinforced soil under the road construction will prevent from undesirable pavement cracks, ruts, potholes or bumps that could be caused by the existing partially consolidated organic soil layers.

This type of use of geosynthetics is not regulated either in Lithuania or in normative documents of other countries (A. Vaitkus et al. 2014). For this reason, an experimental study is carried out to determine whether this type of road strengthening is suitable or not, looking from long term perspective, for the road to maintain its strength and quality requirements (A. Šiukščius et al 2017).

2. Tested road sections

Two different categories of roads were chosen in order to get the best results when estimating the geogrid influence on the pavement evenness. The first road is the main road A6 Kaunas–Zarasai–Daugavpils, the second is the country road No. 131 Alytus–Simnas–Kalvarija. The main road A6 was reconstructed in sections from Kaunas to Zarasai, and the reconstruction works began more than 10 years ago.

We were looking for a geological situation as described above, where there are filled soils over the peat layers and where the geogrid reinforcement was used under the road construction. After studying the whole available information it was found that main road A6 had those conditions in the sections of 66,20–68,76 km (the end of construction at 2007), 137,35–142,00 km (the end of construction at 2011) and 142,00–150,10 km (the end of construction at 2016). Altogether in these sections there are 5 places with a geogrid reinforcement over the organic soils that lays deeper in the strata. On the same principle the country road No. Nr. 131 was reconstructed and it also has already 10 years passed from the beginning of the reconstruction works. The local road No. 131 had required conditions in the sections of 44,80–50,80 km (the end of construction at 2010) and 52,45–57,12 km (the end of construction at 2009). Those two sections has 8 places with a geogrid reinforcement over the organic soils that lays deeper in the strata.

3. Designed solutions in the projects

3.1. *Main road A6 Kaunas–Zarasai–Daugavpils (section 66,20–68,76 km)*

Geogrid was installed under the designed road construction without any additional soil replacement in section 67,26–67,45 km. The geosynthetics installed was a 30/30 kN/m tensile strength polypropylene (PP further in the text) geogrid and a GRK3 (according MN GEOSINT ŽD 13) class non-woven geotextile. According to the project, in this section, the road goes over a 0,9–1,4 m thick peat layer, which is compressed till 30% and it lays in 2,3–3,4 m depth. Under the peat layer, in a 2,3–4,8 m depth, there is a mud layer with the thickness of 0,7–1,5 m. There is no geological investigations given in the design project. Construction year 2006—2007.

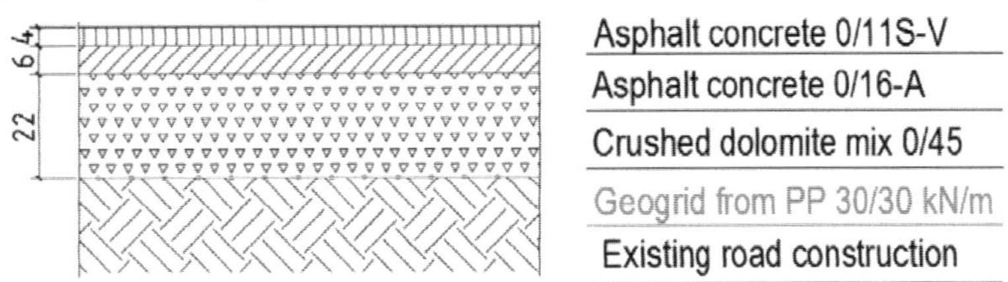

Figure 1. Main road A6 section 67,26–67,45 km designed road construction.

IOP Conf. Series: Materials Science and Engineering **356** (2018) 012020 doi:10.1088/1757-899X/356/1/012020

3.2. Main road A6 Kaunas–Zarasai–Daugavpils (section 137,35–142,00 km)

Geogrid was installed under the designed road construction with an additional 0,3 m soil replacement in section 141,047–141,123 km. The geosynthetics installed was a 400/40 kN/m tensile strength polyester (PET further in the text) geogrid in crosswise direction and a 200/40 kN/m tensile strength PET geogrid along the road, a GRK3 (according MN GEOSINT ŽD 13) class non-woven geotextile was also used. The geological investigations shows that in this section, there is a 0,4 m thick well-decomposed peat layer in a 2,7 m depth. Not only is the peat layer is soaked, but also the part of the loamy soil part above it. Below peat there is a soft plastic clay. Construction year 2010–2011.

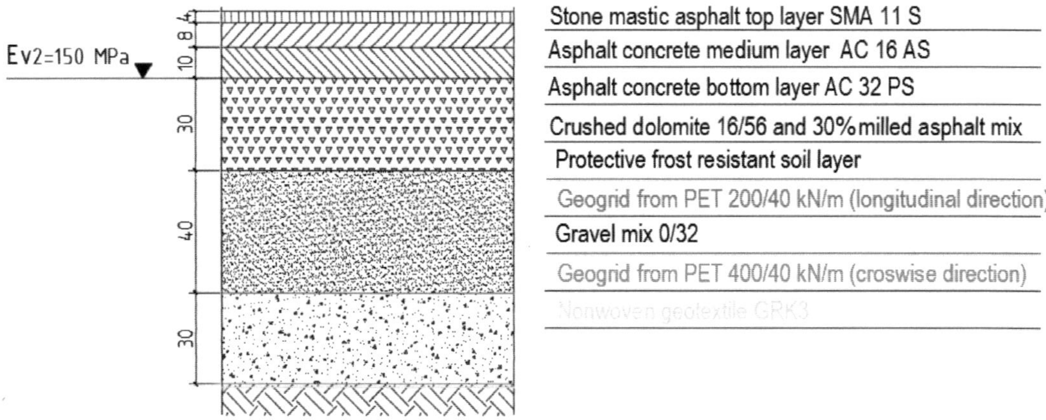

Figure 2. Main road A6 section 141,047–141,123 km designed road construction.

3.3. Main road A6 Kaunas–Zarasai–Daugavpils (section 142,0–150,1 km)

Construction year 2015–2016. Geogrid was installed under the designed road construction with an additional 0,3 m soil replacement in sections 146,21–146,44 km; 147,65–147,80 km; 148,94–149,26 km and 149,73–149,83 km. The geosynthetics installed was a 120/120 kN/m tensile strength PET geogrid and a GRK3 (according MN GEOSINT ŽD 13) class non-woven geotextile. The geological investigations shows that:

3.3.1. Section 146,21–146,44 km. At a depth of 3,9–6,3 m, there is a 0,3–2,0 m thick well-decomposed peat layer. Below it, at a depth of 5,2–8,3 m there is a 0,3 m thick mud layer. Not only peat layer is soaked, but also a part of the dusty sand above it. Below mud there is a liquid plasticity dust.

3.3.2. Section 147,65–147,80 km. At a depth of 14,0 m, there is a 0,5 m thick well-decomposed peat layer. Below it, at a depth of 14,5 m there is a 3,5 m thick mud layer. Not only peat layer is soaked, but also a part of the dusty sand above it. Below mud there is a medium roughness gravel.

3.3.3. Section 148,94–149,26 km. At a depth of 1,0 m, there is a 0,4 m thick filled softly plastic clay layer which is with organics. Below it, there is a filled softly plastic clay layer and deeper there is natural softly plastic clay layer. Ground water level is in 2,2 m depth, under the clayey soils.

3.3.4. Section 149,73–149,83 km. At a depth of 3,1 m, there is a 0,6 m thick well-decomposed, soaked peat layer. Above it, there is a 1,5 m thick filled dusty sand layer. Below peat there is a natural low plasticity, softly plastic clay.

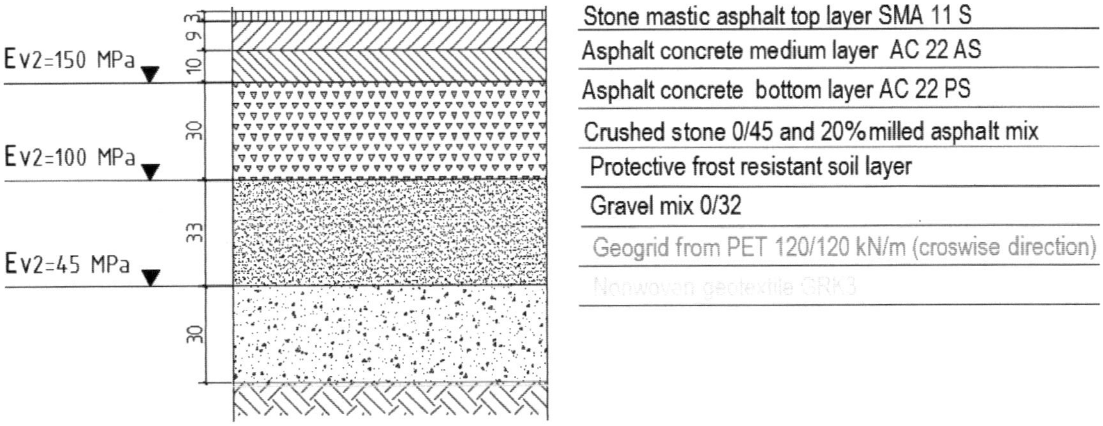

Figure 3. Main road A6 sections 146,21–146,44 km; 147,65–147,80 km; 148,94–149,26 km and 149,73–149,83 km designed road construction.

3.4. Country road No. 131 Alytus–Simnas–Kalvarija (section 44,80–50,8 km)

Construction year 2009–2010. Geogrid was installed under the designed road construction with an additional 0,6 m soil replacement in sections 45,965–46,14 km; 46,45–46,72 km; 46,865–46,95 km; 47,32–47,425 km and 49,22–49,295 km. The geosynthetics installed was a 60/60 kN/m tensile strength PP geogrid (crosswise in two layers at every 0,3 m) and a GRK3 (according MN GEOSINT ŽD 13) class non-woven geotextile. The geological investigations shows that:

3.4.1. Section 45,965–46,14 km. At a depth of 2,2 m, there is a 0,8–0,9 m thick well-decomposed soaked peat layer. Above it, there is a filled low plasticity clay, in some places with organics. Below peat there is a natural low plasticity clay.

3.4.2. Section 46,45–46,72 km. At a depth of 2,0–2,5 m, there is a 0,5–1,2 m thick medium-decomposed soaked peat layer. Above it, there is a filled low plasticity clay, in some places with organics. Below there is a natural low plasticity clay.

3.4.3. Section 46,865–46,95 km. At a depth of 2,6 m, there is a 0,8 m thick well-decomposed, unsaturated peat layer. Above it, there is a filled low plasticity clay, in some places with organics. Below there is a natural low plasticity clay.

3.4.4. Section 47,32–47,425 km. At a depth of 3,0–3,1 m, there is a 0,8 m thick well-decomposed, unsaturated peat layer. Above it, there is a filled low plasticity clay. Below there is a natural dusty clay.

3.4.5. Section 49,22–49,295 km. At a depth of 2,3 m, there is a 0,5 m thick well-decomposed soaked peat layer. Below it, at a depth of 2,8 m there is a 0,8 m thick mud layer. Above it, there is a filled dusty clay. Below there is a natural dusty sand and clay.

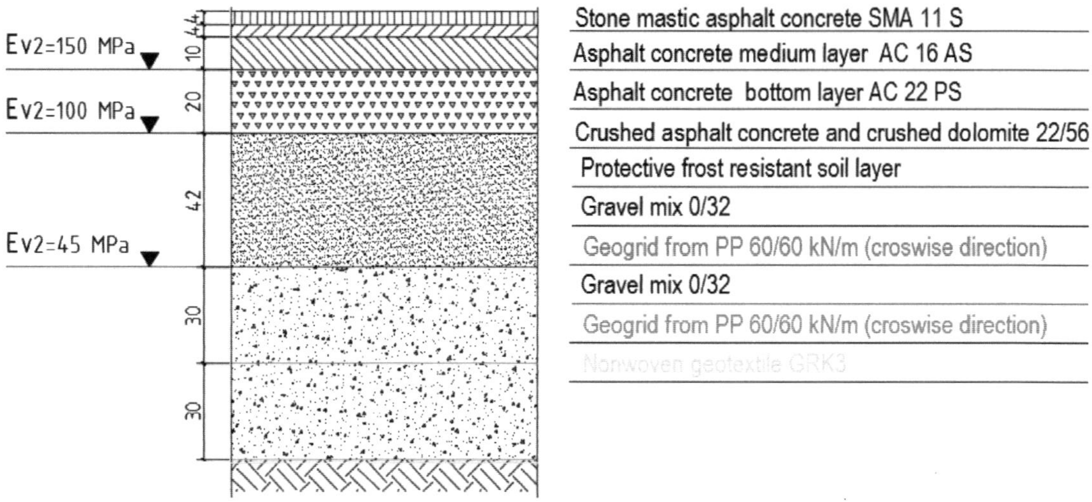

Figure 4. Country road No. 131 sections 45,965–46,14 km; 46,45–46,72 km; 46,865–46,95 km; 47,32–47,425 km and 49,22–49,295 km designed road construction.

3.5. Country road No. 131 Alytus–Simnas–Kalvarija (section 52,45–57,12 km)
Construction year 2007–2009. Geogrid was installed under the designed road construction with an additional 0,3 m soil replacement in sections 53,91–54,09 km and 56,74–56,94 km. The geosynthetics installed was a 60/60 kN/m tensile strength PP geogrid (crosswise) and a GRK3 (according MN GEOSINT ŽD 13) class non-woven geotextile. The geological investigations shows that:

3.5.1. Section 53,91–54,09 km. At a depth of 2,2 m, there is a 0,8 m thick well-decomposed soaked peat layer. Above it, there is a filled dusty clay. Below peat there is a soaked natural dusty sand.

3.5.2. Section 56,74–56,94 km. At a depth of 3,0 m, there is a 0,7 m thick mud layer and in 3,7 m depth there is 0,6 m thick peat layer. Above them, there is a 0,3 m thick organic clay layer and filled dusty sand. Not only peat and mud layers are soaked, but also a big part of the dusty sand above them. Below peat there is a soaked natural dusty sand.

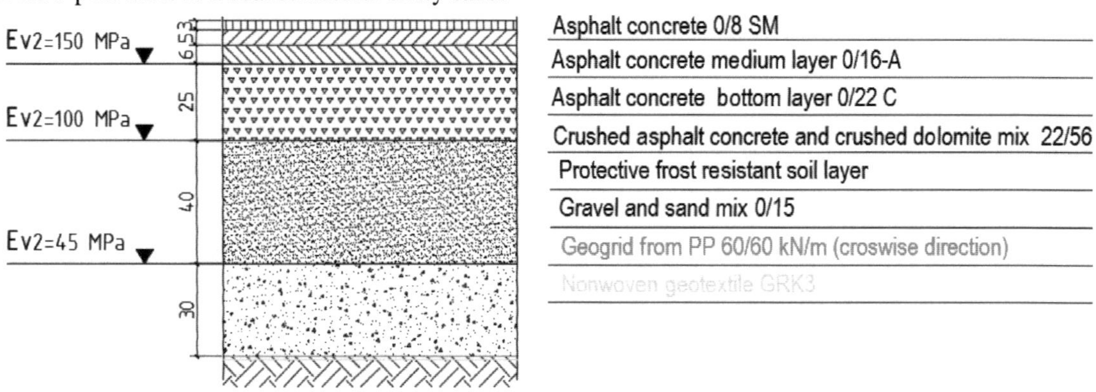

Figure 5. Country road No. 131 sections 53,91–54,09 km and 56,74–56,94 km designed road construction.

4. Research

The road surface evenness is one of the main parameters that describes the ride comfort and it is the indicator of the pavement condition. That can be described by the International Roughness Index (IRI further in the text) which is most commonly obtained from measured longitudinal road profiles. It is calculated using a quarter-car vehicle math model, whose response is accumulated to yield a roughness index with units of slope (m/km). When trying to evaluate pavement vertical displacements comparing the geogrid reinforced and unreinforced road sections from a long term perspective, the IRI measurements were taken in a different periods of the road service.

Periodic IRI measurements were carried out by the Road and Transport Research Institute. The mobile road research laboratory RST-28 was used to take the IRI measurements. RST-28 laboratory is a mobile, multi-component pavement surface quality measuring equipment, which uses the technology to ensure the precise and comprehensive test results of the road condition. To collect the data, it uses 19 laser sensors, 2 accelerometers, 2 gyroscopes and 2 inclinometers, which are mounted on the front measuring beam (laser frame). The car has road surface photographing equipment, consisting of two video cameras, aligned vertically to the road, pavement surface illuminators, an air compressor for cooling the equipment, a control unit, an equipment start/stop switch. The cabin of the car has a signal processing system, main computers and other devices. The entire system is controlled by the system switches, keyboard and monitor from the operator's workplace. An air compressor and generator, which is mounted in a special box, is located on the rear of the cabin (separated from the front by the device cabinet).

5. Results of the research

After IRI measurements were completed, all 5 test sections were evaluated throughout their whole length where some trends can be observed. The tested roads were divided in to the sections with geogrid reinforced subgrade and the remaining parts with various subgrade conditions. Statistical analysis with average IRI values of the sections shows that in general, when assessing the results of IRI measurements, it should be noted that the pavement evenness is not in any case much lower or higher comparing with unreinforced sections pavement evenness. However, looking at each tested section it can be stated that geogrids has a positive effect on the pavement evenness.

The first good example can be main road A6 Kaunas–Zarasai–Daugavpils section 66,20–68,76 km, where we can see that even the 1,6-2,9m thick peat and mud layer in the subgrade can't do no clear negative effect on only one geogrid layer reinforced section 67,26–67,45 km (see figure 6). Geogrid reinforced section shows average IRI value of 1,68 m/km after 4-5 years after reconstruction, when remaining parts shows average IRI value of 1,40 and 2,16 m/km. Unfortunately the result for 2013-2014 year test is not clear because of the installed 6 speed bumps in that section, where they are clearly visible in the whole section (figures 6 and 7).

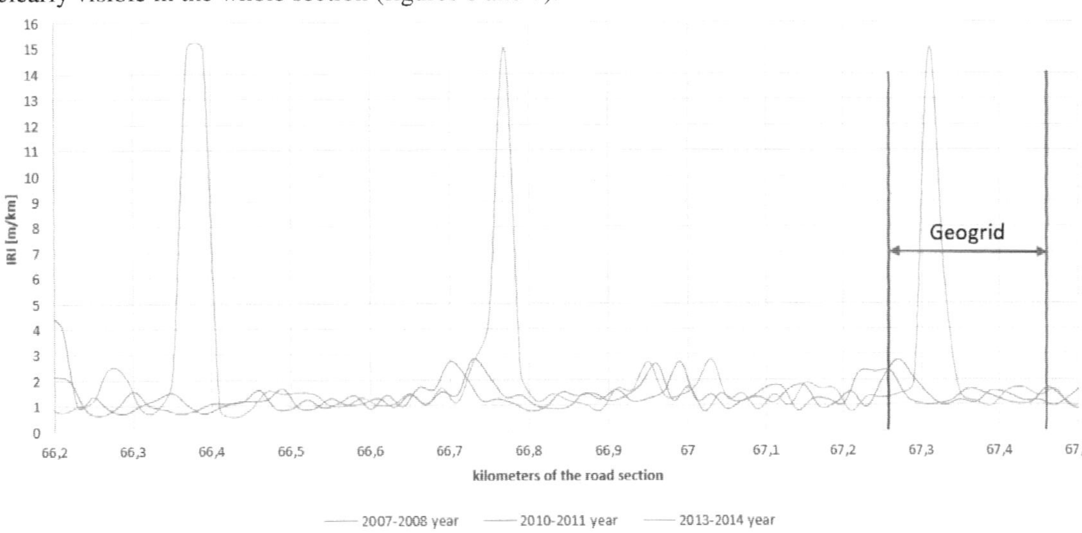

Figure 6. Main road A6 section 67,26–67,45 km pavement IRI results (first half of the section).

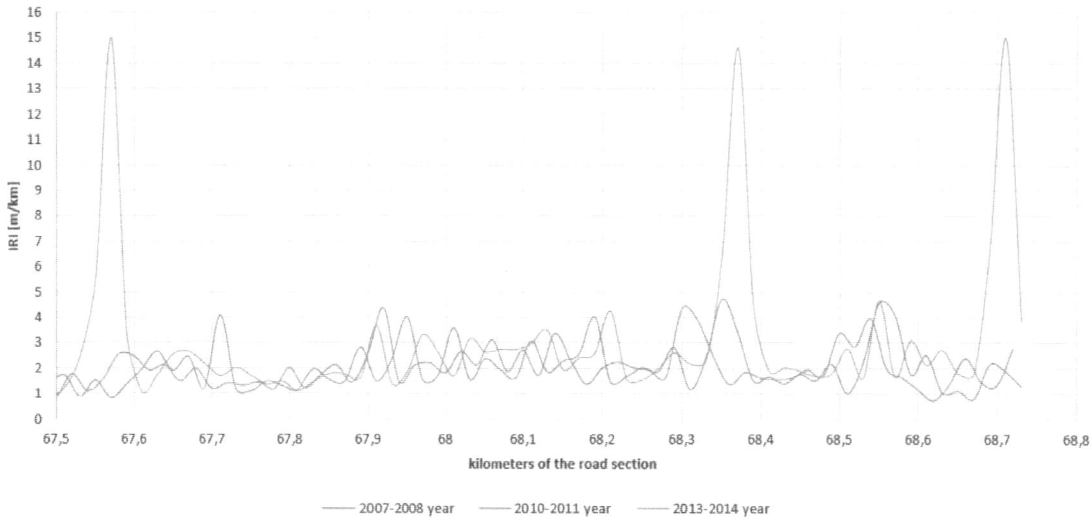

Figure 7. Main road A6 section 67,26–67,45 km pavement IRI results (second half of the section).

Anyway, comparing the average IRI vales we can see the same tendency as for the year 2010-2011. Geogrid reinforced section shows average IRI value of 3,20 m/km after 7-8 years after reconstruction, when remaining parts shows average IRI value of 2,39 and 3,34 m/km. Unfortunately there is no data how IRI looked before the road reconstruction, so there is no possibility to see how this reinforced section improved the road quality.

The main road A6 Kaunas–Zarasai–Daugavpils section 142,0–150,1 km has 4 sections reinforced with geogrids and there is lots of IRI data before the road reconstruction was finished. Unfortunately there is only one IRI measurement that was taken at 2016, few months after the road was finished so there is no comparable data as this section should be monitored for more years. However, there is one place, where we can see a benefit from the geogrid reinforced section, when we compare the IRI at the end of the section where the reinforcement ends at 149,73–149,83 km. There we can clearly see much higher average IRI values (2,95 m/km) which means that the road is reflecting the situation that was earlier (3,56 m/km), which is not the case with reinforced section (0,98 m/km), and this is measured only a few months after the reconstruction finished.

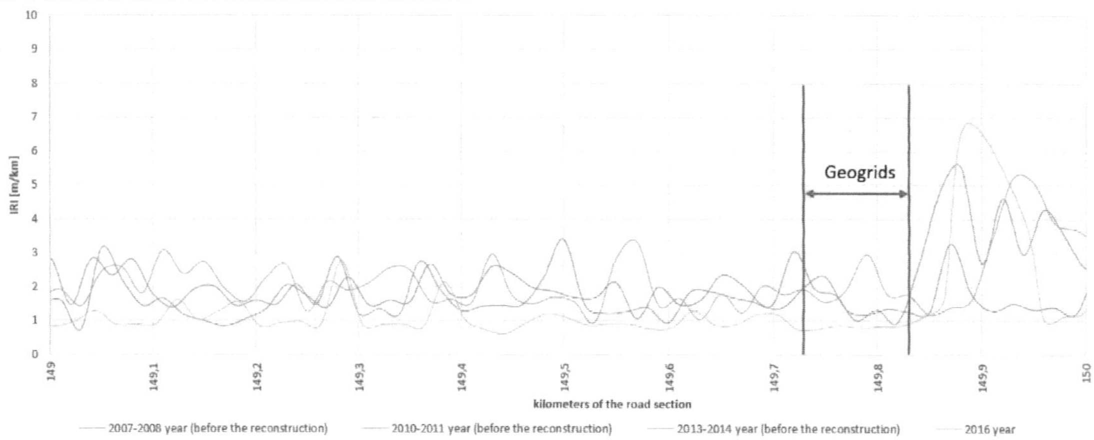

Figure 8. Main road A6 Kaunas–Zarasai–Daugavpils section 142,0–150,1 km, reinforcement at 149,73–149,83 km.

Country road No. 131 Alytus–Simnas–Kalvarija section 52,45–57,12 km has 2 geogrid reinforcement sections and the data of IRI measurements is available a year before the reconstruction and 5 years after the reconstruction. This example shows the general view of the pavement evenness. The IRI values before the road reconstruction was really high, the highest value of 14,75 m/km. In the reinforced sections the average IRI values before the reconstruction were 4,70 and 5,49 m/km and the first measurements after the reconstruction which were performed 2 years after the reconstruction shows the average IRI values of 0,81 and 0,87 m/km. After two more years another measurements were performed and they showed average IRI values of 0,82 and 0,85 m/km (see table 1 and fig. 9 and 10). Comparing average IRI results for this section it can be stated that sections with geogrid reinforcement has not higher IRI values than sections without geogrid reinforcement and in fact slightly lower, which shows that geogrid has long term influence. High IRI values at 56,55 km also shows speed bump, so this place is giving a negative effect on the average value. Having IRI not higher than 2 m/km for a country road is a really good result. Another point to note is that sections before and after the geogrid reinforced sections are stabilized by lime and if we look at the charts, we can see that the pavement with lime stabilized subgrade is not as smooth as geogrid reinforced section.

Table 1. Average IRI values for the road No. 131 Alytus–Simnas–Kalvarija section 52,45–57,12 km.

Section	Reinforcement	Year tested		
		2007-208 (before reconstruction)	2010-2011	2013-2014
52,45–53,91 km	NO	4,21 m/km	0,85 m/km	0,91 m/km
53,91–54,09 km	YES	4,70 m/km	0,81 m/km	0,82 m/km
54,09–56,74 km	NO	4,29 m/km	1,06 m/km	1,12 m/km
56,74–56,94 km	YES	5,49 m/km	0,87 m/km	0,85 m/km
56,94–57,12 km	NO	4,98 m/km	N/A	1,78 m/km

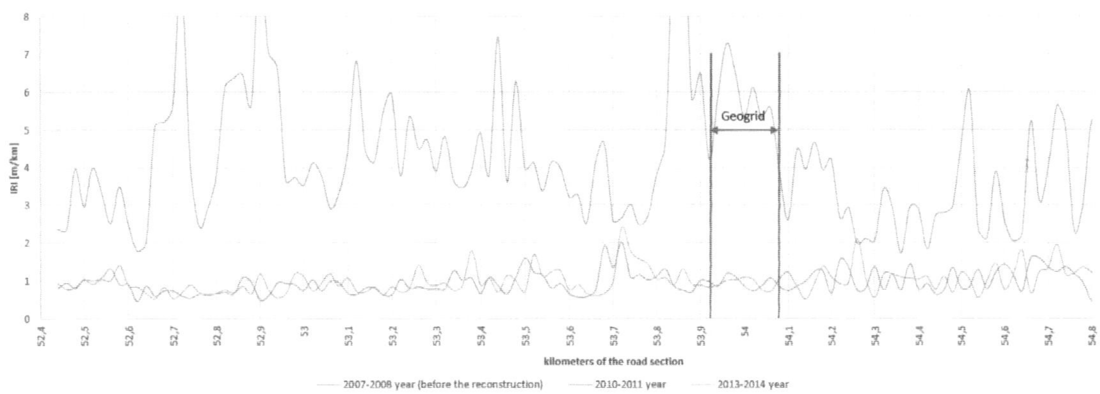

Figure 9. Country road No. 131 Alytus–Simnas–Kalvarija section 52,45–57,12 km pavement IRI results (first half of the section).

IOP Conf. Series: Materials Science and Engineering **356** (2018) 012020 doi:10.1088/1757-899X/356/1/012020

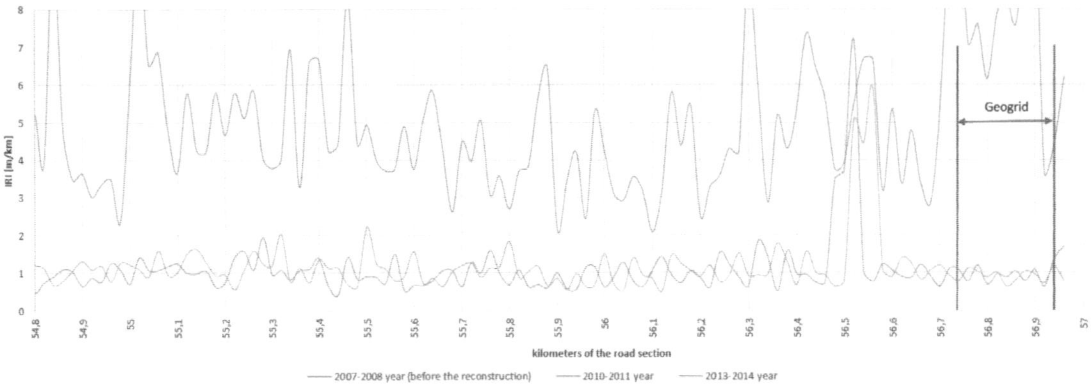

Figure 10. Country road No. 131 Alytus–Simnas–Kalvarija section 52,45–57,12 km pavement IRI results (second half of the section).

6. Conclusions

After examining the results of the International Roughness Index (IRI) for the tested road sections and observed tendencies, we can draw the following conclusions:

All geogrid reinforced subgrade sections performed well, showing average IRI values not higher compared to other road sections, excluding A6 Kaunas–Zarasai–Daugavpils section 66,20–68,76 km, which needs further investigation. This leads to a positive geogrid effect on the pavement evenness not even when there are soft soils deeper in the subgrade, but it also shows that in long term perspective geogrid reinforced subgrade can handle unfavorable conditions better then lime stabilized soil;

In some places, tested few months after the road reconstruction, IRI results are starting to show high values and it needs more investigation. This could lead to a necessity of Falling Weight Deflectometer (FWD) test;

The geogrids used for the road reconstruction is really different, starting from one layer of 30/30 kN/m and ending with two layers of 200/40 kN/m and 400/40 kN/m tensile strength. All of them are stiff geogrids with low elongation. However, the place which is starting to show high IRI values is reinforced with the 120/120 kN/m tensile strength geogrids. This also leads to the necessity of additional investigation of soil bearing capacity;

The geogrid reinforcement application for paved roads over the peat and mud layers, which are deeper in the subgrade, could be explained in a national methodical instructions for usage of geosynthetics MN GEOSINT ŽD 13.

References

[1] Vaitkus A et al 2016 *Improvement of road pavement maintenance models and technologies,* The Baltic Journal of Road and Bridge Engineering 11(3): 242–249.

[2] Vaitkus A and Šiukščius A 2014 *Regulations for use of geosynthetics for road embankments and subgrades,* The Baltic Journal of Road and Bridge Engineering 9(2): 88–93.

[3] Šiukščius A et al 2017 *Geogrid reinforced subgrade influence to ensure paved road durability,* 10th International Conference "Environmental Engineering": 1–7.

[4] Geosintetikos naudojimo žemės darbams keliuose metodiniai nurodymai *MN GEOSINT ŽD 13.*

IOP Conf. Series: Materials Science and Engineering **356** (2018) 012021 doi:10.1088/1757-899X/356/1/012021

Repeatability of road pavement condition assessment based on three-dimensional analysis of linear accelerations of vehicles

Marcin Staniek

Silesian University of Technology, Faculty of Transport, Department of Transport Systems and Traffic Engineering, Krasińskiego 8, 40-019 Katowice, Poland

marcin.staniek@polsl.pl

Abstract. The article provides a discussion concerning a tool used for road pavement condition assessment based on signals of linear accelerations recorded with high sampling frequency for typical vehicles traversing the road network under real-life road traffic conditions. Specific relationships have been established for the sake of road pavement condition assessment, including identification of road sections of poor technical condition. The data thus acquired have been verified with regard to repeatability of estimated road pavement assessment indices. The data make it possible to describe the road network status against an area in which users of the system being developed move. What proves to be crucial in the assessment process is the scope of the data set based on multiple transfers within the road network.

1. Introduction

Up-to-date information about technical condition of road pavement is the key element of the Pavement Management System (PMS) In accordance with the PMS standards, the technical and operating parameters established while road inspections are conducted include load bearing capacity of structures, pavement evenness in both longitudinal and transverse profile, pavement coarseness conditioning anti-slip properties as well as identified and recorded pavement defects. In order to produce a complete description of the state of road infrastructure, it is often required that highly specialised measuring instruments be used. Choosing from among such solutions, one may use mobile laboratories which enable identification and recoding of many technical and operating parameters in a single measuring run. What results from their application in road inspections is the assessment of longitudinal and transverse evenness as well as the said identification and recording of pavement defects [1]. The evenness measurement is conducted on a continuous basis using a laser surface analyser which allows for the results thus obtained to be analysed against a theoretical road profile. This tool ensures highly accurate road cross-section mapping as measurements are performed [2]. The pavement defect identification is mainly based on image processing and recognition technologies. By means of extraction of characteristic features of an image as well as by thresholding, segmentation or morphological transformations of image objects, one can detect the defects and classify them according to pre-assumed dependences of their description [3]. Image processing technologies require ongoing verification of the research material quality, i.e. precision of road mapping in a sequence of images. This, in turn, often leads to the necessity of noise reduction and correction of image defects resulting from the setup of the camera's optical components, but it also requires calibration of the measuring system itself [4]. Application of image processing technologies enables detection and recording of such defects as bumps, patches, deformations, chipping defects and fractures of various types: longitudinal, transverse, alligator-type and technological cracks [5]. Moreover, test vehicles record the roadway surroundings in a sequence of photographs for purposes of recording and assessment of the condition of individual infrastructure elements in roads' direct vicinity [6,7].

Highly specialised measuring devices are also used to determine anti-slip properties of road pavement. Their operating principle is generally based on measurement of the forces that affect a standard test wheel installed in a test vehicle once the wheel has been set in parallel and at a constant pre-set angle to the direction of motion [8]. A literature review confirms that one can also successfully use visual solutions based on application of light patterns to estimate the level of road pavement coarseness [9]. There are also methods which enable road pavement texture testing by means of calibrated sand [10].

The load bearing capacity of the road structure is typically estimated by the pavement deflection method under pre-set load. Tools which enable such a measurement include Benkelman deflectometers [11] or falling weight deflectometers [12]. However, the latest solution of this kind is the traffic speed deflectometer (TSD) which does not require application of any additional weight (e.g. a concrete slab) to load the road structure, since the thrust applied on the road structure in the process is induced by wheels of a test trailer. Measurements of road pavement displacements due to loading are conducted by means of laser sensors [13]. In order to describe the arrangement of the road's structural courses, and particularly to identify internal voids in the road structure, one can apply a number of solutions based on a ground penetrating radar [14].

Another solution applied to assess the state of road infrastructure is to measure the linear accelerations recorded while the test vehicle is moving. According to the basic approach of this method, a data logger unit records accelerations induced by vibrations of the driven wheel moving on the road subject to examination, while at the same time it is loaded by the mass applied. Paper [15] describes the system architecture proposed, comprising a data collector, a car mounted terminal and a processing system. Field tests of the prototype system were conducted in the Chinese province of Zhejiang. Results of the tests show that, compared to the method of unevenness assessment by means of laser tools, the relative error of the system in question is below 10%, which confirms the accuracy, efficiency and reliability of the solution proposed. What the analysis of road description signals requires is to filter out the noise emerging as an outcome of the vehicle mass system damping. Sample solutions to this problems have been discussed in papers [16] and [17], describing a quarter-car model used in the road unevenness identification process. Paper [18] addresses a case where, in order to determine parameters of the vehicle motion description, drive tests were performed over a portable hump with a known size. The proposed transfer function, established in the process of the half-car model simulation, reflects the vehicle tilt as well as the measuring device installation point. Results of the tests discussed in paper [19] have confirmed the linear dependence of the accelerations recorded (taking average speed into account) on the road pavement unevenness. It has also been proved that data provided by MEMS sensors best describe the magnitude of unevenness in the measuring frequency range from 40 to 50 Hz.

The method described in this article, used for recording and analysis of acceleration signals as well as for implementation of the data processing tool, is referred to as Road Condition Tool (RCT), and it has been devised under the S-mileSys platform being developed as a means to support supply chain links involved in the sector of freight transport. The platform's main goals include efficient transport route planning over the first/last mile by taking the technical condition of the road infrastructure into consideration. The idea behind the system is pursued under the international project entitled "Smart platform to integrate different freight transport means, manage and foster first and last mile in supply chains (S-mile)" under the "Sustainable Logistics and Supply Chains" call within the framework of the ERANET Transport III programme [20]. There are six institutions involved in the project, including businesses and higher technical schools from three countries: Spain, Turkey and Poland. A study concerning acquisition of data for the sake of description and assessment of the road infrastructure condition is the task assigned to the Silesian University of Technology. The assumed concept underlying the development of the RCT tool stems from a review of the available solutions for identification of road traffic parameters and road pavement condition assessment as well as from analysis of their comprehensiveness and simplicity of the measurement procedure involved.

2. Mobile solutions for acquisition of data on the state of road infrastructure

According to the concept underlying the S-mileSys platform, carriers' vehicles are equipped with mobile devices ensuring connection between the dispatcher at the transport company's headquarters and the driver transporting cargo. Besides the functions typically dedicated to urban freight transport, mobile devices featuring the application in question offer feedback capabilities, such as acquisition of digital data describing the state of road infrastructure. They describe motion dynamics of a vehicle as it traverses the chosen area of the road network. The solution proposed is a mobile device in the form of a tablet or a smartphone running on the Android operating system and featuring the RCT mobile application developed under the S-mile project.

According to the pre-assumed concept, the solution is intended to enable recording of linear accelerations for purposes of identification of the state of road infrastructure. The acceleration data are obtained from a microelectromechanical system (MEMS) and combined with location data obtained from the GPS receiver. The mobile devices to be used must feature both these components as well as a GSM unit. After a part of the route (transport process) is completed, an output file is transmitted from RCT mobile to RCT server via the FTP protocol. Due to communication using different transmission technologies and at different levels of the GSM signal coverage, successful retrieval of data is acknowledged by an ACK signal generated by RCT server. Fig. 1 shows a device with the RCT mobile application installed and running in the manual mode.

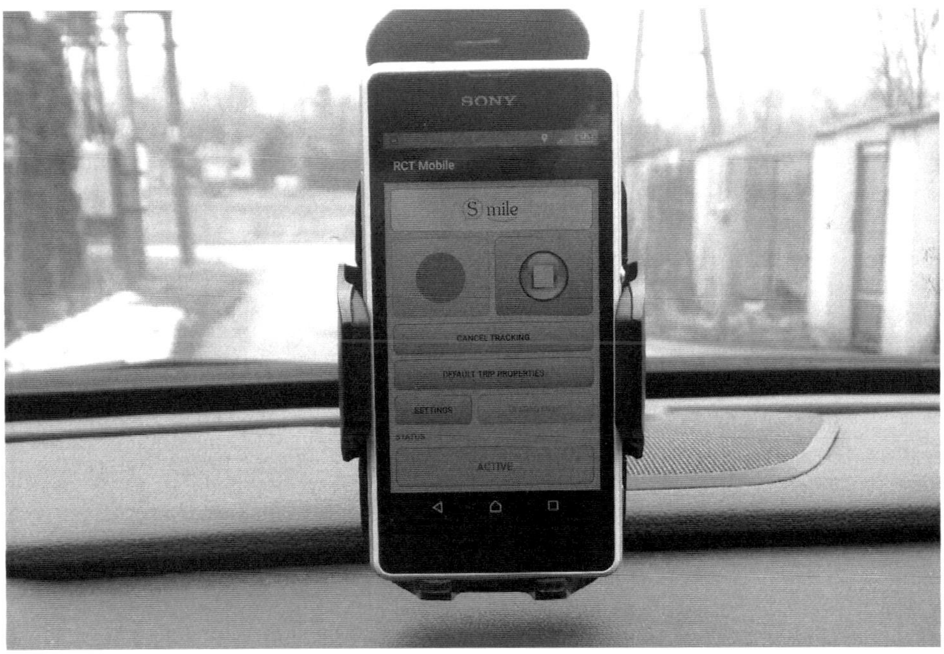

Figure 1. Mobile device with the RCT mobile application installed.

The available graphical user interface comprises six buttons used by the driver and an application status window. Respective functions enable individual application features, including start and stop of the vehicle motion dynamics recording procedure, determining identifiers of the current transport route, setting frequency of data retrieval from the MEMS and GPS modules as well as data recording accuracy.

In the remote operating mode, RCT mobile is handled by Main Mobile Application for Smartphone with the driver being prevented from any interference. The data required for identification of the route, the company and the vehicle as well as the order currently being processed are transferred from S-mile Freighter Tool to Main Mobile Application for Smartphone [20].

An output file is created as a result of measurements, and it consists of two parts. The first part comprises a header with the following parameters that identify the route covered:

- Carrier ID,
- Route ID,
- Vehicle ID,
- Vehicle type,
- Vehicle weight,
- Cargo weight,
- Data acquisition frequency.

The second part is a series of data of the recorded vehicle motion dynamics, including:

- Measurement time stamp,
- GPS geographic coordinates GPS (longitude and latitude),
- Linear accelerations in a three-dimensional system,
- Vehicle speed at the time of measurement.

Below is a sample fragment of an RCT file containing a series of data describing the vehicle motion dynamics.

```
1508043927692;50.265592;19.018385;5.2480;10.1130;-1.6560;92.91;
1508043927704;#;#;-3.7920;9.0210;-3.2840;#;
1508043927713;#;#;4.3380;8.2450;0.4110;#;
1508043927722;#;#;1.0910;9.8540;-3.6770;#;
1508043927731;#;#;-0.9190;10.1610;-4.7500;#;
1508043927741;#;#;3.6680;10.0940;0.3440;#;
```

The # symbol designates the data retrieved from the GPS module. On account of the GSM module's lower frequency of data acquisition compared to the MEMS module as well as the memory occupancy minimisation criterion, data of duplicated values are not repeated. Most typical GPS receivers/modules enable acquisition of data concerning location at the frequency of 1 per second. From the perspective of route mapping, such a record is sufficient. The linear acceleration data retrieved from the MEMS module, on the other hand, can be acquired with the frequency of up to 1,000 per second.

The data recorded in the RCT mobile application file are then sent to RCT server where they are verified in order to check correctness of the data retrieved and filter out irrelevant or failed data. What the server checks is the data series continuity and correctness of the values obtained by analysing affiliation with pre-defined value ranges. In the event of data discontinuity and high deviations of the data recorded from mean values, the filtering procedure is initiated with regard to the given data series. Once the verification procedure is complete, a route recorded as a series of GPS coordinates is converted into a set of the OSM map sections which reflect the actual route. The data conversion proceeds in stages, while the assignment algorithms are based on the following criteria:

- criterion of the GPS track coverage by OSM sections,
- criterion of conformity between directionality of the GPS track and the route of the OSM sections,
- criterion of continuity of mapping of the GPS track against the direction of motion in the OSM sections.

With the foregoing criteria in mind, it is possible to accurately map a vehicle's route recorded as the GPS positions over a set of the OSM sections forming a complete connection between the point of origin and that of the transfer destination. On account of the precision of aggregation of data which describe vehicle motion dynamics, and consequently also the road infrastructure condition assessment, for the sake of the measurements, a fixed measurement section length has been assumed at 10 m (part).

Aggregated data of linear accelerations, initially subject to filtering and assigned to individual parts of the road network, provide the baseline for estimation of the road pavement condition indices. Such an index assumes a value ranging between 0 and 1, where 0 corresponds to a road part of good technical condition (technical class A as per the applicable guidelines [21]), while 1 defines a section of poor technical condition (technical class D as per the applicable guidelines [21]). For the given road part, the road pavement condition assessment index is estimated in accordance with the following relationship:

$$\delta\left(d_{id\_part}\right) = \left[1 + e^{-\beta(d-T)}\right] \tag{1}$$

where:
d – sum of absolute differences between the recorded linear accelerations (2),
β – parameter defining the function slope, where a high value leads to stepwise behaviour of the function,
T – parameter defining the function shift in a direction that matches value d,

$$d_{id\_part} = \sum_j \left|a_j^y - a_{j-1}^y\right| \tag{2}$$

where:
a^y – recorded linear accelerations assigned to the given road part.

Parameters β and T, which determine the function behaviour, were estimated with reference to the linear accelerations recorded in tests of the RCT tool. Their values depended on the category of the given vehicle whose motion dynamics was recorded.

Fig. 2 illustrates the linear acceleration data recorded as well as the results obtained in the road pavement condition assessment for two selected road sections. Section 1 is a road fragment where the surface course of the pavement is on the good condition, while section 2 is a road fragment in poor technical condition.

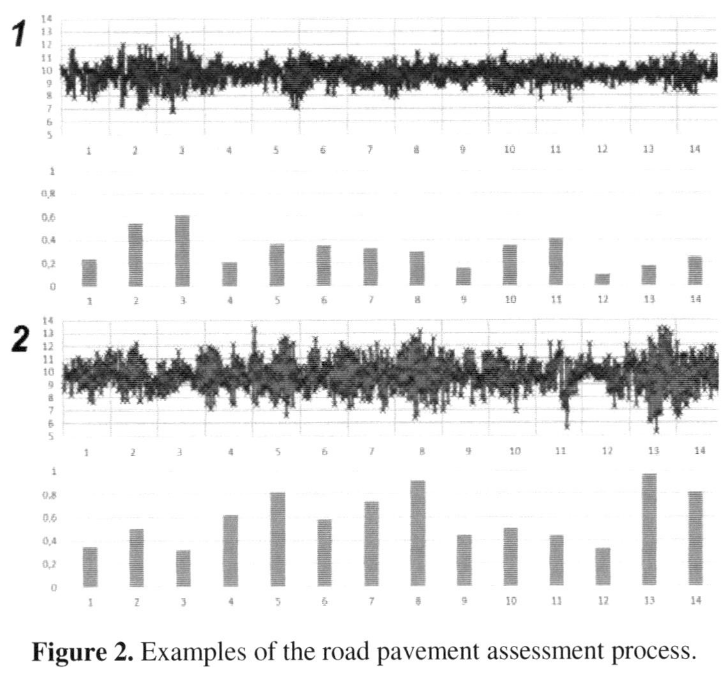

Figure 2. Examples of the road pavement assessment process.

3. Tests of the measurement method proposed

The implemented RCT solution was verified in terms of repeatability of the road pavement assessment indices being established by two kinds of analysis. One of them required tests to be conducted in a pre-set road network fragment by repeated transfers over road parts with a vehicle of specific category and at pre-set constant running speeds. The vehicle used for testing purposes was a passenger car with the RCT mobile data recording tool on board. All measurements were performed by one driver, which was assumed to ensure consistent driving style. Following the measurements, the recorded data of the motion dynamics description were uploaded to RCT server. By extracting road pavement condition indices of individual road parts from the database, it was possible to analyse the repeatability of verified results that described the road infrastructure. Table 1 provides mean values of the pavement condition assessment indices for the selected 10 road parts against specific ranges of the test vehicle driving speed. Both the mean values and the standard deviations have been calculated for all pavement condition assessment indices of respective road parts. The analysis conducted under the study has been illustrated in Fig. 3, where standard deviations of the pavement condition assessment indices have been additional marked for individual ranges of speed at which measurements were taken.

Table 1. Verification of repeatability of the δ index depending on the vehicle running speed.

Road parts	Mean values of the pavement condition assessment index δ for constant speeds			Mean values of δ index	Standard deviations of δ index
	~20 km/h	~45 km/h	~70 km/h		
1	0.405	0.358	0.399	0.387	0.040
2	0.381	0.437	0.381	0.400	0.062
3	0.410	0.422	0.441	0.424	0.053
4	0.433	0.450	0.514	0.466	0.058
5	0.447	0.492	0.415	0.451	0.057
6	0.481	0.477	0.499	0.486	0.042
7	0.413	0.366	0.449	0.410	0.046
8	0.482	0.468	0.546	0.499	0.071
9	0.450	0.445	0.502	0.466	0.060
10	0.472	0.462	0.407	0.447	0.057

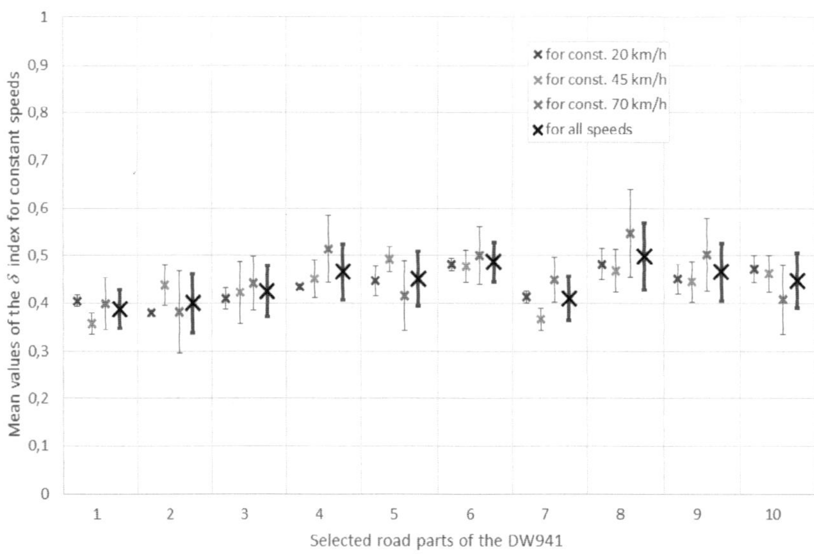

Figure 3. Graphical analysis of the obtained results for constant speeds.

The measurements repeatedly conducted in a 2.12 km long section of the DW941 road, constituting an analysis of 212 road parts, clearly imply that the standard deviation of the road pavement condition assessment indices increased as the test vehicle increased its speed. The highest standard deviation value for the indices analysed came to 0.071. Having assumed a double standard deviation and normal distribution of the values obtained for the road pavement condition assessment indices, one may presume that roughly 95% of the indices being established will display repeatability of values in the range of ± 0.142.

The second type of the analysis performed consisted in isolating from the database the road pavement condition assessment indices of the given route where measurements were performed by different drivers using vehicles of different categories. The analysis covered transfers made by a passenger car SO, a delivery van SD and heavy goods vehicle SCbP. With regard to the specific design of the vehicle, and particularly its suspension system, an off-road car ST was also isolated from the passenger car category [22]. On account of the driving styles and the recorded running speeds, individual sections of the S52 expressway were chosen for purposes of the analysis, since that is where the smallest speed differences were observed.

Table 2 provides mean values of the pavement condition assessment indices for the selected 10 road parts, as they were established based on transfers made by vehicles of different categories. Fig. 4 provides graphical representation of this analysis.

Table 2. Verification of repeatability of the δ index depending on the vehicle category.

Road parts	Mean values of the pavement condition assessment index δ for vehicles of different categories			
	SO	SD	SCbP	ST
1	0.177	0.133	0.223	0.171
2	0.176	0.179	0.135	0.165
3	0.354	0.393	0.368	0.362
4	0.423	0.428	0.439	0.409
5	0.445	0.493	0.446	0.403
6	0.357	0.354	0.311	0.337
7	0.793	0.770	0.754	0.778
8	0.759	0.757	0.781	0.712
9	0.692	0.739	0.648	0.718
10	0.780	0.816	0.751	0.732

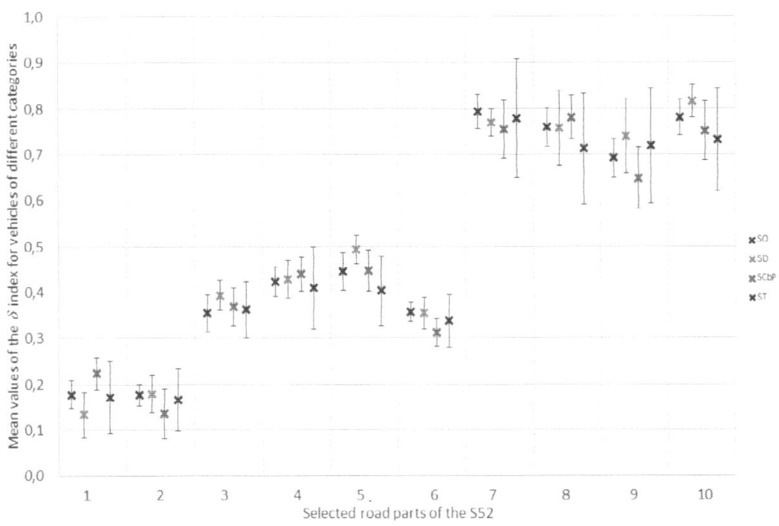

Figure 4. Graphical analysis of the obtained results for vehicles of different categories.

Based on the analysis of the road pavement condition assessment indices established for the S52 expressway section of 14.95 km comprising 1495 road parts, one could observe an increase in the level of standard deviation for the road pavement condition assessment indices established in measurements performed by means of a heavy goods vehicle. For higher values of the road pavement condition assessment index (i.e. road of poor technical condition), the standard deviation observed is higher by 0.087 on the average. For the road section of good technical condition [21], where values of the road pavement condition assessment indices ranged between 0.1 and 0.5, standard deviation values were insignificantly higher, i.e. by 0.043 on the average. In the delivery van and off-road car categories, no considerable change to the standard deviation value of the road pavement condition assessment index was observed.

4. Conclusions

The solution discussed in the paper makes it possible to assess technical condition of road pavement based on the description of vehicle motion dynamics. What proves to be of key importance in this respect is the linear accelerations recorded in a three-dimensional system, associated with the vehicle driving speed at the time of data readout from the MEMS module. Having analysed these data, one can determine specific relationships enabling identification of the current road condition, and once they are correlated with the GPS location – the indices thus obtained can be entered into records.

As discussed in this paper, the verification of repeatability of the assessment indices established has confirmed that analysis of linear acceleration signals can be successfully used to describe the condition of road pavement. The study in question has revealed low correlation between the values of the assessment indices and the speed of test vehicles used to conduct measurements as well as their category. An exception to the foregoing regularity is the category of heavy goods vehicles, which results from the specific design of the shock absorbing system of the driver's cab. Further research is assumed to eliminate this problem.

From the perspective of road infrastructure administration bodies, the solution proposed in the paper may facilitate decision making in the scope of transport system management and road infrastructure maintenance. It provides overall information on the preliminary assessment of technical condition of road pavement in the part of the road network subject to measurements, and enables identification of points in need of immediate or comprehensive repairs.

Bibliography

[1] Technical specification 2017 Pavement Monitoring by the ARGUS® Survey System by TÜV Rheinland Schniering GmbH. Available schniering.com
[2] McGhee K 2004 *TRANS RES B* **334**
[3] Varadharajan S et al 2014 Applications of Computer Vision - IEEE Winter Conference
[4] Chambon S and Moliard J 2011 *International Journal of Geophysics*
[5] Oliveira H and Correia P L 2013 *IEEE T INTELL TRANSP* **14** 1 pp 155-168.
[6] Rogers M and Enright B 2016 *Highway engineering*
[7] Forczmański P and Małecki K 2013 COMM COM INF SC **395**
[8] Wasilewska M, Gardziejczyk W and Gierasimiuk P 2016 *TRANSP RES PROC* **14**
[9] El Khoury j et al 2014 *INT J PAVEMENT ENG* **15** 4
[10] Hall J. et al 2009 NCHRP Report 01-43
[11] Baran E 1988 Australian Road Research Board (ARRB) Conference **14** 6
[12] Sybilski D, Mechowski T, Harasim P 2007 *Roads and Bridges-Drogi i Mosty* **6** 2
[13] Zofka A et al 2014 *TRANS RES B* 2457
[14] Graczyk M et al 2016 *TRANSP RES PROC*, **14**
[15] Du Y et al 2016 *INT J DISTRIB SENS N* **12** 8413146.

[16] Ghose P et al 2012 Pervasive Computing and Communications Workshops - IEEE International Conference
[17] Agostinacchio M, Ciampa D and Olita S 2014 *EUR TRANSP RES REV* **6** 3
[18] Zhao B and Nagayama T 2017 *PROCEDIA ENGINEER* **188** pp 9-16
[19] Douangphachanh V and Oneyama H 2014 *EURASIP Journal on Wireless Communications and Networking* **1** 114
[20] Staniek M and Sierpiński G 2016 *DOI: 10.20858/sjsutst.2016.92.14*
[21] Regulation of General Director of National Roads and Motorways on the diagnosis of the road pavement condition and its elements (in Polish) **34**
[22] Wolshon B and Pande A 2016 *Traffic engineering handbook*

ACKNOWLEDGEMENTS
The present research has been financed from the means of the National Centre for Research and Development as a part of the international project within the scope of ERA-NET Transport III Programme "Smart platform to integrate different freight transport means, manage and foster first and last mile in supply chains (S-MILE)".

era-net TRANSPORT The National Centre for Research and Development S mile

Identification of deep-rooted transverse cracks using Ground Penetrating Radar

Jacek Sudyka[1], Lech Krysiński[1], Adam Zofka[1], Tomasz Mechowski[1], Przemysław Harasim[1]

[1] Road and Bridge Research Institute, Warsaw

E-mail: jsudyka@ibdim.edu.pl

Executive summary. Transverse cracking is a surface distress resulting from various mechanisms appearing in the pavement and their occurrence proves improper action of the structure, which, in a long run will be subject to accelerated degradation. The proper evaluation of the progress of cracking is certainly helpful in finding out the origin of this type of damage; therefore, also in the correct selection of the repair technology.

The use of GPR (Ground Penetrating Radar) as an auxiliary tool in assessing the structural condition of the pavement in terms of identification of transverse cracking is not a new practice. However, it is still a complicated task due to many external factors which may distort the result or even prevent analysis of data, especially in terms of interpretation of the obtained radar images.

This article presents the case study for the practical use of the method of GPR inspection of the transverse cracks structure to analyse the process of restoring transverse cracks in bituminous pavement. This procedure is based on the earlier catalogue of the GPR manifestations of the elements of deep cracking structure and requires precise synchronization of the GPR measurement with cracks observed on the pavement. Explicit manifestations are only visible in expanded cracks, where the zone of material changes in the vicinity of the cracking surface has a width of at least one decimetre. Such defects usually result from the long-lasting operating process, and their presence in the deeper layers of the structure is an evidence of breaking continuity of mechanical consistency of the foundation under the cracking line.

Keywords: road pavements, cracking, GPR radar, identification of cracks

1. Method of GPR evaluation of intensity and advancement of the cracking process

The GPR diagnostics of the pavement cracks is a discipline arousing great interest due to practical significance resulting from identification of characteristics of the process of formation and development of cracks. The possibilities of the GPR (Ground Penetrating Radar) method apply to identification of the detailed structure of cracks i.e. their geometrical composition together with certain material characteristics of that structure, which are demonstrated in the electromagnetic properties [1, 2, 3].

One of the practical benefits of the development of the methodology for the GPR identification of cracks is the possibility of the evaluation of the rooting depth of the cracking structure and the extent of their development. These characteristics are treated in the analysis in a semi-quantitative and descriptive manner as the characteristics of the degradation process. This procedure aims, inter alia, at finding the possible evidence for the presence of deep-rooted cracks understood as weakening of the pavement and its subgrade in the crack vicinity, resulting primarily from long-term mechanical activity of the cracks, though perhaps inspired by earlier local features of the subgrade. The negligible visibility of the underdeveloped initial cracks in the GPR method is a feature so favorable in the practical assessment that the acquisition of visible traits of the cracking structure proves that the defect is very developed and is related to its long-term, intense mechanical activity.

The GPR traits of the cracking structure are intensively tested in terms of phenomenology [4] as well as in terms of laboratory and numerical [3, 5] modelling. The practical interpretation of echograms may be in particular assisted with the catalogue of typical elementary traits and the notes concerning the diagnostic significance of synchronous occurrence of those traits at the various levels or in approximate parallel measuring profiles [4]. The key examples of elementary traits include *diffraction hyperbola*, the *Chi pattern* together with their reduced variants, such as nearly *point form* (a spot of local scattered emission). Among the particularly spectacular traits is the *Christmas tree form* composed of several hyperbolas or chi patterns occurring one above another. The catalogue of traits also includes the significant considerations ad observations concerning the typical structures of the medium generating those traits. The GPR image is not a simple representation of the geometrical structure, because its nature is diffractional, which is particularly visible in the case of edges, linear or nearly-point elements, or other step changes of the structure. What's more, this image indirectly expresses the contrasts of certain electromagnetic properties of the material, the interpretation of which requires knowledge in typical properties of the medium encountered in the specific test area (for example in the area of pavement testing). The interpretation of echograms may be, in particular, assisted with the list of frequently encountered architectural arrangement of the cracks' surroundings generated in relation to the activity of the cracks, as the features suggesting local subsidence of the structure (collapses, breaks). The list is accompanied by vast reflections on the typical changes in materials (mostly in the main crack crevice and its close vicinity) and on the phenomena (processes) responsible for those changes [6].

In numerous cases that method may be used in a completely non-invasive variant for testing of a large number of cracks in the selected section subjected to an intense cracking process. Testing requires a thorough surveying of the cracks visible on the surface, as well as arduous synchronization of their visual position with the GPR image. The presence of numerous traits of cracking visible on the echogram allows to reliably interpret them, for example as the hallmarks of long-term mechanical activity leading to local material changes of the medium. The assessment may include the depth of occurrence of the elementary traits, their intensity (mostly if these are clearly typical traits) and the manifestations of the horizontal stretch of the damage zone. It should be noted whether there are material changes in the base course, which are strictly correlated with the cracking position. The echogram is analysed also in terms of presence of the evident traits of cracking which are not so evident on the surface. The summary of that procedure includes the quantitative characterization of the tested group of cracks to answer the question what is the share of cracks of a very well developed structure in the population of the visible cracks, are most of them deep-rooted and whether more or less all the cases of the evident GPR traits of cracking are also already visible of the surface. Such a method of characterization of the population of cracks allows for formulating the realistic view on certain characteristics of the course of the cracking process, which are significant from the technological point of view.

2. Methodology of measurements and preparation of data for interpretation

The below-described project uses the GPR method (a GSSI device, an air-coupled antenna with the central frequency of 2 GHz) and two visual observation methods: surveying of cracks with manual measurements of cracking positions (measuring wheel) and marking of that position on the roadway edge, as well as continuous photography of the surface using the SPDE measuring system (Fig. 1). The perfect synchronization of all three measurements is a key element of the procedure presented in this paper and these works usually require major efforts. The procedure leads to removal of large errors of the single observations and the often encountered faults of the individual distance defining methods (such as skips or changes in distance calculation trend). The essence of these effects is the relative identification of cracks, i.e. making sure that in the specific case all those observations apply to exactly the same crack. In practice, this means that the precision of the mutual synchronization of distance from the various methods is situated at the level of single decimetres.

Figure 1. Multifunctional SPDE measuring system for measuring surface characteristics of the pavement and photographic recording of damages.

Marking of positions of cracks on the roadway edge during preliminary visual surveying is a very important part of the synchronization procedure. It allows for accurate location of manifestations of specific cracks on an echogram (initial identification) through an arduous correlation of the distance of GPR traits with visual observations. This procedure will allow for a precise synchronization of the measuring distance with an official chainage of the road or other working chainages, or other local methods used for defining the distance. Direct observation also facilitates the initial identification of cracks on the photographs taken with a surface analyser and further synchronization of the measuring distance of SPDE with the measuring distance of GPR.

The photographs taken during the continuous measurement provide a very good documentation of the position and the exact route of the cracking line which could not have been obtained using other methods of field measurements. It is possible or even mandatory to use the graphical transformations highlighting lineations and edges (Fig. 2D) during layout. Then these images reveal their huge diagnostic potential. A precise synchronization of photos with measurements of other diagnostic methods present new possibilities in the field of analysis of pavement structure and in particular the structure of defects.

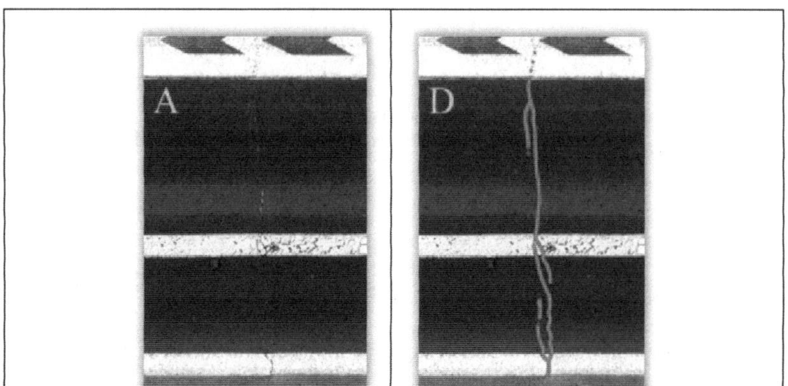

Figure 2. Identification of transverse cracking based on the photo. A) Original form of the photo, D) course of the cracking line marked on the original photo.

3. Characteristics of the population of cracks and characterization of the degradation process

Approximately 85 discontinuities in the form of transverse cracks were found in the tested road section with a length of 1,000 m About 65 of them may be characterized as intersecting the profile of GPR measurement, presenting interpretable GPR traits of the presence of cracking. For practical reasons, 56 cracks intersecting the entire width of the tested traffic lane were selected for further consideration as a group representative for characterization of the cracking process.

Among this test group, only two cracks did not have a clear GPR response, six from the other cracks provided a weak or a very weak response, while the remaining ones, i.e. 48 cases (86% of the test group) included cracks with a well developed (8 cracks) and a very well developed structure (40 cracks), providing a clear or even spectacular GPR response. Nearly all of their GPR images indicate the presence of changes in the medium related to the advanced cracking evolution process up to the depth above 50 cm (and even up to 80 cm in extreme cases), that is to the diagnostic limits of the used equipment. More than 30 cracks are accompanied by spectacular GPR manifestations (Fig. 3). In many cases (22 cracks), the changes in the deeper layers of the structure, i.e. in the sub-base (aggregate "Kr") and in the mechanically stabilised layer ("Stb") have a significant transversal size (i.e. in the profile direction), covering the strip along the crack with a width of above 1 m (rarely above two metres). At the same time the tendency to increase the width of the zone covered by changes in the structure and the material together with increase in depth is very visible.

The cracks identified in the photos were marked with a violet line and were afterwards compared with a fragment of the echogram corresponding to that section, on which spectacular and well-located manifestations of cracks inside the pavement structure are visible (Fig. 3). They have very well developed forms that are variants of the Christmas tree layout [4]. This combination shows that the GPR manifestations of the cracking structure, and more precisely their axial planes, occur synchronously with their surface manifestations, with accuracy to one decimetre. It is worth mentioning the crack marked with number 02, that line is observed on the surface during the measurement and which does not intersect the line of the measuring profile yet, but very intense internal manifestations of the crack appeared below the profile at the point where it intersects with the extension of the axial plane of the crack. This is characteristic of very developed cracks recently covered with a new layer of asphalt, when this restoration process is not yet very advanced. The cracks visible in that figure have a very developed structure. Their presence in the structure is obvious based only on echograms, without confirmation with visual observations. The intensity of manifestations and their width (stretch along the profile) increase together with the depth, so that their presence in the sub-base and deeper courses is beyond doubt. The width of the zone with structural and material changes in these deeper levels reaches values greater than one metre. The forms of scattering caused by the structure of the cracks at the levels of inter-layer borders differ considerably from the pattern of the hyperbola, and take the shape of a Chi pattern, which proves wide zones of structural change.

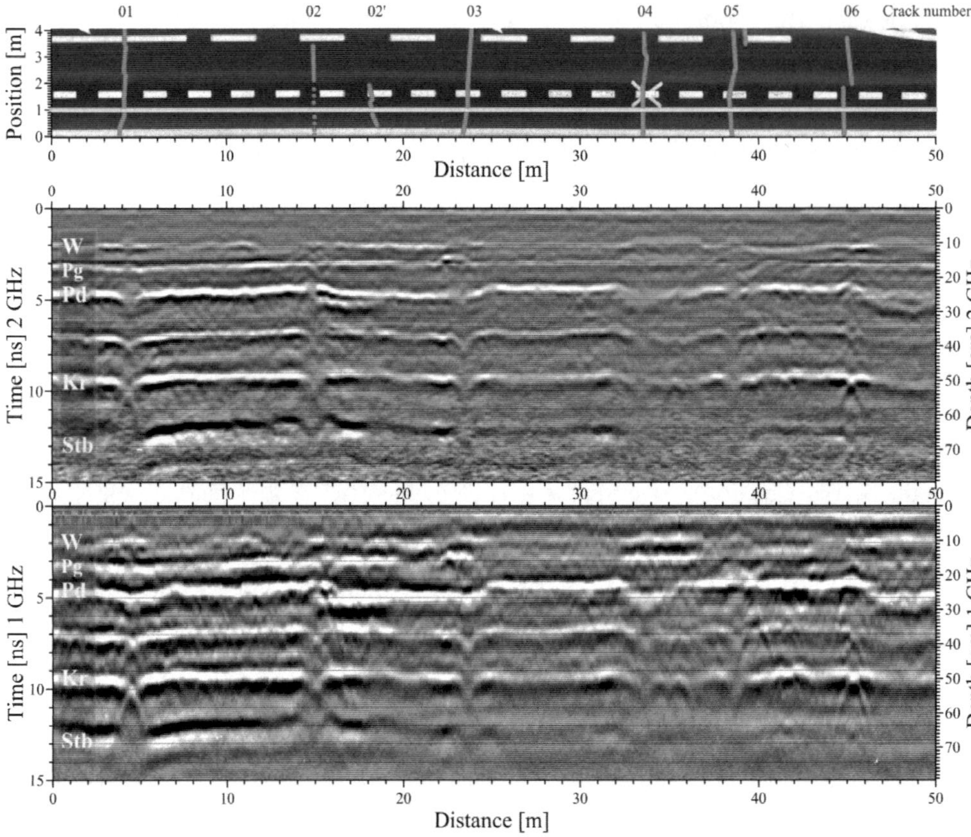

Figure 3. An example of an excellent correlation of the position of cracks visible on the surface with their very distinct and well located manifestations on a GPR image. The above figure presents the photographic map of the pavement with the cracking lines visible on the surface marked with a violet line and the route of the measuring profile marked in green. Synchronously compared echograms obtained using antennas with the central frequencies of 2 GHz and 1 GHz are presented below. The reflection levels corresponding to the lower surfaces of the structural layers are designated as W (binder course), Pg (asphalt roadbase), Pd (asphalt sub-base), Kr (sub-base), Stb (mechanically stabilised layer).

Another part of the tested pavement section (Fig. 4) may be indicated as a comparative example presenting less obvious cases requiring a more complicated discussion. The image of the layered structure should be in this case defined as heavily distorted. The base course structure is especially distorted. Here one can see numerous elementary characteristics of potential presence of cracks, but they are not arranged in vertical groups clearly suggesting the specific locations of cracks. What is more, the defects and disorders of the layered structure of the packet are not numerous (three short ranges) and do not present conclusive evidence of piercing of the packet by cracking. Three cracks found on the surface of this section are accompanied by certain GPR traits of the potential presence of cracks, providing a chance to determine the probable position, but which alone are not the basis for suggesting its probable presence. In the case of crack 11, with precise synchronization with photography it is possible to recognize which elementary traits are the manifestation of the crack and what's more they form a typical Christmas tree pattern; however, it is masked by a multitude of elementary figures in its surrounding. Crack 12(S) and the group 07, 08 and 09 constitute two examples of branched cracking systems, which suggest the early stage of development of alligator cracks. Those areas in the asphalt packet present defective interlayer connections, whose range of occurrence strictly corresponds to the area subject to cracking. In this regard particularly significant are defects in asphalt base. These defects have a form characteristic for debonding, but a chaotic scattering pattern is visible in those areas in the packet, which even leads to weakening and partial masking of structures located below, which is visible in the echogram in a form of a vertical zone called the *masking umbrella*. This irregular scattering suggests that the lower layers of the packet may be significantly mechanically degraded. An accurate view of the image of layers underneath the asphalt packet allows for determining that Chi patterns forming a Christmas tree arrangement are visible there and the axial plane of that arrangement corresponds to one of the transverse cracks (449.5 m in the case of network 12(S), while the cases of group 07, 08, 09 correspond to crack 08 of 410 m). The initial crack 10 which appeared in the vicinity of the construction joint (Tech, 419.2 m) may be regarded as not having the clear manifestations of deep-rooted cracking apart from the generally disturbed structure of the base course.

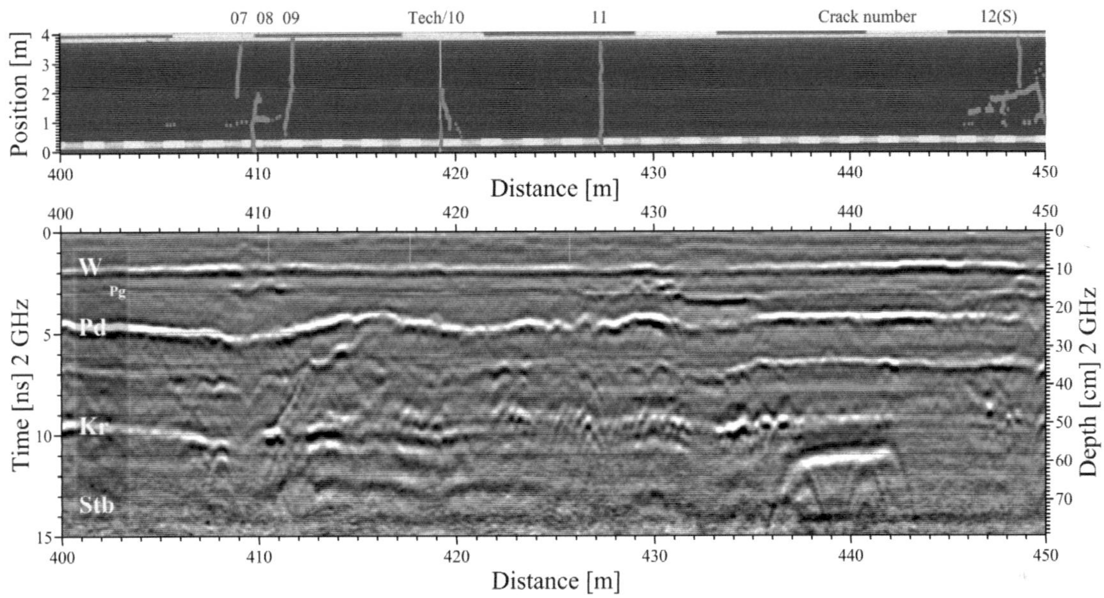

Figure 4. An example of a section with a disturbed structure suggesting occurrence of cracks, the precise location of which in the echogram is however difficult. The above figure presents the photographic map of the pavement with the cracking lines visible on the surface marked with a violet line, and the construction joint marked with a blue line. An echogram obtained using an antenna with the central frequency of 2 GHz and synchronously compared with the map is presented below. The reflection levels corresponding to the lower surfaces of the structural layers are designated as W (binder course), Pg (asphalt roadbase), Pd (asphalt sub-base), Kr (sub-base), Stb (mechanically stabilised layer).

Due to the varying degree of severity of the interpretable GPR manifestations of cracks and the varying degree of visibility of the cracks on the surface during identification of the extent and intensity of the phenomenon, it is good to describe the size of the individual groups of the process manifestations. This initial presentation make it possible to indicate the typical traits of cracking (depth of rooting, width of the zone of structural and material changes, tendency to intersect upper layers and other structural characteristics) which are the premises to draw a conclusion about the route of the process and the level of its advancement.

The echograms also show the traits of cracks not belonging to the defined test group. Those include cracks (about 20, including many with strong GPR manifestations) which, although intersecting the measurement line, have not yet intersected the entire traffic lane and have not been included in the test group. There are also less numerous manifestations of the possible cracks — generally weaker and

IOP Conf. Series: Materials Science and Engineering **356** (2018) 012022 doi:10.1088/1757-899X/356/1/012022

without the apparent deep rooting, where it is difficult to assign to them the clear traits of surface cracking or where inconclusive initial characteristics occur on the surface in those locations. The process of piercing of the new overlay by the underneath, old yet active weak zones, is by no means complete, but it may be assumed that the majority of the most developed among those structures is already visible on the surface.

The described deep traits are characteristic for the very well developed cracks, which are accompanied by material changes of a width of 1 dm (or more) in the vicinity of the cracking gap and much more extensive structural changes (to the more than a dozen of decimetres in width) visible in the disturbed system of layers, mostly the layers of the sub-base. The cracks intersect into the whole structure; in some cases suggesting the presence of material changes also under the structure. In many cases, the form of deformation of the layered system points to the loss of the bearing capacity of the structure under the cracking.

4. Summary

A common use of photographs of the pavement taken using the surface analyser and the GPR inspection of its structure is a combination indeed providing new possibilities of structural diagnostics of cracking. Great potential of this method is revealed during the synchronous comparison of photos with echograms at the precision of synchronization at the level of single decimetres. The primary benefit of this method is the possibility of a very precise and reliable assignment (identification) of the corresponding transverse cracks visible on the surface to the GPR characteristics of cracks. The second benefit of this combination is the ability to develop a structural discussion in situations more complicated than a transverse crack (e.g. alligator cracks), when a single measuring profile does not provide sufficient insight into the complexity of the structure. Particularly convenient is the analysis of the relationship between the cracks not intersecting the measuring profile on the surface and the deep manifestations of cracking, which may appear under the profile at the extension of the cracking plane.

Moreover, the presented case study shows practical possibilities of the presented method used for characterization of the pavement cracking process. This method allows for inspection of the structure of cracks and assessment of the depth of their rooting, as well as for the assessment of the process advancement. This insight is possible in the case of very well developed cracks.

Bibliography

[1] M. Solla, S. Lagüela, H. González-Jorge, and P. Arias, "Approach to identify cracking in asphalt pavement using GPR and infrared thermographic methods: Preliminary findings," *NDT E Int.*, vol. 62, pp. 55–65, 2014.

[2] J. Pedret Rodés, V. Pérez-Gracia, and A. Martínez-Reguero, "Evaluation of the GPR frequency spectra in asphalt pavement assessment," *Constr. Build. Mater.*, vol. 96, pp. 181–188, 2015.

[3] N. Diamanti and D. Redman, "Field observations and numerical models of GPR response from vertical pavement cracks," *J. Appl. Geophys.*, vol. 81, pp. 106–116, 2012.

[4] L. Krysiński and J. Sudyka, "GPR abilities in investigation of the pavement transversal cracks," *J. Appl. Geophys.*, vol. 97, pp. 27–36, 2013.

[5] F. M. Fernandes and J. C. Pais, "Laboratory observation of cracks in road pavements with GPR," *Constr. Build. Mater.*, vol. 154, pp. 1130–1138, 2017.

[6] L. Krysiński and J. Sudyka, "Identyfikacja pęknięć nawierzchni z wykorzystaniem techniki radarowej," *Mag. Autostrady*, vol. 7, pp. 44–50, 2012.

IOP Conf. Series: Materials Science and Engineering **356** (2018) 012023 doi:10.1088/1757-899X/356/1/012023

The application of non-destructive methods in the diagnostics of the approach pavement at the bridges

M Miskiewicz, J Lachowicz, P Tysiac, P Jaskula and K Wilde

Faculty of Civil and Environmental Engineering, Gdansk University of Technology, Gdansk, Poland

Email: mikolaj.miskiewicz@pg.edu.pl

Abstract. The article presents the possibility of using non-destructive methods of road pavement diagnostics as an alternative to traditional means to assess the reasons for premature cracks adjacent to bridge objects. Two scanning methods were used: laser scanning to measure geometric surface deformation and ground penetrating radar (GPR) inspection to assess the road pavement condition. With the use of a laser scanner, an effective tool for road deformation assessment several approach pavement surfaces next to the bridges were scanned. As the result, a point cloud was obtained including spatial information about the pavement deformation. The data accuracy was about 3 mm, the deformations were presented in the form of deviation maps between the reference surface and the actual surface. Moreover characteristic pavement surface cross-sections were presented. The in situ measurements of the GPR method were performed and analysed in order to detect non-homogeneity in the density of structural layers of the pavement. Due to the analysis of the permittivity of individual layers, it was possible to detect non-homogeneity areas. The performed GPR measurements were verified by standard invasive tests carried out by drilling boreholes and taking cores from the pavement and testing the compaction and air voids content in asphalt layers.. As a result of the measurements made by both methods significant differences in layer compacting factor values were diagnosed. The factor was much smaller in the area directly next to the bridgehead and much larger in the zone located a few meters away. The research showed the occurrence of both design and erection errors as well as those related to the maintenance of engineering structures.

1. Introduction

Nowadays, non-destructive diagnostic methods are used increasingly in structural health monitoring and technical state evaluation. The possibilities of using non-invasive methods were described in [1]-[10]. The article presents road pavement [11] diagnostics of transverse cracks and surface deformation [12] adjacent to the bridge joint area of a bridge structure. The GPR scan [10] and the laser scanning technology [13]-[19] the chosen methods, the results were compared to the traditional ones performed by core drilling [20],[21]. Non-destructive diagnostic methods are commonly used [22],[23], however the novel approach is to combine them and use both. This allows a broad spectrum of inferences about the state of the analysed structure, comparable to destructive methods. Therefore this solution is able to more widely implement.

Problem description

The structure of the approach pavement has been designed and made as susceptible. Asphalt layers are laid on the base course of the unbound mixture and the subbase of the improved subgrade as a cement treated soil. In the binder course and base course, high modulus asphalt concrete (HMAC) was used. The pavement was designed for heavy traffic load. Details regarding the pavement structure are shown in Fig. 1b.

During the construction of the road nine bridge structures were built. There were two solutions for the carrying system joints. One-module leak-proof expansion joints were used, but airtight bitumen expansion joints were used as well. Two types of transition plates and their backs were designed depending on the type of a load- carrying structure applied on the back wall of the abutment. The length of the transition plates was 5 m, the thickness of them both was equal to 30 cm. The differences occurred in the shape of the transition plate fragment at the bridge joint.

In a relatively short time after the road started operating transverse cracks appeared during the first winter at the majority of expansion joints of modular bridge structures. The cracks occurred 30 cm from the steel profile of the expansion device, at the abutment of the transition plate at the bridgehead. Moreover, the cracks did not appear with bituminous expansion joints. At that time, damage was secured by flooding with a flood mass. It was a provisional only solution. No major actions were taken to remove the cracks from the moment they arose during the measurements based on documentation analysis. During the measurements it was noticed that damage was often accompanied by a visible lack of pavement surface pieces just before the cracking (Fig. 1a). In some circumstances only vertical deformation of the pavement surface appeared without cracks. Lowering the pavement surface usually occurs in the vehicle tracks, also in the areas not subjected to direct impact of the wheels, e.g. in the area of emergency lane or at the inner edges of the wings of abutments.

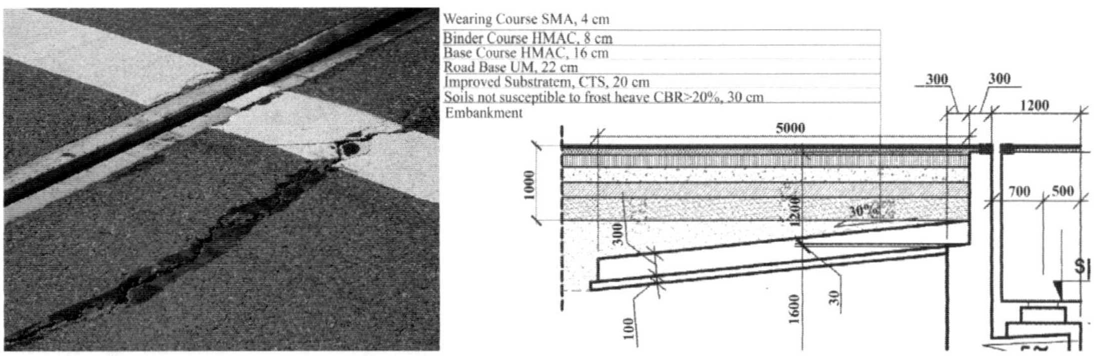

Figure 1. Analyzed detail, a) approach pavement surface condition b) pavement structure details.

2. Non-destructive testing

2.1. Ground penetrating radar method

The road surface research was carried out with the use of the IDS GPR set, including: a DAD 400 kHz control unit, a portable computer and two antennas including two transmitter-receiver pairs with 400 MHz and 900 MHz frequency. The entire system is powered by a battery. Moreover, the choice of the antenna frequency acts strongly on measurements results. The higher frequency applied, the smaller penetration depth of the electromagnetic wave is obtained, but the resolution of the radargram is higher. During the measurement mission, two antennas were used, but the paper presents the 400 MHz frequency case only, due to a greater penetration depth.

In the region of the transition bridge plate transverse and longitudinal profiles were conducted on the most-loaded lane and on the emergency lane adjacent to the bridge joint of the bridge structure. A general view with the arrangement of scanning routes on the bridge object is shown in Fig. 2.

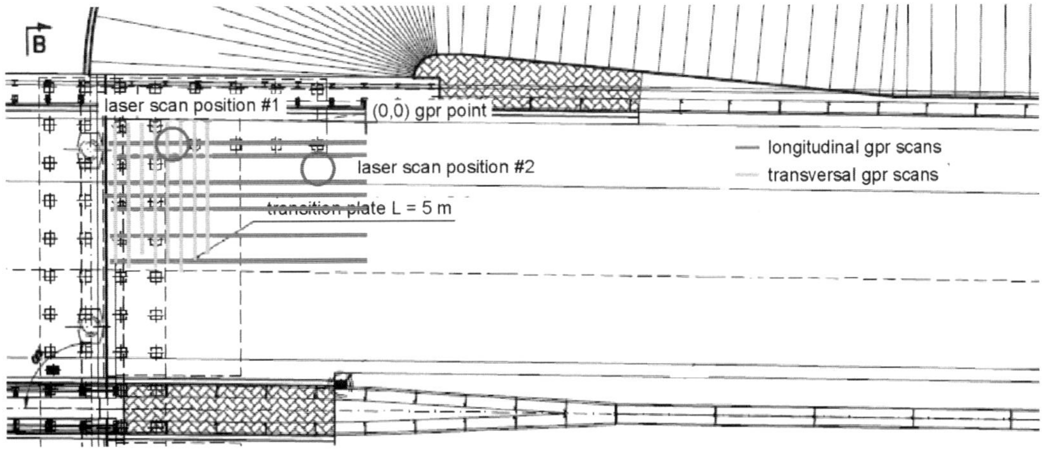

Figure 2. A general view with the arrangement of scanning routes on a bridge object and positions of the Terrestrial Laser Scanner.

Longitudinal GPR maps of the most loaded lane and the emergency lane are shown in Fig. 3 and Fig. 4a. It is worth mentioning that below each map a graph of estimated effective values of permittivity for each detected pavement structure layer is presented. Based on the thickness of successive layers and reflection times of the electromagnetic wave information, the proprietary GPR data processing algorithm estimated changes of this electrical parameter along the profile.

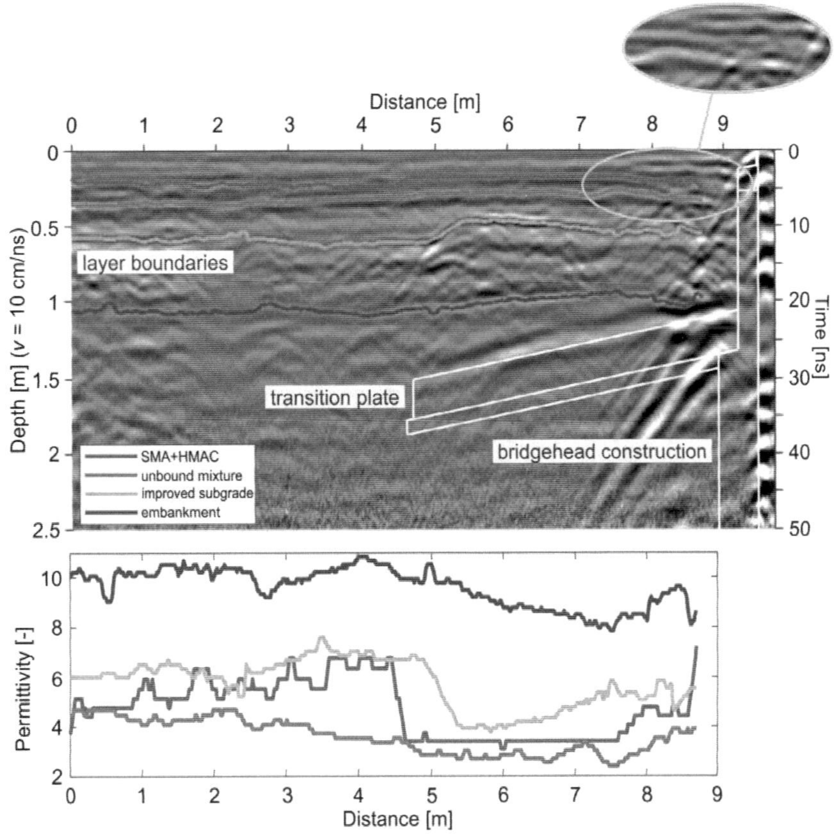

Figure 3. The slow lane (most- loaded lane) longitudinal GPR map.

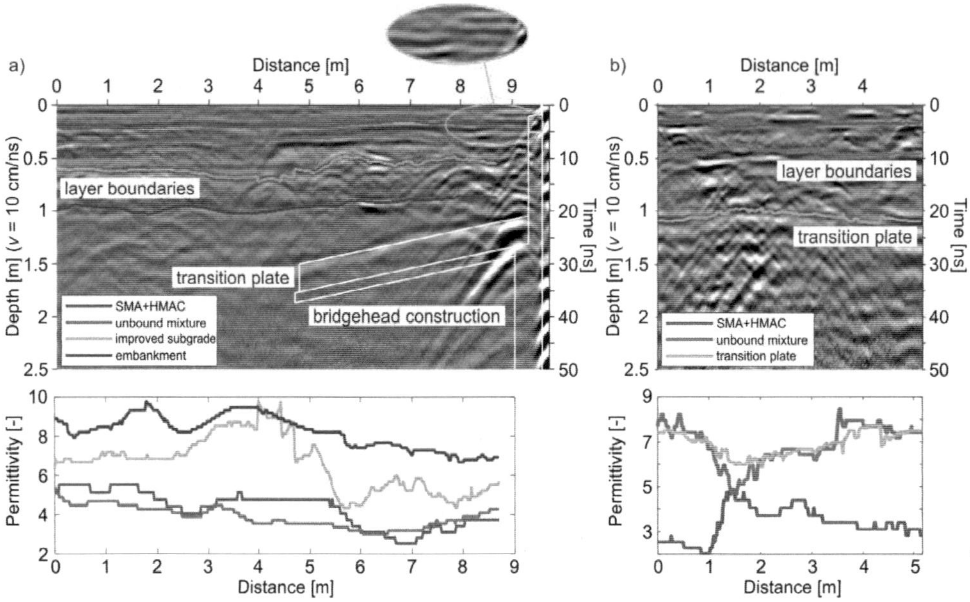

Figure 4. GPR maps a) The emergency lane longitudinal GPR map b) Transversal GPR map of approach pavement at bridge.

In addition, Table 1 lists values of the permittivity depending on the distance from the bridge joint. This parameter could be varied for one material depending on many factors including moisture and density. In the case of the analysed object it is clearly visible that the electrical permittivity changes significantly in the vicinity of the bridge joint over the transition plate, i.e. up to 4 m from the bridge joint. This may indicate areas of increased humidity or zones of inhomogeneous compaction in particular construction layers (with increased air capacity). In Table 1 these dielectric constant values are marked in red, they differ by more than 20% from the average value of a given layer.

Table 1. Effective value of the relative permittivity of the analyzed layers.

Lane	Layer	Distance from the bridge joint [m]						
		1	2	3	4	5	6	7
Emergency	1	3.73	3.23	2.82	4.34	4.77	4.92	4.06
	2	4.28	3.54	3.19	3.36	3.54	4.09	3.90
	3	5.53	4.94	5.69	4.71	8.38	8.74	7.18
	4	6.88	7.18	7.67	8.23	8.78	9.49	8.23
The slow lane (most-loaded)	1	4.69	3.41	3.41	3.41	6.69	5.53	5.69
	2	3.90	2.56	2.71	2.80	3.36	3.72	4.17
	3	4.74	5.19	4.25	3.98	6.69	7.15	5.83
	4	9.57	8.07	8.69	9.60	10.47	10.19	10.22

Legenda:
Layer 1 – Asphalt layers SMA +HMAC,
Layer 2 – Road base from the unbound mixture,
Layer 3 – Improved subgrade made of cement-treated soil,
Layer 4 – The top layer of the embankment.

Anomalies near the bridge joint were observed, i.e.: change of the reflected signal phase, increased reflected signal power and multiple longitudinal reflections (Fig. 3 and Fig. 4a). These anomalies may indicate structural disturbance, increased humidity or air voids. In order to interpret the observed anomalies correctly, numerical calculations of electromagnetic wave propagation were carried out by means of the finite-difference time-domain method (FDTD). Transverse maps with permittivity diagrams are shown in Fig. 4b. It can be noticed that differences in the dielectric constant occur across the pavement, possibly caused by the low traffic load of the emergency lane.

Diagnostic tests using the GPR method showed changes in the value of permittivity of individual structural layers of the pavement in a zone around the bridge joint. Detected changes in the dielectric constant in the area of four meters from the bridge joint trigger increased porosity or humidity of the tested medium. Moreover, in case of the elevated humidity, the value of permittivity is higher than the average value due to the value of water permittivity (ε_r = 81). However, in the case of increased porosity of the material, the permittivity value is lower than the average due to the value of air permittivity (ε_r = 1). In conclusion, the occurrence of inhomogeneous compaction zones in the expansion zone is considered highly probable.

2.2. Laser scanning

The surface of the pavement was measured with use of the Leica P30 laser scanner. The device is capable to collect information on a scanned object at a speed of 1 million points per second with the accuracy of 2 mm at 50 meters. A huge advantage of the scanner is the noise reduction of points, at the level of 0.4 mm.

In order to process the acquired data, the ICP algorithm for scan positions alignment was used firstly. The next steps were surface extraction and noise reduction from the scan. Based on the acquired data, the most important step was the plane fitting because the reference surface of the experiment was the approximate one. After choosing points, belonging to the surface, the plane was estimated using a least- square method.

The accuracy of the point cloud resulting from scanning is equal to the approximate value of 2 mm. This refers to the accuracy of the measured distance by the scanner (1 mm by 10 m), the noise level of the point cloud (0.4 mm by 10 m) and the position alignment, where the standard deviation did not exceed 1 mm. As a result, the isometric map and cross- sections marked in Figure 5 were created.

Based on a number of the selected bridge objects, the measurements were performed on the slow lane (most-loaded lane) and on the emergency lane near the bridge joint of the bridge structure. A general view of the location of the scanner positions on the object is shown in Fig. 2. The isometric map of the scanned pavement is shown in Fig. 5. It is clearly visible that the right, more loaded lane is lowered by about 1 cm related to the reference surface. In turn, the edge of the left lane is about 1 cm above the reference position. In the immediate vicinity of the bridge joint, on the right lane zone is lowered by about 0.5 cm- 1.5 cm in relation to the reference position and by 2-3 mm in the left lane.

In addition, Fig. 6 shows cross-sections of the surface: across the carriageway in close neighborhood to the bridge joint (A-A), across the road in a section away from the bridge joint by 10m (B-B), along the right edge of the right lane (C-C). As a reference line, the points on expansion joints were taken into account in the case of cross-sections and two points at the edges of the study in the case of longitudinal sections (one on a bridge joint, the other located 10 metres away).

The cross-sections indicate a significant reduction in the right lane of the road in relation to the reference surface. In the lane adjacent to the bridge joint, the measured distance from the reference plane to the real state is on the approximately equal of 1.5 cm. On the fast lane (left lane), elevation of the pavement surface was observed in relation to the reference plane. The elevation value reaches 5 mm. In the cross-section, 10 meters away from the bridge joint, an increase in the area at the edge of the left lane was observed (1.5 cm), without noticeable changes on the right lane. In the longitudinal cross-section case, in the area of a bridge joint lowering the area was observed at the level of 1.5 cm. At a distance of about 10 m from the bridge joint the surface height oscillated near the reference values.

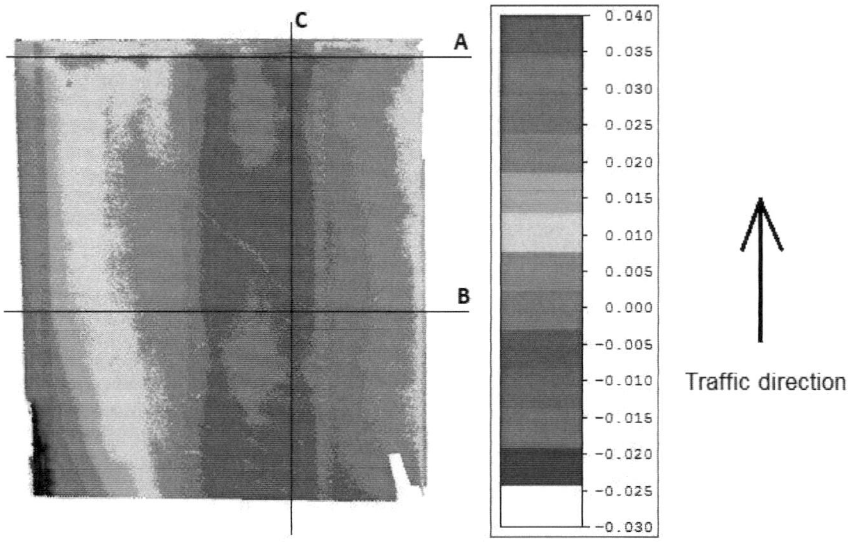

Figure 5. Isometric map of the scanned pavement surface with marked cross-sections, on the right - the distance scale between the reference surface and the real one.

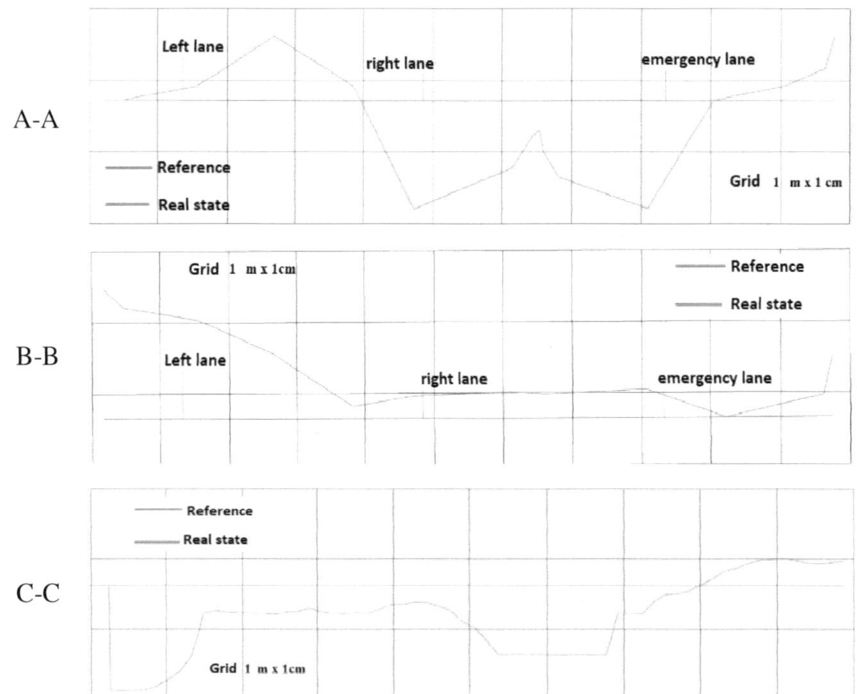

Figure 6. Cross sections of the scanned pavement (scale of 1m along the section x 1 cm after height).

The measurements show that in the close neighborhood of the bridge joint on the slow lane (most-loaded) lane there are significant deformations of the pavement surface. The maximum deformation values of the pavement in comparison to the theoretical reference level were about 1.7 cm. The recorded deformation values of the pavement indicate an uneven settlement of the surface at the entrance to the object and a resulting threshold on the edge of the wall of the outboard bridgehead.

3. Reference invasive test

In order to verify the observed zones of inhomogeneous compaction from non-destructive methods, destructive tests were performed (Fig. 7). The cores were used to determine the air void content in the asphalt layers and to determine the range of crack propagation. In addition, information on cores made it possible to determine the thickness of the each layers.

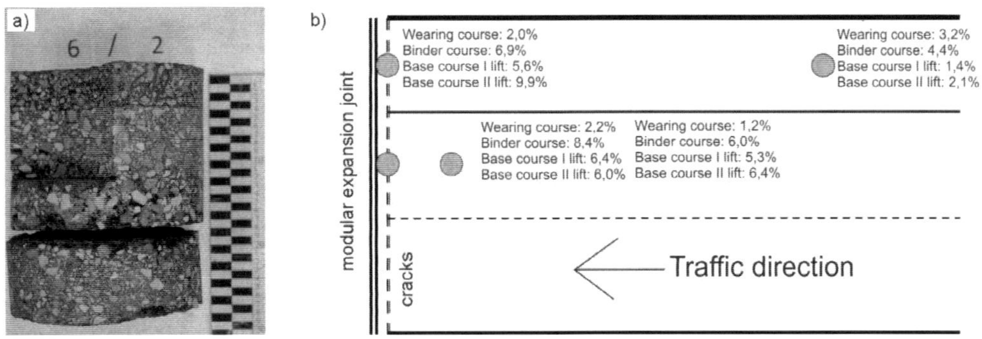

Figure 7. Invasive test results a) core picture b) results of the air voids content in the asphalt layer.

The content of the air voids in the binder course HMAC and two lift layers of the base course from HMAC in samples taken directly at joints, in the vicinity of cracks, was about 2-3 times greater than in the place located about 15 meters from the bridge joint. This would indicate a worse compaction of these layers next to the bridge, compared to the compaction on the regular section. This is confirmed by the appearance of the samples themselves. The samples taken directly at the bridge joint were heterogeneous and porous in some fragments. Tests performed in the laboratory using cores, confirmed the conclusions of non- destructive testing.

4. Conclusions

The paper presents diagnostic methods used to assess the condition of the approach pavement surface in the area of transverse cracks at the bridge joint of selected bridge objects. The measurements were performed using the GPR method and the laser scanning technology. In order to confirm the diagnosis, destructive tests were carried out.

Based on the measurements of the GPR method, as longitudinal reflections observed on radargrams, it was possible to determine relative values of electrical permeability for individual structural layers. The variations of this parameter along the scan length allowed to state that there may be zones of inhomogeneous compaction on particular layers. In addition, it can be concluded that differences in permeability values were greater for the slow lane (most-loaded lane). The collected results obtained by the terrestrial laser scanner results made a conclusion that in the close neighborhood of the bridge joint on the slow lane (most-loaded lane) significant deformations of pavement surface occur, showing the maximum pavement deformations in comparison to the theoretical reference level about 1.7 cm. The cores obtained during the destructive tests allowed to state that the air void content in the samples taken at bridge joint is much larger than in the samples taken 15 m from the bridge joint. The analysis carried out confirms that the GPR method and laser scanning technique can be successfully applied analysis to diagnose the condition of road surfaces.

This paper focuses only on presentation of the final results of research. For brevity the details of numerical simulations are omitted in this paper. These will be presented in the forthcoming papers.

References

[1] Lachowicz J and Rucka M 2016 Experimental and Numerical Investigations for GPR Evaluation of Reinforced Concrete Footbridge *16th International Conference of Ground Penetrationg Radar (GPR)* (Hong Kong) pp 1–6

[2] Rucka M et al 2016 GPR investigation of the strengthening system of a historic masonry tower *J. Appl. Geophys.* 131 94–102
[3] Owerko T et al Investigation of displacements of road bridges under test loads using radar interferometry - case study *Sixth International IABMAS Conference* 181-188.
[4] Kuras P et al Advantages of radar interferometry for assessment of dynamic deformation of bridge *Sixth International IABMAS Conference* 885-891
[5] Mazurkiewicz L et al 2017 Experimental and numerical study of steel pipe with part-wall defect reinforced with fibre glass sleeve *Int. J. of Pressure Vessels And Piping* 149 108-119
[6] Miskiewicz M et al 2016 Structural Health Monitoring of Composite Shell Footbridge for Its Design Validation *Baltic Geodetic Congress (Geomatics) IEEE* 228-233. DOI: 10.1109/BGC.Geomatics.2016.48
[7] Miskiewicz M et al 2017 Technical monitoring system for a new part of Gdańsk Deepwater Container Terminal *Pol. Maritime Res.* S1 (93) 24 149-155, DOI: 10.1515/pomr-2017-0033
[8] Miskiewicz M et al 2017 Monitoring system of the road embankment *Baltic Journal of Roads and Bridge Engineering* 12 4 218-224
[9] Chroscielewski et al 2014 Assessment of tensile forces in Sopot Forest Opera membrane by in situ measurements and iterative numerical strategy for inverse problem *Shell Structures: Theory and Applications* 3 499-502, DOI: 10.1201/b15684-125
[10] Saarenketo T and Scullion T 2000 Road evaluation with ground penetrating radar *J. Appl. Geophys.* 43 119–38
[11] Judycki J et al 2017 New polish catalogue of typical flexible and semi-rigid pavements *MATEC Web of Conferences* 122
[12] Judycki J et al 2017 Field investigation of low-temperature cracking and stiffness, moduli on selected roads with conventional and high modulus asphalt concrete *Bestinfra 2017, IOP Conference Series: Materials Science and Engineering* 236, 012002
[13] Zai D et al 2016 3D road surface extraction from mobile laser scanning point clouds, *2016 IEEE Int. Geosci. Remote Sens. Symp.* 1595–1598 DOI:10.1109/IGARSS.2016.7729407
[14] Szulwic J and Tysiąc P 2017 Searching for road deformations using mobile laser scanning *MATEC Web of Conferences 122 04004* DOI: 0.1051/matecconf/201712204004
[15] Bobkowska K et al 2017 Bus bays inventory using a terrestrial laser scanning system *MATEC Web of Conferences* 122 04001 DOI: 10105104001
[16] Janowski A et al 2016 Remote sensing and photogrammetry techniques in diagnostics of concrete structures *Comput. Concr.* 18 3 405–420
[17] Janowski A et al 2015 Airborne and mobile laser scanning in measurements of sea cliffs on The Southern Baltic *15th Multidisciplinary Scientific GeoConference SGEM 2015, Vol 2*
[18] Nagrodzka-Godycka K et al 2014 The method of analysis of damage reinforced concrete beams using terrestrial laser scanning *14th SGEM GeoConference on Informatics, Geoinformatics and Remote Sensing* 3 335-342
[19] Janowski A et al 2014 Modes of Failure Analysis in Reinforced Concrete Beam Using Laser Scanning and Synchro-Photogrammetry. How to apply optical technologies in the diagnosis of reinforced concrete elements? *Second Int. Con. on Advances in Civil, Structural and Environmental Engineering - ACSEE 2014* 16-20
[20] Judycki J et al 2015 Investigation of low-temperature cracking in newly constructed high modulus asphalt concrete base course of a motorway pavement *Journal of Road Materials and Pavement Design* 16 362-388 DOI:10.1080/14680629.2015.1029674
[21] Judycki J and Jaskula P 2010 Diagnostyka i modernizacja konstrukcji nawierzchni drogowych *Autostrady* 12/2010 24-31
[22] Graczyk M et al 2014 Zastosowanie georadaru w diagnostyce stanu konstrukcji nawierzchni i ocenie bezpieczeństwa eksploatacyjnego lotnisk *Logistyka* 3 2128–39
[23] Krysinski L and Sudyka J 2013 GPR abilities in investigation of the pavement transversal cracks *J. Appl.* Geophys. 97 27–36

IOP Conf. Series: Materials Science and Engineering **356** (2018) 012024 doi:10.1088/1757-899X/356/1/012024

Usage of digital image correlation in assessment of behavior of block element pavement structure

M Grygierek[1] B Grzesik[1] P Rokitowski[1] T Rusin[2]

[1] The Silesian University of Technology, Gliwice, Poland
[2] Dantec Dynamics GmbH, Ulm, Germany

marcin.grygierek@polsl.pl

Abstract. In diagnostics of existing road pavement structures deflection measurements have fundamental meaning, because of ability to assess present stiffness (bearing capacity) of whole layered construction. During test loading the reaction of pavement structure to applied load is measured in central point or in a few points located along a straight on a 1.5 ÷ 1.8 m distance (i.e. Falling Weight Deflectometer) in similar spacing equal to 20 ÷ 30 cm. Typical measuring techniques are productive and precise enough for most common pavement structures such as flexible, semi-rigid and rigid. It should be noted that in experimental research as well as in pavements in complex stress state, measurement techniques allowing observation of pavement deformation in 3D would have been very helpful. A great example of that type of pavements is a block element pavement structure consisting of i.e. paving blocks or stone slabs. Due to high stiffness and confined ability of cooperation of surrounding block elements, in that type of pavements fatigue life is strongly connected with displacement distribution. Unfortunately, typical deflection measurement methods forefend displacement observations and rotation of single block elements like paving blocks or slabs. Another difficult problem is to carry out unmistakable analysis of cooperation between neighboring elements. For more precise observations of displacements state of block element pavements under a wheel load a Digital Image Correlation (DIC) was used. Application of this method for assessment of behavior of stone slabs pavement under a traffic load enabled the monitoring of deformations distribution and encouraged to formulate conclusions about the initiation mechanism and development of damages in this type of pavement structures. Results shown in this article were obtained in field tests executed on an exploited pavement structure with a surface course made of granite slabs with dimensions 0.5x1.0x0.14 m.

1. Introduction

Block element road pavement structures were well known in Roman Empire. Importance of this type of road pavement structures occurred in the time of vibrant development of industry and metropolitan areas. Communication passages with block pavement structures were very popular then, because of high endurance and fatigue life, and respective blocks were easy to assemble. Nowadays block elements are not so popular and not so widely applied, being used only as a surface layers of pedestrian footpaths, cycle paths, squares, in representative city districts or low volume roads [13]. Block element pavement structures have layered construction. The role of sub-base or base is frequently performed by unbound granular layers on which a thin cement-sand laying course and block elements are settled. Joints between block elements are typically filled with fine-graded granular material (i.e. sand) or grout.

Research on behavior of block element pavements were carried out in many works [3][4][10][11][12]. The most important aspect of proper work of block element pavement is cooperation between respective elements. Mampearchchi and Senadeera [8] described so called

IOP Conf. Series: Materials Science and Engineering **356** (2018) 012024 doi:10.1088/1757-899X/356/1/012024

blocking mechanism in which block elements block each other and make impossible a respective element to rotate or translate (Figure 1).

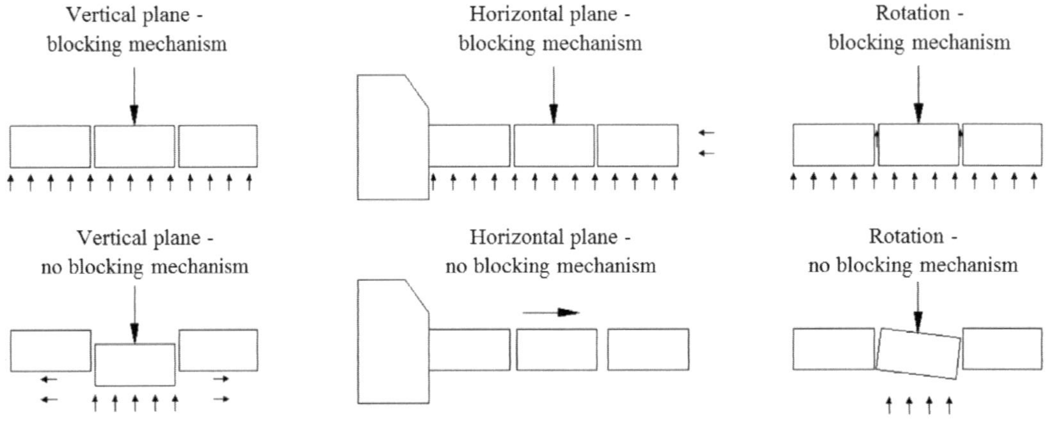

Figure 1. Blocking mechanism in vertical and horizontal plane and rotation.

Behavior of block element pavements was also carried out in numerical analyses [6][7][9]. Hengl and Füssl [7] point out variable mechanisms of block pavement destruction. On the basis of numerical model composed of 14 block elements limited on both edges by curbs authors gained three different mechanisms of destruction - through buckling, kinematic chain formation and mixed (Figure 2).

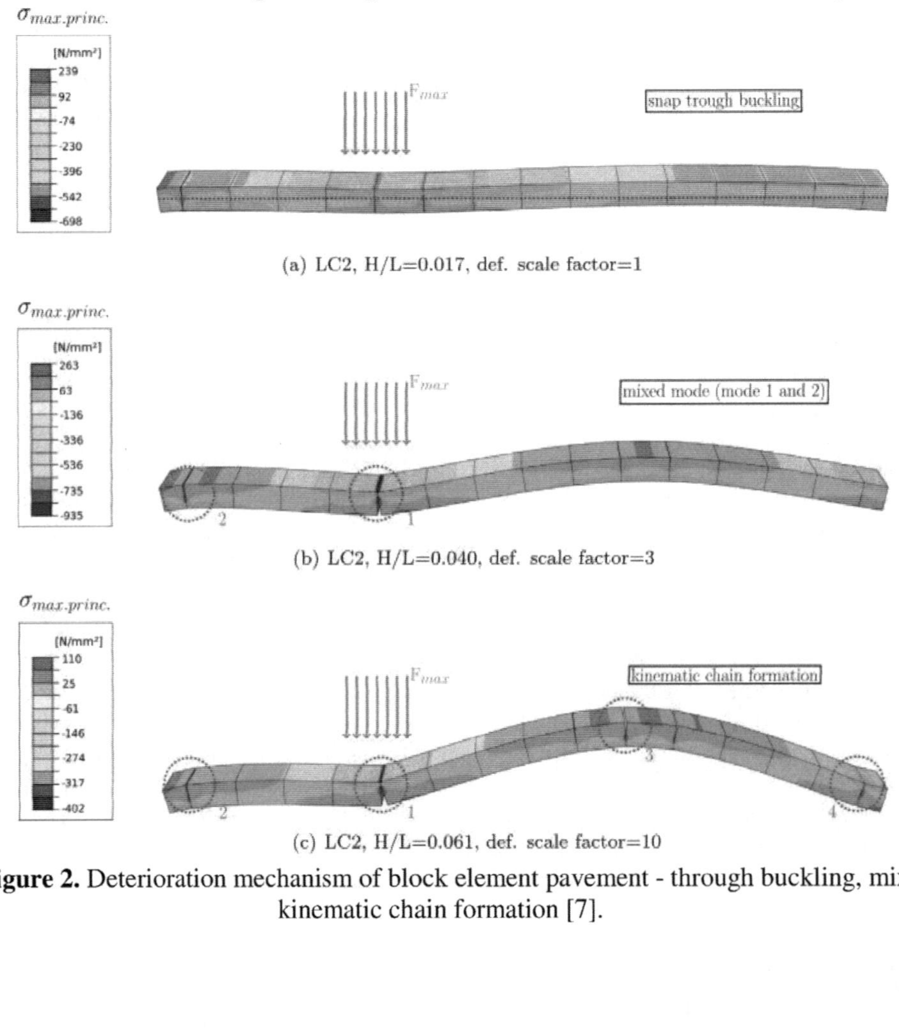

Figure 2. Deterioration mechanism of block element pavement - through buckling, mixed, kinematic chain formation [7].

For small values of superelevation and high values of applied loads pavement structure was destructed through buckling, because in block elements high values of compression forces were locally obtained. For high values of superelevation the destruction of pavement was caused by kinematic chain formation due to ability of single element to rotate or translate under a low value of applied load. In this destruction scheme block elements were also peeled away causing improper behavior of pavement structure.

This issue was also noticed as an important problem by Polish researchers. Koba [5] in his work showed that paving made of concrete blocks transfers loads on lower layers by two mechanisms: effect of distribution and effect of plate. Especially important is effect of plate (Figure 3) which can cause various pavement damages such as characteristic upheaval of whole pavement structure in the loading area. This situation is connected with block pavements laid on weak subgrade and is expressed in modulus of elasticity. On the sections with observed upheaval tensile stresses were obtained which can cause damages to joints located between the edges of lifted elements. Mechanisms of destruction of block element pavements mentioned in the literature are difficult to observe in situ. The main problem is to measure displacements of respective elements with satisfactory accuracy and in all characteristic points at one time.

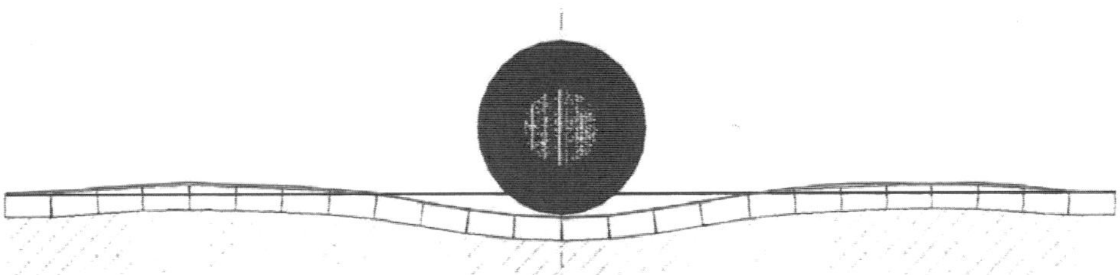

Figure 3. „Effect of plate" in pavement with block elements in surface course [5].

The Digital Image Correlation method mentioned in this article and used in field tests on pavement with surface course made of granite slabs with dimensions 0.5 x 1.0 x 0.14 m allowed to attain adequate area of measured displacements. Due to DIC method it was possible to explain causes of existing damages of pavement, which were strongly connected with mechanisms described in other research.

2. General characteristic of Digital Image Correlation

The Digital Image Correlation system used to measure pavement's displacements in 3D is a measuring technique widely used in aviation, space and automotive industry, nuclear and wind energy and in laboratories to investigate mechanical parameters of high-tech and advanced materials (i.e. composites, CFRP, GLARE etc.). Nowadays DIC systems are increasingly used in civil engineering [1][2], especially in laboratories. DIC 3D measurement technique is based on stereoscope vision systems in which at least two cameras record image of tested object from two different perspectives. Due to prior calibration of system, dedicated software analyses the image and obtain information about the distance between cameras and tested object, the angle between both cameras and distance between them. Additionally the calibration process allows to correct errors resulting from imperfection of optical devices such as spherical aberration. Tested object has to be marked with special markers to gain proper results. Markers are also reference points in the test. The algorithm of software finds the markers and assigns them three dimensional coordinates x, y,z. Systems makes measurements of displacements of marked points and due to comparison of gathered data in every step of loading it is possible to show the relationship between load and deformation or displacement. Stereoscope cameras system and 3D space calibration enables the measurement of 3D displacements. Results are shown in a graphic manner similarly as it is in Finite Element Method, therefore system can be used to validate FEM models.

Due to multiple options offered by hardware and software the whole system can be used to investigate variable objects and specimens. DIC system can analyze complete 3D surface in real time with frequency up to 5 Hz (depends on maximum data acquisition frequency of camera and on maximum data transfer link between PC or laptop) and in frequency up to one million frames in post processing [14]. This innovative technique gives also new possibilities in road engineering.

3. Road pavement characteristic

Tested pavement structure was a pathwalk with permission of traffic of heavy trucks and lorries servicing local public facilities and stores. Surface course was made of granite slabs with dimensions 0.5 x 1.0 x 0.14 m. The pavement structure also consisted of cement-sand laying course (4 cm), unbound granular base 0/31,5 mm (25 cm), unbound granular sub-base 0/31,5 (30 cm) and geotextile layed on subgrade.

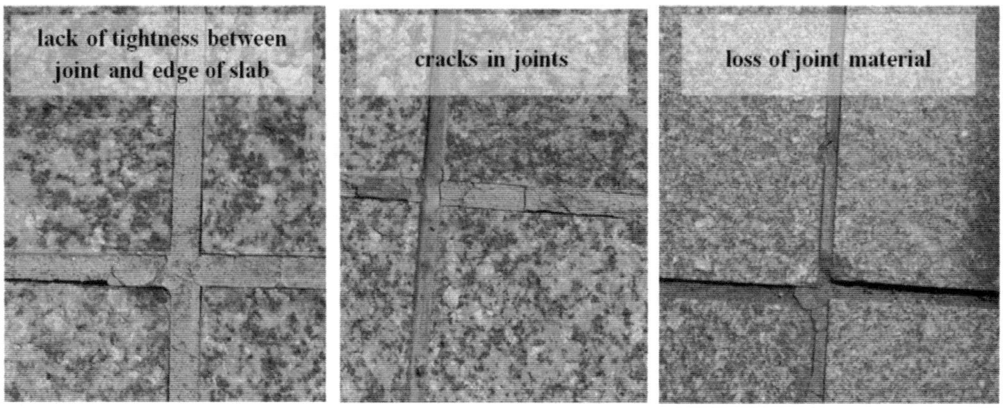

Figure 4. Stages of joint's deterioration.

Characteristic feature of granite slabs pavement were damages of joints between slabs. Several types of this damages could be distinguished such as lack of tightness between joint and edge of slab, cracks in joints and loss of joint material (Figure 4). Dilatation joints were exceptional, because of the elastic material used to fill joint which was not damaged (Figure 5).

Figure 5. Example of dilatation joint in good condition (vertical), grout joint with some damages (horizontal).

4. Test stand

Tests were performed on a selected area with dimensions 2.3 x 1.4 m (Figure 6). During the tests a heavy truck with single axes and twin wheels on rear axis was used. Heavy truck meets the requirements for vehicles used to beam deflectometer measurements. The pressure on the rear axis was equal to 99.8 kN.

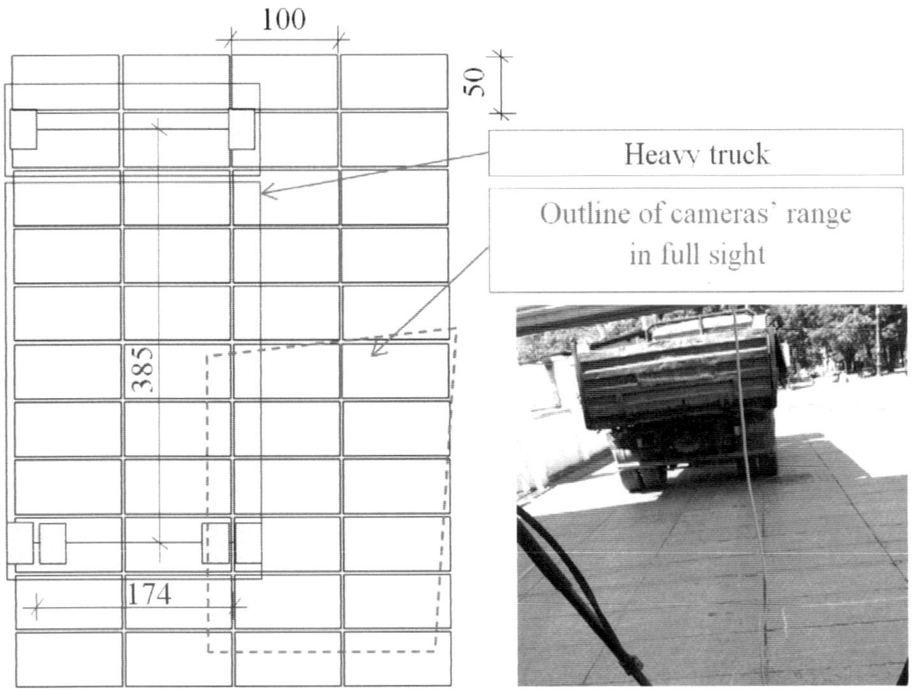

Figure 6. Scheme of tested pavement and heavy vehicle.

Figure 7. Test stand during DIC deflection measurements of block element pavement structure.

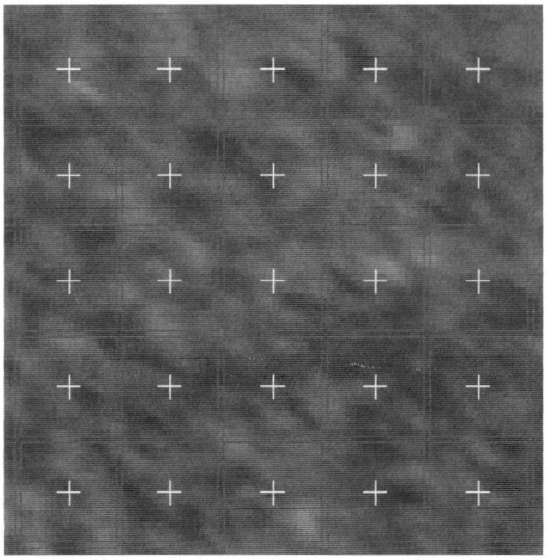

Figure 8. Magnified structure of pavement surface. Natural markers (grains) in gray-scale are visible.

Displacements of pavement structure under a load generated by heavy truck were measured using Digital Image Correlation system. Test stand was shown in the Figure 7. The test stand consisted of three cameras located linearly on a stiff aluminum profile with length of 2.0 m. Monochromatic cameras with 5 Mpx resolution and lenses with focal length equal to 12 mm were used. The central camera was located around 3678.32 ± 0.10 cm from the central point in the field of view (FOV). For granite slabs there were no necessity to mark the surface with stochastic markers, because granite has a natural grained structure with different gray scale tones (Figure 8). Analysis of deformation was executed using a mesh with more than 13 000 elements with size of single facet equal to 17 px. In the principal point (central point of the FOV) resolution is equal to 1Px/1mm, but resolution of DIC measurement is different in different areas of the FOV. In the areas covered by three cameras resolution is better than in areas covered just by two cameras. Optical axis of cameras was inclined at 45 degrees to the measured surface, what means that in the area closer to the cameras resolution was better than for area further away from the cameras. Resolution in horizontal and vertical direction in Px/mm is the same. Calibration was executed using square calibration board with a 750 mm long edge. The X and Y axes in coordinate system were set in plane of tested surface.

5. Test and analysis of results

During the tests pavement surface was loaded by heavy truck which was moving with low velocity along the shorter edge of slab (Figure 6) which is along the OY axis (Figure 8). Vehicle started passing in point y = 5.0 m (rear axis location) and moved slowly to point y = 0.75 m. During the passing joints between the slabs were located between twin wheels. Measurements were registered only for the rear axis, because of vanishing field of view during the movement of heavy truck. Map of vertical displacements of observed area of pavement in several positions of rear axis is shown in Figure 8. Maximum measured and calculated with DIC deflections of pavement structure are equal to 0.500 mm. It is worth to say that this value is not the maximum vertical displacement value, because it was unable to measure displacements of pavement between twin wheels. The analysis of vertical displacement maps (Figure 8 - step 25 and step 30) shows that along the loaded edge (along OY axis) negative values of displacements were noticed ("top down") and along the non-loaded edge positive values of displacements were noticed ("bottom up"). In order to further analysis of vertical displacements the distribution of vertical displacements along OY (Figure 8) and perpendicular OX (Figure 9) axis was made. The distribution of vertical displacements along the loaded edge (Figure 8) shows only the negative value of displacement. It is worth emphasizing the strong influence of rear axis on the 1.0 m distance.

Figure 9. Maps of vertical displacement of pavement surface due to different location of load (rear axis of vehicle).

On the following 1.5 m the reaction of pavement structure is also observed, however displacements registered on this section are definitely lower (around 0.04 ÷ 0.06 mm) and they are slightly higher than error (around 0.03 mm) of their calculation (Figure 10). The distribution of vertical displacements along the perpendicular OX axis shows characteristic upheaval of outer edge of slab (slab A) near the joint (Figure 10). Also the slab B's outer edge is uplifted. Vertical displacement in the „upper direction" reach values even equal to 0.200 mm (Figure 9) and it is higher value than the error of calculation of DIC equipment. It can be said that during the loading of pavement non-loaded edge is uplifting. Near the edge of uplifting slabs tensile stresses appear between edge of slab A - material of joint - edge of slab B. Having considered that tensile strength of edge - joint - edge configuration, appearance of tensile stressess causes deterioration of pavement structure - rupture the connection between elements and deterioration of material in joints. This model of destruction is similar to mechanism showed by Hengl and Füssl [7] and called kinematic chain formation (Figure 2).

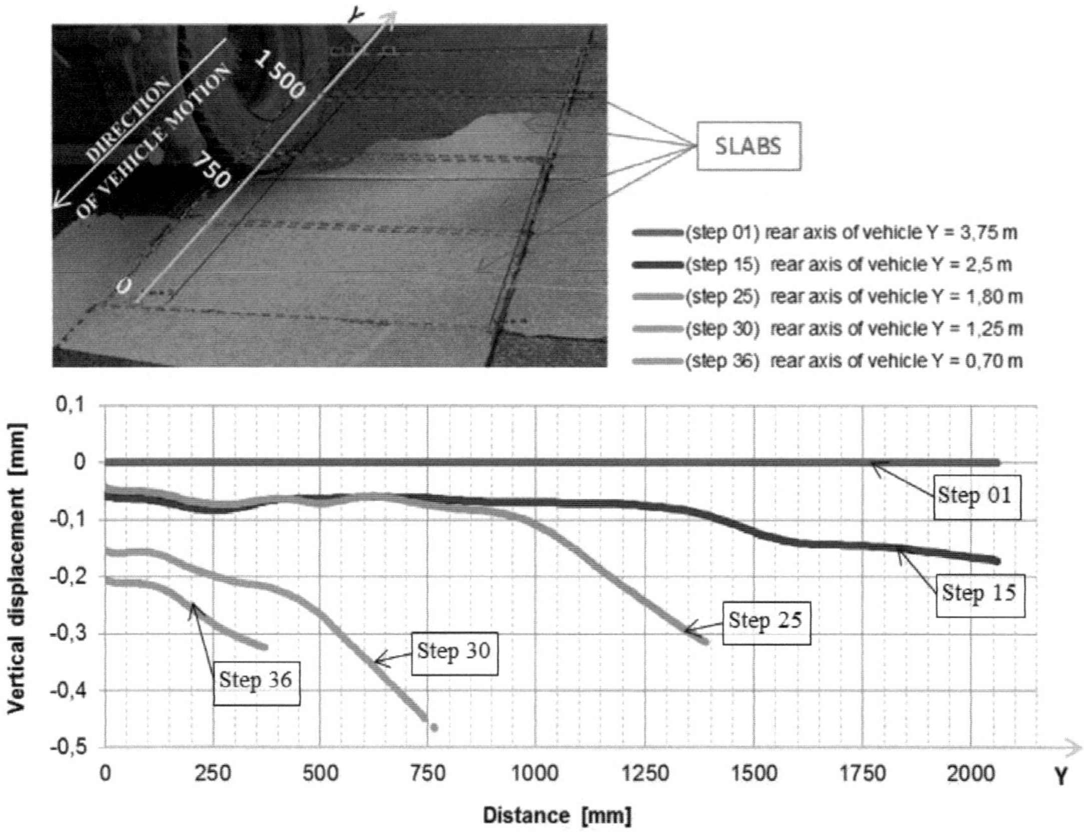

Figure 10. Distribution of vertical displacements of pavement surface along Y axis due to different location of rear axis of moving vehicle.

It is worth noticing that upheaval of elements only appears on the shorter edge of slab - opposite to the loaded edge. When load was applied along the longer edge of slab, the counter edge was not being uplifted (Figure 11). The results allow to formulate thesis - vertical displacement "bottom - up" will appear when one of the edges will be loaded and the subgrade beneath the block element pavement structure is too flexible and weak. The value of vertical displacement (uplifting) will depend on the value of loading, weakness of subgrade and distance between the loaded edge of slab and opposite edge.

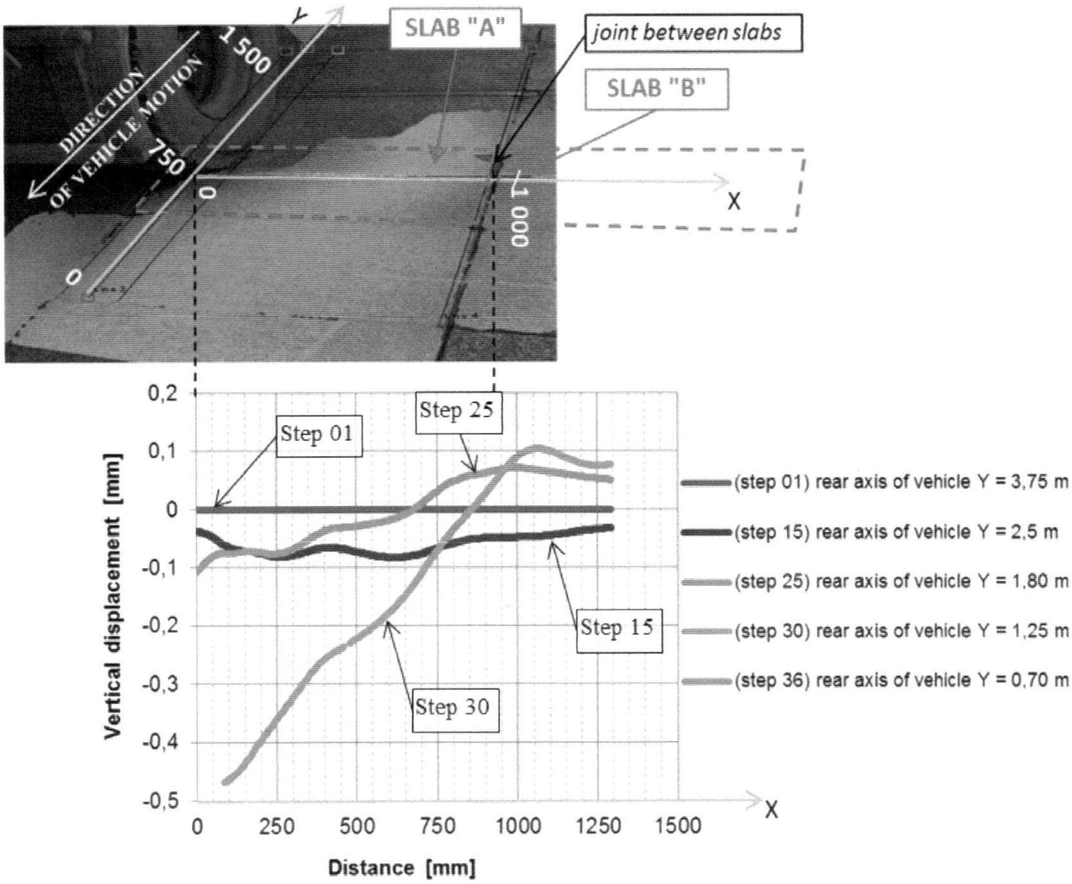

Figure 11. Distribution of vertical displacements of pavement surface along X axis due to different location of rear axis of moving vehicle.

6. Conclusions

Executed tests and analysis allow formulation such conclusions:

Pavement structure with surface course made of natural stone slabs was tested. On the surface of pavement there were noticed damages such as damages and lack of tightness between slabs.

Digital Image Correlation system enabled showing the distribution of displacements of pavements with block natural stone elements (dimensions around 2.0 x 1.3 m) in surface course. Due to generated vertical displacement map the distribution of displacements in random profile set in the area of measurement. DIC system allows to analyze complex displacement state of pavement i.e. with surface course made of block elements with load applied by traffic.

It was stated that loading shorter edge of slab causes rotation of the slab and upheaval of opposite edge. This type of deformation of pavement made of block elements causes tensile stresses in area of edge o slab A - joint - edge of slab B to result in break off the connection between side edge of slabs and joint. Vertical displacement in the "bottom - up" direction has to be negative, because it leads to destruction of joint and further leads to reduction of fatigue life and aesthetic values of pavement structure

Further research will be focused on analysis of loading position on distribution of displacements. Furthermore, gathered DIC data enable validation of physical model describing behavior of pavement from block elements.

References

[1] Ramos T, Furtado A, Eslami S, Alves S, Rodrigues H, Arede A, Tavares P, Moreira P 2015 2D and 3D Digtial Image Correlation in Civil Engineering – Measurements in a Masonry Wall. Procedia Engineering 114 (2015) 215-222

[2] Malesa M, Szczepanek D, Kujawska M, Kołakowski P 2010 Monitoring of civil engineering structures using Digital Image Correlation technique. The European Physical Journal Conferences 6, June 2010, DOI 10.1051/epjconf/20100631014

[3] Working Group D3 1988 Design of small element pavements. Netherlands Centre for Research and Contract Standarization in Civil and Traffic Engineering, Ede.

[4] Rollings R, Rollings M, Sharp K 1992 Evaluation of thickness design concepts for block pavements. International Concrete Block Paving Conference, 4th, 1992, Auckland, Nowa Zelandia.

[5] Koba H 2001 Nośność nawierzchni z betonowej kostki brukowe. VII Międzynarodowa Konferencja „Trwałe i bezpieczne nawierzchnie drogowe", s. 65-72, Kielce, 8-9.05.2001.

[6] Huurman M, Houben L, Kok A, 1992 Development of a three-dimensional finite element model for concrete block pavements, Delft University of Technology.

[7] Hengl H, Fussl J 2016 The influence of superelevated profiles of paving block structures on their load-bearing behaviour, Engineering Structures 117, s. 195-203.

[8] Mampearachchi W, Senadeera A 2014 Determination of the most effective cement concrete block laying pattern and shape for road pavement based on field performance. Journal of Materials in Civil Engineering 02.2014, s.226 - 232,

[9] Mampearachchi W, Gunarathna W 2010 Finite-element model approach to determine support conditions and effective layout for concrete block paving. Journal of Materials in Civil Engineering 11.2010, s. 1139 - 1147.

[10] Panda B, Ghosh A 2002 Structural behavior of concrete block paving. I:Sand in bed and joint. Journal of Transportation Engineering 03-04.2002 s. 123 – 129.

[11] Knapton J, O'Grady M 1983 Structural behavior of concrete block paving. International Journal of Concrete Structures and Materials 1983, s.17-18.

[12] Mokaddes A, Binod S 2013 Overview on structural behaviour of concrete block pavement. International Journal of Scientific & Engineering Research, Volume 4, Issue 7, 2013.

[13] Stypułkowski B, Koba H 1999 Konstrukcja kamiennych nawierzchni historycznych ulic i placów. Drogownictwo 7/99, s. 208 - 213.

[14] http://www.dantecdynamics.com

IOP Conf. Series: Materials Science and Engineering **356** (2018) 012025 doi:10.1088/1757-899X/356/1/012025

Evaluation of Homogeneity of Thickness of New Asphalt Layers Using GPR

Jacek Sudyka[1], Lech Krysiński[1]

[1] Road and Bridge Research Institute, Warsaw

E-mail: jsudyka@ibdim.edu.pl

Executive summary. The use of GPR (Ground Penetrating Radar) as an auxiliary tool in assessing the structural condition of the road pavement is now a common practice. Such equipment is employed by both private consulting companies and main road management laboratories. The field of application of GPR is constantly growing. Radar techniques are combined with sophisticated methods of data analysis, which facilitates the process of assessment of damage to the pavement structure, such as cracks, inhomogeneity of compaction or the condition of interlayer connections. The primary role of GPR operators is to determine the thickness of layers, including new ones.

This article discusses the main elements of the method of measuring the thickness of new asphalt layers. This method can support the acceptance testing phase. Particular attention is attached to the methods for the calibration of the wave travel velocity, including the role of reference drill holes, their precise positioning and determination of layer thickness. The article revolves around the dispersion of measurements and uncertainty of average and local results, as well as pointing to the key factors having an influence over accuracy, which is of importance in the consecutive phases of acceptance testing.

Keywords: measurement of layer thickness, GPR, acceptance tests, measurement accuracy.

1. Introduction

The accurate measurement of thickness of pavement structure layers basically occurs in the assessment of roads intended for renovation and during the phase of acceptance tests of new pavements. While the former case does not pose any special challenges and falls within the designer's responsibility, the latter is more demanding, since the results may entail financial penalties imposed on contractors.

In assessing the thickness of asphalt layers, the most common approach is to collect a core sample from the pavement, as provided by the road manager. The thickness measurement can be carried out in accordance with the PN-EN 12697-36 standard. In recent times, the Ground Penetrating Radar (GPR) has been increasingly used to support the inspection process carried out as part of acceptance tests. Although the use of GPR clearly improves the quality of assessment of layer thickness homogeneity, still the sole application of this technique does not necessarily guarantee a high quality of data. No less important is the choice of methodology that takes into account strict data synchronization and the correct algorithm for determining the velocity of electromagnetic waves (related to electric permittivity value). Obtaining thickness results of satisfying quality from the GPR measurement depends, first and foremost, on the correct determination of electrical permittivity of the examined layer. There are four basic methods for assessing permittivity: reflectometric (described in this article) reference drill holes [1, 2], laboratory measurement of electrical permittivity [3, 4, 5] and the common mid-point method (CMP) [6, 7].

This article discusses the methodology of measurement and data processing relying on data from core samples collected in locations selected on the basis of the initial analysis of radar data. This method requires a careful approach to the phase of field measurements; at the same time, it does not require complicated calibration methods or very expensive laboratory equipment. The described method can be successfully used in tests of asphalt and concrete pavements supporting the acceptance testing phase required in technical specifications, standards and national regulations. As regards concrete pavements, the specific nature of placed material should be considered along with the associated phenomena, such as significant microwave absorption and dispersion.

2. Typical measuring equipment and profiling

GPR measurements of the thickness of asphalt layers or of the entire package of asphalt pavement layers are usually performed by means of profiling (continuous scanning) along a selected line. Profiling is generally carried out along the road axis using a mobile GPR system intended for extensive road surveys (Fig. 1). The scanning system allows the use of various types of antennas, yet, in practice, the air-coupled impulse antennas with the central frequency f about 2 GHz are most advantageous, as they allow high mobility of the scanner and offer a very clear and high-resolution picture of the structure of asphalt package. Ground antennas with a similar central frequency have superior sensitivity and horizontal resolution, and they offer interesting results in detecting defects [8] but are used only for local tests with closed traffic due to mobility constraints. In overview surveys, it is possible to perform the measurement at a speed of 50 km/h or higher, but this is definitely inadvisable for measurements that require high accuracy in thickness determination. In such a case, the recommended speed is between 10 and 20 km/h, and individual profiling is done in subsequent passes. With such an organization, it is possible to ensure synchronization of the measurement distance with reference objects or other measurements with an imprecision of no more than 1 meter, which permits, for example, correct calibration of the wave velocity by drilling.

Figure 1. Mobile scanning system.

If the aim of measurement is to obtain a representative image of the layer thickness, as in acceptance testing, the so-called mapping is performed in the form of several scans (profiles) along the road axis, parallel to and equidistant from each other (Fig. 2). In order for the result of mapping to be considered reliable, the distance between the profiles should not be greater than 2 meters, and the profiles themselves should run on the center lines of belts of no more than two meters wide and covering the entire width of the examined road. On the other hand, the distance between the profiles should not be less than 50 cm because the main antenna radiation cone covers the oval road surface with a diameter of about 1 meter and produces, as a result of measurement at a given point, an image of the averaged (in some sense) package structure beneath this area. Maintaining the distance between the profiles of significantly less than 1 meter is also difficult to control in the field without drawing profile lines on the surface.

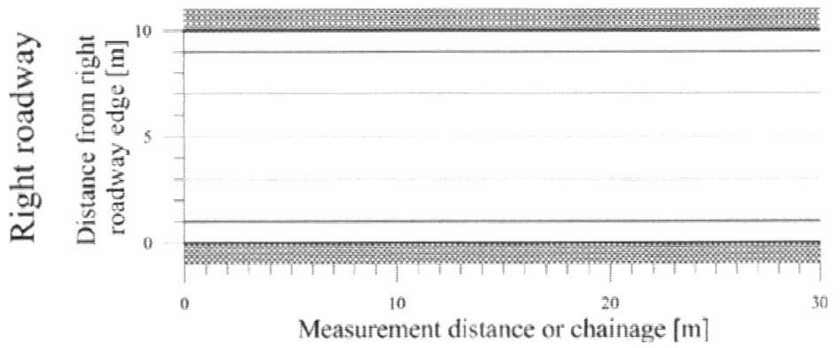

Figure 2. An example of the organization of a measurement profile system when mapping surface thickness.

3. Method of thickness measurement

The scanning process involves the transmission of a microwave pulse towards the pavement and recording the returning response signal. The scans are made at successive profile points at fixed intervals (e.g. every 10 cm). As a result of scanning and after the numerical processing of measurement data, an echogram is obtained (Fig. 3), i.e. a graphic representation of subsequent scans one next to another, where the amplitude of the return signal is shown as the shades of gray, the distance from the upper edge of the echogram corresponds to the time of return, and the numbers of subsequent scans reflect the measuring distance. The echogram is an image of the layered package structure, but it must be remembered that the vertical dimension is not a direct and simple equivalent of depth. This image is also partially diffractive, which is clearly seen in the areas of sudden changes in the layer structure and may cause shifts in the time of return in the case of significantly inclined boundaries. However, in the case of asphalt package layers (apart from area of local structural disturbances), the interlayer boundaries usually have very small inclinations in relation to each other within a one-meter distance scale, so that no adjustments (migration) are needed to compensate for the inclination.

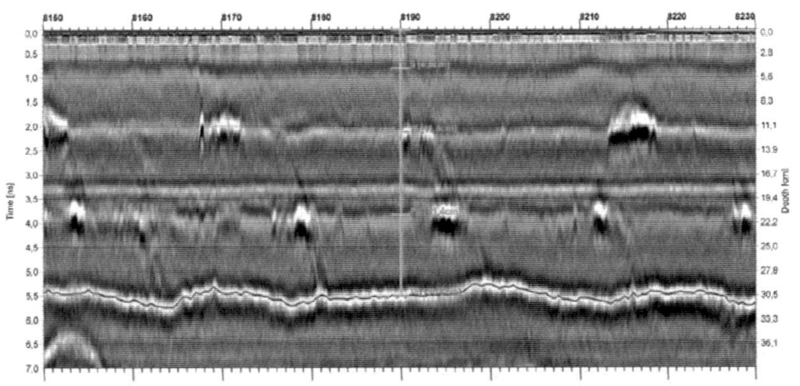

Figure 3. Example of half-velocity calibration $v_{1/2}$ of a package of asphalt layers. At value $v_{1/2}$ = 5.55 cm/ns, a perfect match was achieved between the double travel time and the thicknesses marked in the drill hole.

Measurement of the thickness ***d*** of a layer package or a single layer involves the reading from the echogram the time ***t*** of wave travel through the package (or layer) and converting this time into thickness:

$$d = v_{1/2} \cdot t \tag{1}$$

Consequently, the main components of error in the determination of layer thickness come from the uncertainty Δt of double travel time ***t*** (DTT) and from the uncertainty $\Delta v_{1/2}$ of adopted half-velocity value $v_{1/2}$:

$$\Delta d \approx \sqrt{(\Delta v_{1/2} \cdot t)^2 + (v_{1/2} \cdot \Delta t)^2} \approx \Delta v_{1/2} \cdot t + v_{1/2} \cdot \Delta t \tag{2}$$

Although the scanning systems have a very good temporal resolution due to the signal sampling frequency, which potentially allows the relative positioning of reflecting boundaries with submillimeter accuracy, a limitation of the real temporal resolution is the width of the microwave pulse $\tau \approx T = 1/f$ and the ambiguity of interpretation of the actual shape of the return pulse. In the case of a simple single reflection [9] corresponding to the boundary where the material contrast occurs expressed in a stepped change of electrical permittivity, the basic uncertainty in determining the travel time comes from the difficulty of determining the time position of the interlayer reflection peak. Both manual and automatic procedures are employed to track this peak. The typical uncertainty of such a determination in a single scan is about one eighth of the wavelength (0.06 ns at $f = 1/T = 2$ GHz, i.e. about 3 mm of the determined layer thickness) or smaller. As long as the procedure does not involve a systematic shift of the determined time, the impact of this uncertainty is usually reduced when determining the average package thickness in a given road section.

Another vital and even much more important reason for uncertainty Δt is the deformations in the shape of the reflection [9]. The bottom reflection of the package is usually a strong single reflection (from the boundary with the contrast of electrical permittivity) and its peak is usually a good representation of the bottom package even if the signal is extended due to the small-scale fluctuation of bottom shape in centimeter and decimeter scales of horizontal variation. In contrast, the forms of interlayer reflections are often severely deformed. They take the form of double or mixed reflections [9, 10] and to identify them requires an experienced interpreter. Ambiguity in the interpretation of the shape of reflection may lead to a systematic, even quarter-wave error of $\Delta t \approx 0.125$ ns (2 GHz antenna), i.e. about 0.6 cm ($v_{1/2}\ \Delta t$) of layer thickness, and this error can increase (can be even doubled) if the mistake occurs both at the top and bottom of the layer. This component of error usually accounts for the uncertainty of measurement of thin, few-centimeter thick layers. In the case of very thin layers, limitation of the resolution of close horizons must be take into account because the upper reflection begins to merge with bottom one when the thickness of a layer is less than half the wavelength in the material ($d < v_{1/2} \cdot T \approx 2.5$ cm at $f = 2$ GHz). In 2 GHz central frequency antennas, it is difficult to distinguish (convincingly) between the upper and lower boundaries of layers thinner than 3-4 cm. In some cases, when the layer presence is certain, the numerical signal modeling may prove helpful in determining its thickness [11, 12].

4. Main factors determining precise velocity calibration using reference drill holes

In the case of relatively thick packages (several decimeters), the effective thickness error Δd depends on the uncertainty $\Delta v_{1/2}$ of determining the wave half-velocity. In principle, this value (according to the definition) is a material property, expressed as half the velocity of travel of an electromagnetic wave of a specific frequency (the case without dispersion). Its value is determined primarily by the petrophysical type of rock factions, but also the content of the binder, the presence of voids, humidity and the spatial organization of aggregate grains have some influence [2, 13]. In the GPR method, however, this velocity is regarded as an effective property of material in a particular measurement (observed typical values range from 4.7 to 7.2 cm/ns). It is a formal factor for converting the time of double travel into layer thickness taking into account the various deficiencies of the measurement system and of the entire method of processing and interpretation. It is assumed that this is an average layer or package attribute, approximately constant, on a certain road section that is considered homogeneous. Its minor fluctuations are ignored. It is approached as an effective value (averaged vertically), also trying to cover the vertical changes of velocity most often resulting from the differences of compaction at different layer depths.

In the adopted method, drill holes are made in selected locations of the profile line, in places that can be precisely located on the echogram. The determined package (or layer) thickness d_r is divided by the time of wave double travel t_r recorded at this very location, thus obtaining a half-velocity estimate:

$$v_{1/2} = d_r / t_r \tag{3}$$

Hence, the uncertainty of such an estimate is expressed as:

$$\Delta v_{1/2} \approx \sqrt{\left(\frac{\Delta d_r}{t_r}\right)^2 + \left(d_r \cdot \frac{\Delta t_r}{t_r^2}\right)^2} \approx \frac{\Delta d_r}{t_r} + d_r \cdot \frac{\Delta t_r}{t_r^2} \approx v_{1/2} \cdot (\Delta d_r/d_r + \Delta t_r/t_r) \quad (4)$$

The calibration function of reference drill holes should be distinguished from their role in identifying a layered structure or in recognizing other package attributes (e.g. defects).

4.1. Mislocation of the reference drill hole

The key challenge in calibration is the accuracy of synchronization between the location of the drill hole and the corresponding formal GPR measurement distance on the echogram. The reason for that is the relatively high variability of package thickness that usually occurs along the measurement profile. In the measurement practice, to synchronize the distance on the echogram with profile locations, real and well-identifiable field reference features are used (chainage, lamp posts, geodetic markers, makeshift surface markers, intersections, etc.). After gaining more practice, it is possible to master the methods that help remove errors resulting from parallax observations, system latency, mistakes of individual observations, turns along the traveling route, etc. It is usually possible to achieve positioning accuracy better that 1 m.

One of well-proven and recommended practices is the drilling in locations carefully selected during the earlier analysis of echograms, i.e. in areas where the thickness variation visible on the echogram (in time domain $t_r(x)$) is negligible, and, thus, a two-meter mistake in the location (along the profile) does not cause any serious distortion of the reference thickness d_r.

There is an option of collecting a large number of drilled holes (e.g. all available results) and perform a statistical analysis of the determined velocities set while rejecting extremely distorted values or using other advanced methods discriminating incorrect measurements, i.e. the modeling of specific statistical properties of such a group.

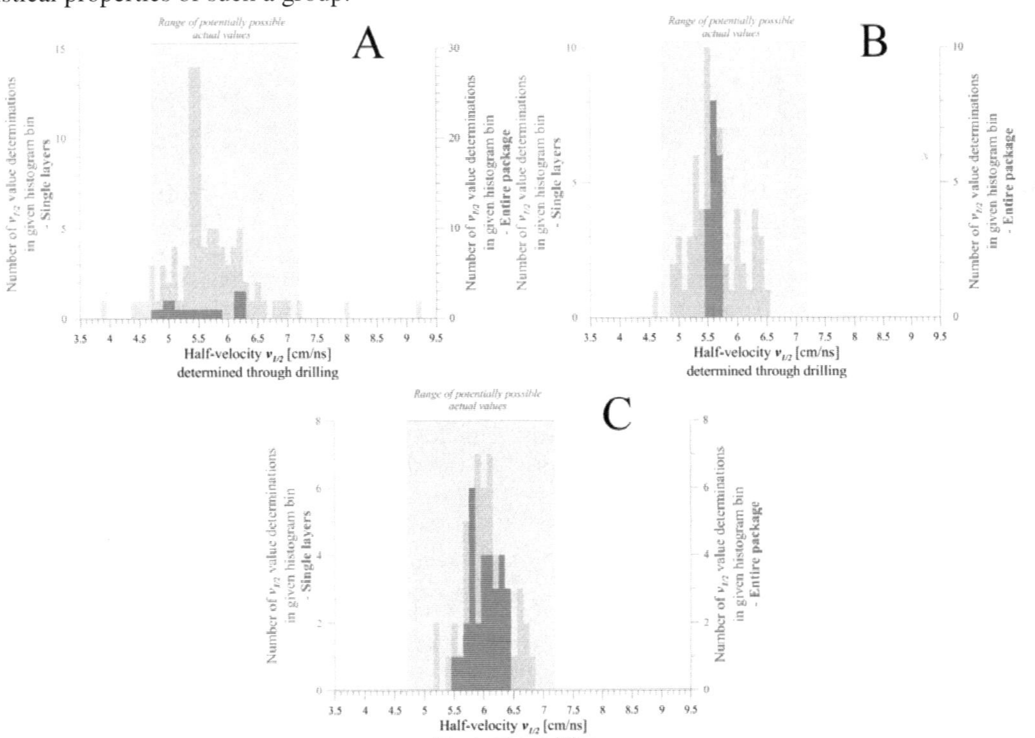

Figure 4. Examples of half-velocity histograms showing distribution of velocity estimated by means of A) data from a large group of previously made holes, i.e. without their precise location based on

echograms, B) a group of holes with a well-controlled positioning and C) a group of holes with a very poor positioning control after several attempts to verify the position, allowing its several-meter modifications.

In the practice of precise determination of thickness of an asphalt layer package, the optimal solution is to drill at least two holes on a 300-meter road section and one hole every 300 meters of its further length. The number of drill holes can be increased if some material differences are detected between the segments of a road section, or this number can be reduced if there are no material differences on a long section. To control the wave velocity effectively, the number of calibration holes should not be reduced to fewer than one per 1 km of the examined road.

4.2. Complications of the shape of the bottom surface of an asphalt layer package

As opposed to the interlayer boundaries, the bottom surface of an asphalt layer package usually reveals significant deformations in the form of undulations in centimeter and decimeter horizontal scales. This shape may be attributed to the unevenness of the previously laid subbase or may result from the interaction of the mineral asphalt mixture with underlying aggregate during incorporation. Such deformations of the boundary surface with scales comparable or slightly smaller than the wavelength in the medium are seen by the electromagnetic wave as a transition zone. When such a zone is encountered, the bottom reflection is extended [10].

Some examples of difficulties in measuring a layer package can be serious deformations and chipping of the bottom, the masking of the bottom by adhering subbase material or the penetration of the bottom of an asphalt package by subbase aggregate grains (Fig. 5). The uncertainty of such determination of package thickness in a single core sample may exceed 0.5 cm, which means that the final examination result is encumbered with significant uncertainty. In addition, in this situation, the uncertainty of a single core sample measurement may not reflect the actual thickness variations in the several-decimeter vicinity around calibration holes. One of the methods to control uncertainty and reduce its impact is to make several calibration drill holes and perform a statistical analysis of the set of determined half-velocities.

Figure 5. Examples of a complicated shape of the bottom surface of an asphalt layer package leading to significant uncertainty in determining package thickness on a core sample: A) significant losses and deformations due to insufficient bottom compaction, B) a well-defined flat surface partially masked by subbase aggregate grains from underneath, C) large aggregate grains (clearly larger than grains embedded in the package) penetrating the package bottom (clearly immersed in the asphalt mix).

4.3. Interlayer delamination

Several-millimeter errors in the determination of package thickness are usually attributable to interlayer delamination. In such cases, it is advisable to measure the thickness of layers after prior matching of collected core parts. It is much easier to perform it in a laboratory than in the field.

4.4. A challenging identification of thin layers

As shown by formula (4), there is a specific challenge in the calibration of velocities by means of drilling in the case of thin packages or single layers (Fig. 4B). The unavoidable uncertainties of reference measurements $\boldsymbol{d_r}$ and $\boldsymbol{t_r}$ translate into the very high uncertainty of velocity determination. For this reason, the method of velocity calibrating by means of drill holes is used primarily in relation to the whole package and without separating it into layers (velocity differentiation between layers).

IOP Conf. Series: Materials Science and Engineering **356** (2018) 012025 doi:10.1088/1757-899X/356/1/012025

Separation into layers is only used for beyond-average velocity differences between layers that can manifest themselves by strong reflections at the interlayer boundary.

5. Other methods of velocity calibration

Among the alternative methods of wave velocity determination there is the analysis of pavement reflectivity [14]; it is also used for effective identification of the material changes in the structure. It is an advanced technique that requires a lot of experience in the necessary measurement calibration procedures and data correction and processing due to the high instability of impulse measuring systems. The simplest and the most precise option in this technique is the analysis of pavement reflectivity which returns the measured wave velocity in the top pavement layer.

A common trend observed in the newly placed packages of asphalt layers is the use of an identical or very similar (when it comes to dielectric properties) set of materials, which can be recognized on the echogram as absent or extremely low amplitudes of reflections on interlayer horizons. In such a situation, the velocity determined in the top layer of the package tends to be regarded as typical also for the whole package. Thus, velocity stratification is somehow neglected, although it occurs, for example, when the lowest package layer is improperly compacted. However, such an approach seems to be very effective in practice if the calibration and adjustment procedures are carefully performed. Reflectometric determinations clearly refer to the calibration relying on drill holes (Fig. 6B), while in the case of thin packages or ones without exact positioning control, they can even be seen as superior to drilling when it comes to dispersion (Fig. 6A and 6C).

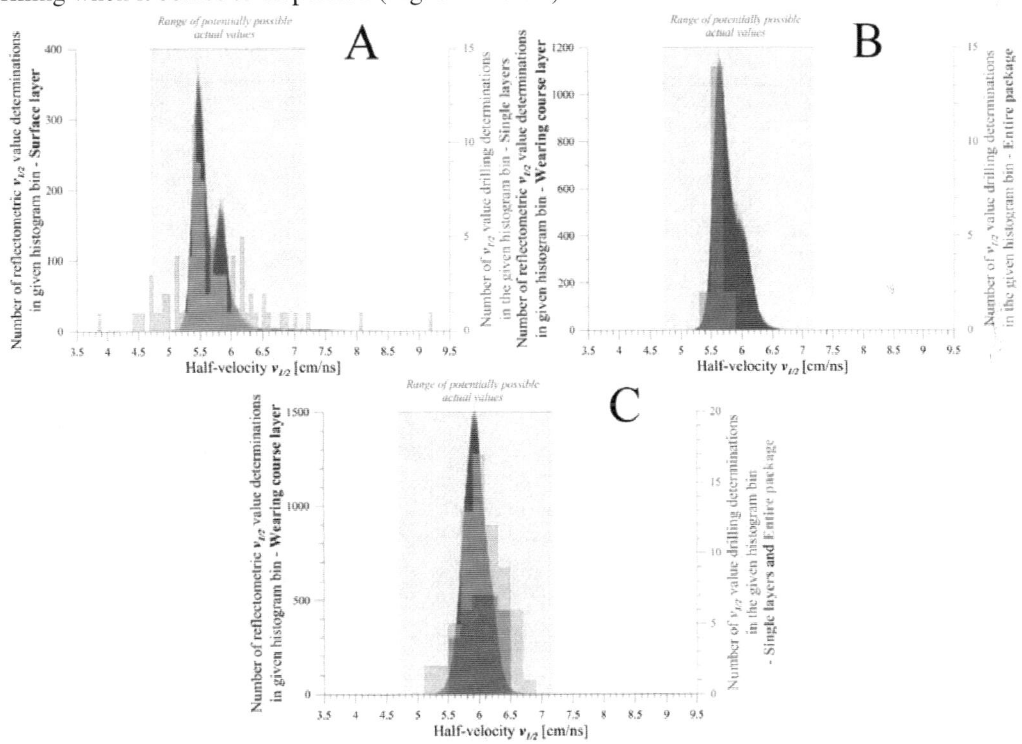

Figure 6. Examples of histograms of half-velocity reflectometric determinations in the upper package layer with the imposed histograms of layer or package half-velocity determination using drillings and wave double-travel times: A) a large set of drill holes in a thin package in the absence of proper control of hole positioning with clearly visible two populations differing in the type of material, B) a set of drill holes with a very good positioning control, C) a large set of drill holes with a very poor positioning control after multiple position verification.

6. Presentation of mapping results

One of the formal ways to document measurements is by means of echograms. They are usually (in the case of mapping or measurements with several antennas) arranged synchronously one over another (Fig. 7). Echograms help imaging a not always unambiguous interpretation of cross-sections and illustrate (in a sketch-like manner) the degree of accordance of thicknesses determined through drillings with a georadar section. When reading echograms, it must be borne in mind that they mostly rely on one optimum velocity determined for a given profile (or the entire road section).

The basic graphic presentation of the results of package thickness mapping is the graphs of variability of the determined thickness along the measurement profiles (Fig. 8). They offer a good insight into the spatial distribution of this parameter.

An important and noteworthy method of visualization of measurement results is the use of thickness histograms of a given road section. Such histograms are helpful in the discussion of the design in the case of new or renovated pavements. In particular, they reveal complications that go with the determination of this parameter in actual structures where the description of thickness with one number and standard statistical concepts does not often reflect the actual situation. Some of these cases shown in Fig. 9B demonstrate the difficulty in measuring thickness in a reliable manner by means of just a few or even single drill holes in randomly selected locations. Such a large span of thickness changes can sometimes lead to misunderstandings.

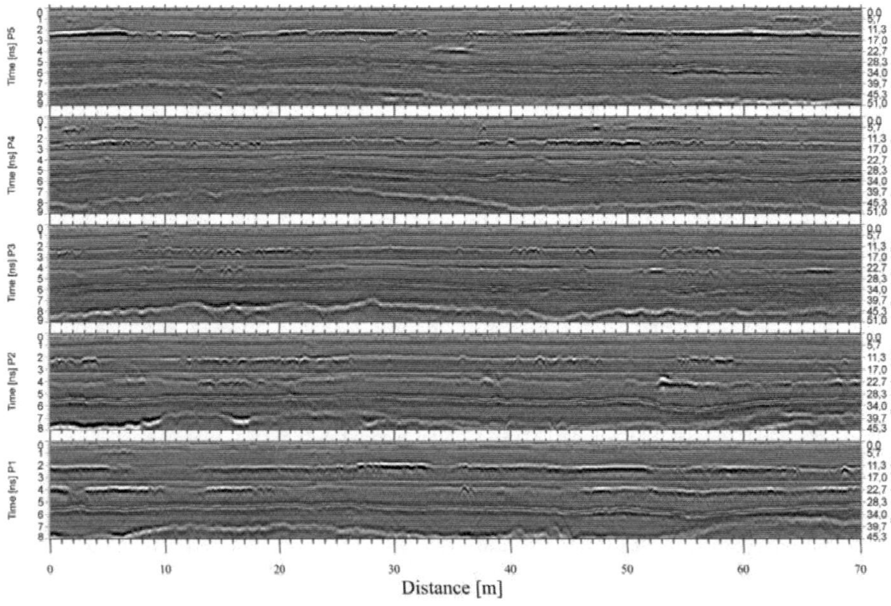

Figure 7. An example of echograms obtained in the mapping of a road section using five parallel GPR profiles.

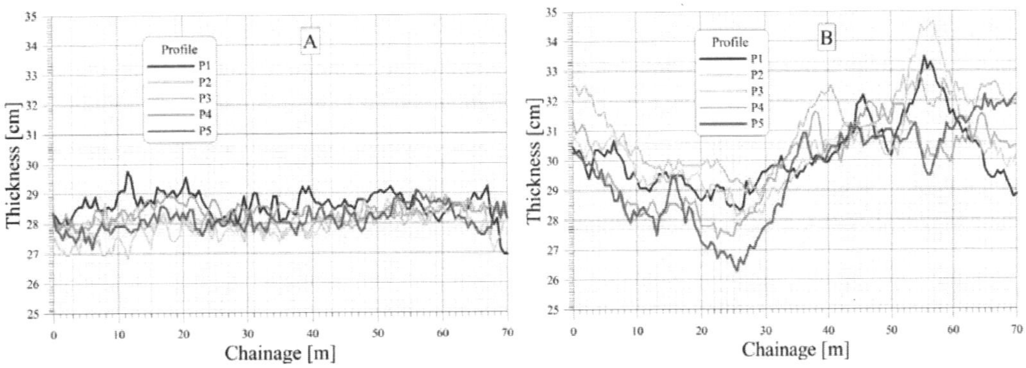

Figure 8. Two examples of collective charts showing the variability of thickness of a package of asphalt layers along mapping profiles.

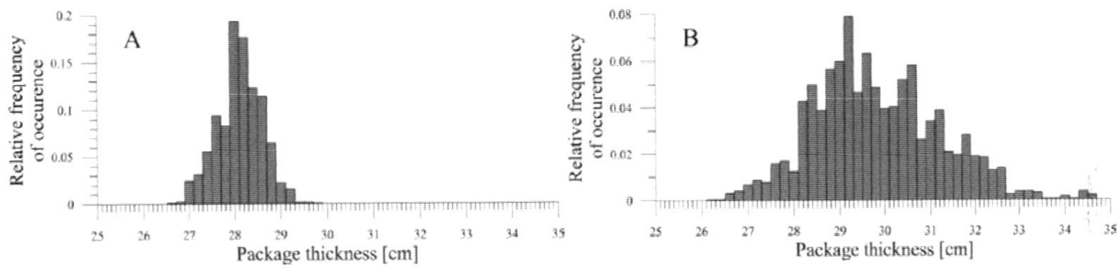

Figure 9. Two examples of histograms of thickness of a package of asphalt layers determined by mapping. A) section with a minor and B) section with significant thickness variation.

7. Summary

The article outlines the methodology of measuring (mapping) the thickness of an asphalt package and its individual layers and discusses the key factors affecting the accuracy and reliability of absolute thickness determination. The following recommendations can be made for such high-precision measurements:

- the use of impulse air-coupled antennas with the central frequency of about 2 GHz,
- calibration of the wave velocity with a strictly controlled location indicated after scanning and initial analysis of the results,
- the application of reflectivity analysis in the preliminary stage of thickness assessment and on sections with a thin asphalt layer package as a substitute method for velocity calibration.

The use of air-coupled antennas with the central frequency of about 2 GHz offers decent capabilities in terms of the depth of penetration, resolution, and accuracy. It also allows the simultaneous application of reflectivity analysis as a preliminary method of thickness assessment.

Based on gather experience, it can be assumed that to calibrate the wave velocity the optimum number of drill holes should be: two on a 300-meter section of a homogeneous pavement and one every further 300 meters of the section length. To control the wave velocity effectively, the number of reference holes should not be reduced to fewer than one per 1 km. The homogeneity of materials forming an asphalt layer package is vital for the proper choice of the number of reference drill holes.

Reflectometric measurements clearly show material differences between road sections. If they are significant, the use of separate velocity calibrations on different sections is recommended. It is therefore a very important control tool in the calibration procedure. However, due to the numerous challenges that go with this method, calibration based on drill holes is preferred. Exceptionally, in the case of very thin packages (less than 10 cm thick) or pavements where drilling is not advisable (e.g. bridges), this method can be used as more effective.

The practice of the discussed measurement method allows the estimation of the thickness with typical uncertainty of thin packets (up to 10 cm thick) at about 0.5 cm and of packages with greater thickness at about 5% in a properly conducted measurement. In the extreme case of a complete absence of estimation of half-velocity, the average value for our region can be referred to, i.e. around 5.5-5.7 cm/ns. However, this can imply an error of even up to 25% of the determined thickness value, also for average values, and an error of 10%+ will be a standard situation.

The described method can be successfully used in tests of asphalt and concrete pavements supporting the acceptance testing phase required in technical specifications, standards and national regulations.

Bibliography

[1] G. Morcous and E. Erdogmus, Accuracy of Ground-Penetrating Radar for Concrete Pavement Thickness Measurement, *J. Perform. Constr. Facil.*, vol. 24, no. 6, pp. 610–621, 2010.

[2] L. Krysiński and J. Hugenschmidt, Effective GPR Inspection Procedures for Construction Materials and Structures, *in Civil Engineering Applications of Ground Penetrating Radar*, A. Benedetto and L. Pajewski, Eds. Cham: Springer International Publishing, 2015, pp. 147–162.

[3] T. Pellinen, E. Huuskonen-Snicker, P. Eskelinen, and P. Olmos Martinez, Representative volume element of asphalt pavement for electromagnetic measurements, *J. Traffic Transp. Eng. English Ed.*, vol. 2, no. 1, pp. 30–39, 2015.

[4] D. Stepins, G. Asmanis, and A. Asmanis, Measuring capacitor parameters using vector network analyzers, *Electronics*, vol. 18, no. 1, pp. 29–38, 2014.

[5] R. A. Tarefder and M. U. Ahmed, Ground penetrating radar for measuring thickness of an unbound layer of a pavement," *in Advances in Intelligent Systems and Computing*, 2018, vol. 598, pp. 160–167.

[6] R. W. Jacob and T. M. Urban, Ground-Penetrating Radar Velocity Determination and Precision Estimates Using Common-Midpoint (CMP) Collection with Hand-Picking, Semblance Analysis and Cross-Correlation Analysis: A Case Study and Tutorial for Archaeologists, *Archaeometry*, vol. 58, no. 6, pp. 987–1002, 2016.

[7] G. Grandjean, J. C. Gourry, and A. Bitri, Evaluation of GPR techniques for civil-engineering applications: Study on a test site, *J. Appl. Geophys.*, vol. 45, no. 3, pp. 141–156, 2000.

[8] F. M. Fernandes and J. C. Pais, Laboratory observation of cracks in road pavements with GPR," *Constr. Build. Mater.*, vol. 154, pp. 1130–1138, 2017.

[9] J. Sudyka and L. Krysiński, Radar technique application in structural analysis and identification of interlayer bonding, *Int. J. Pavement Res. Technol.*, vol. 4, no. 3, 2011.

[10] L. Krysiński and J. Sudyka, Typology of reflections in the assessment of the interlayer bonding condition of the bituminous pavement by the use of an impulse high-frequency ground-penetrating radar, *Nondestruct. Test. Eval.*, vol. 27, no. 3, 2012.

[11] N. Diamanti and A. P. Annan, Characterizing the energy distribution around GPR antennas, *J. Appl. Geophys.*, vol. 99, pp. 83–90, 2013.

[12] K. Belli, H. Zhan, S. Wadia-Fascetti, and C. Rappaport, Comparison of the accuracy of 2D VS. 3D FDTD air-coupled GPR modeling of bridge deck deterioration, *Res. Nondestruct. Eval.*, vol. 20, no. 2, pp. 94–115, 2009.

[13] L. Pajewski, A. Benedetto, X. Derobert, A. Giannopoulos, A. Loizos, G. Manacorda, M. Marciniak, C. Plati, G. Schettini, and I. Trinks, Applications of Ground Penetrating Radar in civil engineering - COST action TU1208, *in IWAGPR 2013 - Proceedings of the 2013 7th International Workshop on Advanced Ground Penetrating Radar*, 2013.

[14] A. De Coster, A. Van der Wielen, C. Grégoire, and S. Lambot, Evaluation of pavement layer thicknesses using GPR: A comparison between full-wave inversion and the straight-ray method, *Constr. Build. Mater.*, vol. 168, pp. 91–104, 2018.

IOP Conf. Series: Materials Science and Engineering **356** (2018) 012026 doi:10.1088/1757-899X/356/1/012026

Assessment of technical condition of concrete pavement by the example of district road

M Linek, P Nita, W Żebrowski and P Wolka

Kielce University of Technology, Tysiąclecia Państwa Polskiego Street 7, 25-314 Kielce, POLAND

E-mail: linekm@tu.kielce.pl

Abstract. The article presents the comprehensive assessment of concrete pavement condition. Analyses included the district road located in the swietokrzyskie province, used for 11 years. Comparative analyses were conducted twice. The first analysis was carried out after 9 years of pavement operation, in 2015. In order to assess the extent of pavement degradation, the tests were repeated in 2017. Within the scope of field research, the traffic intensity within the analysed road section was determined. Visual assessment of pavement condition was conducted, according to the guidelines included in SOSN-B. Visual assessment can be extended by ground-penetrating radar measurements which allow to provide comprehensive assessment of the occurred structure changes within its entire thickness and length. The assessment included also performance parameters, i.e. pavement regularity, surface roughness and texture. Extension of test results by the assessment of changes in internal structure of concrete composite and structure observations by means of Scanning Electron Microscope allow for the assessment of parameters of internal structure of hardened concrete. Supplementing the observations of internal structure by means of computed tomography scan provides comprehensive information of possible discontinuities and composite structure. According to the analysis of the obtained results, conclusions concerning the analysed pavement condition were reached. It was determined that the pavement is distinguished by high performance parameters, its condition is good and it does not require any repairs. Maintenance treatment was suggested in order to extend the period of proper operation of the analysed pavement.

1. The purpose of technical condition assessment of concrete pavements

Taking into consideration safe operation of concrete pavements, diagnostics of pavement structure performance is of crucial importance. Diagnostics of concrete pavements refers to the set of methods and procedures aimed at the assessment of technical condition of the analyzed structures. The diagnostics refers to the anticipation of the pavement changes occurring as a result of external loads and weather conditions and anticipating the period of failure-free pavement performance. Assessment criteria of the approved concrete pavement diagnostics system are the following: load bearing capacity, evenness and coarseness requirements. Due to the traffic safety, the performance criterion referring to pavement evenness and coarseness, along with surface macro texture is of particular significance.

So far, less attention has been paid to structural factors caused by operating conditions concerning the assessment of pavements. In this regard, properties of working surfaces of pavement slabs appear to be particularly significant. The condition of these slabs can be of tribologic and non-tribologic wear and tear nature. Wear of pavement is the damage of its surface consisting in the gradual reduction of surface layer subject to vehicle traffic or chemical reactions, for example: this type of wear is also the result of such phenomena as: corrosion, cracking, material fatigue. The process of tribologic wear refers to all materials occurring naturally and in technology. Intensity of this phenomenon depends on surface layer resistance, type of interaction and its intensity. Within the area particularly exposed to pavement wear, abrasion and slowly increasing flaking occurred within immediate environment, caused by previous structural defects of surface layer, i.e. so called spalling may occur – Figure 1.

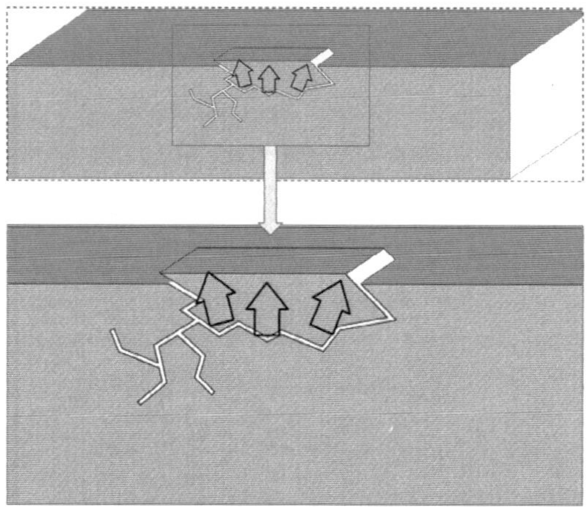

Figure 1. Diagram of spalling destruction of surface layer of concrete pavement.

2. The purpose and scope of the research

The purpose of the research was to assess technical condition of pavement made of cement concrete. The subject of the analysis was the district single-carriageway used for 9 and 11 years. Field research was conducted in spring 2 years apart and was aimed at disclosing the progressive pavement degradation process. Within the area of the entire pavement of the length of 5000m and the width of 5m, within each traffic lane, the research cross-sections located every 50m were determined, where the research was conducted each time. For the purposes of analyses, the entire research section was divided into five uniform sections.

Arrangement was defined and the most frequently occurring damages within the analysed pavement were determined, with reference to the reasons for their occurrence. The area of analyses in this case included the entire road surface within transversal cross-section and profile section. The analysis included the classification of the occurring damages, according to [12]. The following symbols of damages were determined: PL for longitudinal and diagonal cracks, PT for transversal cracks, BS – for broken slab, WU – for defective sealing, UN – for corner damage, WS – crack chipping, UP- surface damages. The occurred damages within the analysed section were defined as portions.

Global assessment of technical condition included assessment criteria of pavement operating condition, due to the traffic safety. The assumed technical and operation parameters in case of concrete pavements each time include the assessment of cracks condition (N_j), occurrence of ruts (K_j), the assessment of pavement condition (Sp_j) and defining anti-skid properties (S_j). The assessment of the pavements was determined on the basis of the global factor according to [11], in compliance with (1).

$$G = 100 - [\, W_N \cdot N_j + W_K \cdot K_j + W_{Sp} \cdot Sp_j + W_S \cdot S_j \,] \tag{1}$$

In formula (1) the following symbols W_N, W_K, W_{Sp}, W_S indicate the weight of the assumed parameters, the sum amounts to 1. While N_j, K_j, Sp_j, S_j indicate crack rate, ruts, pavement condition factor, anti-skid properties – determined in accordance with [12, 13, 14, 15]. Due to the weight value variations, which can be dependent on the assumed strategy of road maintenance, the following assumptions were made for the article purposes:

a) priority of improvement of pavement structural condition (which assumed that $N_j + Sp_j$ are 70%)
b) priority of improvement of traffic safety condition (which assumed that $N_j + K_j + S_j$ are 70%).

Within the scope of current assessment of technical condition of the analyzed pavement, the following indicators, according to [4] were determined: Z_B – pavement damage indicator according to (2), D – pavement performance indicator (3) and N – repair indicator (4).

$$Z_B = \frac{1}{A}[\,1.8\,U_1 + 1.22\,U_2 + U_3 + U_4 + U_5\,] \quad (2)$$

$$D = Z - Z_p - N \quad (3)$$

$$N = \frac{R}{A} \cdot 100 \quad (4)$$

The following symbols were approved in formulas (2), (3) and (4): U_1 – surface of cracks, corners and entire slabs subsidence [m^2]; U_2 – surface of cracking and edge chipping [m^2]; U_3 – flaking off surface [m^2]; U_4 – multiple cracks surface [m^2]; U_5 – deep damage surface [m^2]; A – slab surface [m^2]; R – repaired surface [m^2]; Z – damage indicator in 2017; Z_p – damage indicator in 2015.

Anti-skid pavement properties, identified with the value of friction coefficient, is the ability to generate friction force between the pavement and wheels of vehicles during adverse weather conditions. During field studies, the roughness factor was determined in case of each test section in left tyre track on each of traffic lanes in six measuring points, according to [15].

In case of individual measuring points, the average of 5 measurements was assumed as the value. Tests were conducted by means of Portable Skid Resistance Tester in accordance with test procedure, subject to [10]. Reliable rut depth was determined, according to [14] in case of each section of right tyre track. According to the research, the rut was defined as permanent deformation of cross-section of pavement which occurred along the road within the area of wheels impact. For research purposes, the method of 2-metre-patch and wedge was used. Cracks condition in case of the analysed pavement was determined, according to the guidelines of [12].

Pavement operation influences the change of performance parameters, among others macro texture and friction coefficient [5, 6]. Due to the fact of relatively long pavement operation period and the possibility of uncovering of grit grains through mortar, surface layer texture was defined. It is the parameter which also influences the extent of pavement coarseness. It was defined in three points in case of each traffic lane within test sections. The texture was defined by means of Sand Patch Method, according to the guidelines. [8].

Based on the non-destructive method, the condition of cement concrete in pavement structure was defined. Ground-penetrating radar method allowed to recognize invisible changes in concrete structure, such as: dampness, detailed information concerning expansion gaps, loosening between contact layer and base course or the change of layer thickness and delamination. Research was conducted in line with the direction of the moving vehicles in three profiles on each traffic lane. Measurement profiles were located 50 cm from the road edge. Each measurement sequence included the length of the analysed pavement, together with expansion gaps located between the slabs. The extent of identification of deep structure depends on the antenna frequency [2], therefore in order to obtain more accurate information about structure layers, two antennas of diversified frequency were used (1.20 GHz and 500 MHz).

Using sclerometer method – rebound hammer allowed to define the compressive strength of concrete by means of non-destructive method based on the assessment of surface hardness. In the course of field study, in order to determine the strength, the rebound hammer type Digi Schmidt 2N was used. Measurements were conducted within the area of the selected road, according to the guidelines of PN-EN 12504-2:2002 [7]. In order to define average compressive strength using sclerometer method, in case of the discussed pavement, 5 measurement points were found along 4 test sections, on each traffic lane. In case of each point, 10 indications were made and then average value of rebound number per compressive strength, assuming conversion curve was determined.

Standard tests which determine concrete parameters of pavements should be extended by observations of internal structure of hardened cement concrete. Particular attention should be paid to the surface layer because within this particular area, the destructive influence of environmental conditions, intensified by operational load may increase the occurrence of surface damage. Changes which occur within cement matrix significantly influence mechanical parameters of concrete [1]. These parameters, with reference to pavement concrete, are the basic factors which determine its durability. In accordance with the scientific literature [1, 3], in case of hardened concrete, the hardened cement slurry can be distinguished (also referred to as cement matrix), grains of coarse and fine aggregate (mainly of diversified mineral composition and diversified shapes and sizes), air voids, pores, cracks and various types of reinforcement. Preliminary observations were carried out, which allowed to distinguish individual components. Microscope analysis of concrete sample was conducted by means of Scanning Electron Microscope of Quanta 3D FEG Dual Beam type. Using concrete samples, preparations, subject to SEM observations were made, surface of which was of more than 100 mm2 each time. Using computed tomography allow obtaining layered images of the tested cement concrete sample. By means of tomography method, we can measure precisely individual elements of internal concrete structure in all directions and identify spatially possible defects or discontinuities. [3] Exemplary micro photographs obtained in case of samples collected from the cement pavement were presented on figure 2. In case of concrete, numerous micro cracks were observed. In case of cement matrix, these cracks were distinguished by average aperture of 8 μm. Cracks of interface between aggregate grains and cement matrix occurred mostly within the whole grains area and their maximum aperture was up to 5 μm. One of the most significant elements of internal structure of cement composite, which is considered as structure concrete service life determinant, is porosity characteristics. Diversity concrete porosity has also been observed. The air pores diameters in case of this concrete were dependend of cross section.

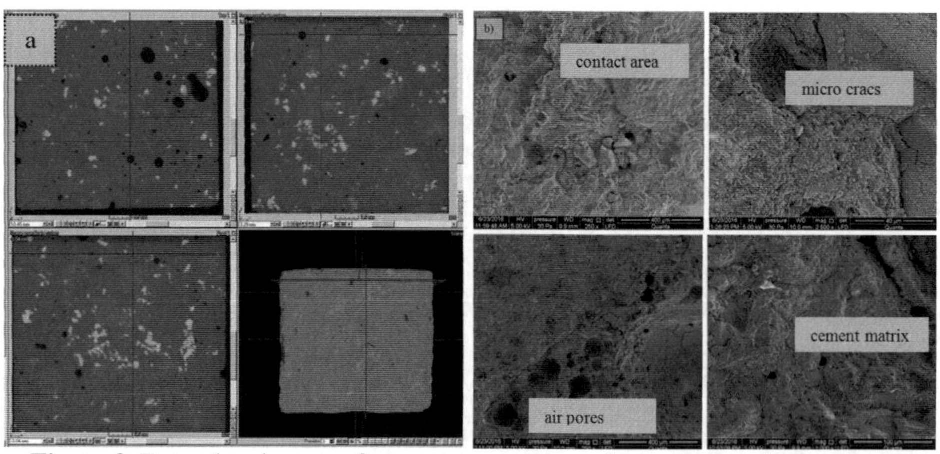

Figure 2. Exemplary images of cement concrete samples together with analysed cross-sections a) using computed tomography method (depth slice, blue and red lines depth section, surface view) b) scaning electron microscopy method (contact area, micro cracs, air pores, cement matrix).

3. Results and discussion

According to the visual assessment of pavement condition, the significant influence of operation on the change of the occurred damage quantity was proved. In case of tests conducted within the first measurement period, in total, 41 damages were classified, out of which the prevailing majority were defective sealing expansion gaps - Figure 3. Subsequent two years of operation contributed to the progressive pavement degradation and the number of the detected damages increased up to 108. Similarly, during the previous period, damages of expansion gaps filling prevailed. Widths of the occurred longitudinal and crosswise gaps ranged from 8 to 11mm.

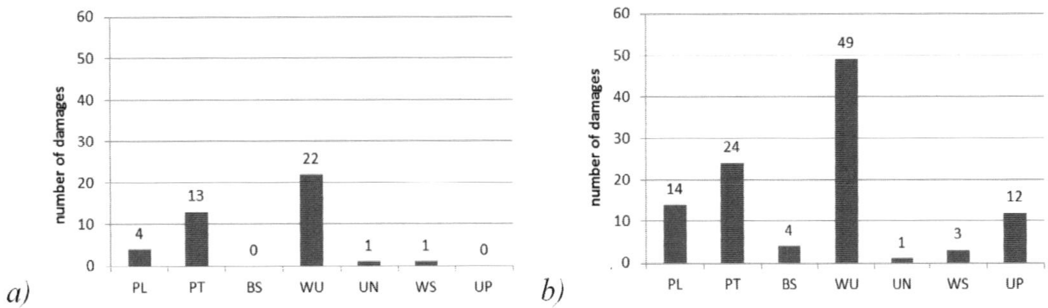

Figure 3. Qantitative specification of the occurred damages within the analysed pavement in 2015 (a) and 2017 (b).

It should be noted that single crosswise, longitudinal and diagonal cracks occurred within the analysed section. All damages of this type, both in case of the first and the second analysed period, were of minor damage potential, according to [11]. In case of analyses conducted in 2017, additionally surface damages in the form of flaking and local defects and 4 broken slabs were found. Predominant amount of defective expansion gaps filling proves omission of proper pavement maintenance. Within the analysed section, the absence of filling and local extrusion of poured sealant were found. Gap filling, in most cases has lost its initial properties and loosened from slab edges, mixture hardened and numerous cracks occurred.

Calculated global pavement condition factor in case of this road reached high values, which proves good pavement condition. The conducted analyses proved that there is no predominant parameter. None of technical and operational parameters was qualified to C or D class, which proves the warning or critical technical condition of the pavement - table 1.

Table 1. Assessment of the analysed pavement for uniform sections, assuming the priority of improvement of pavement structural condition (a) and the improvement of traffic safety condition (b).

Year	W_N priority		N_j	W_K priority		K_j	W_{Sp} priority		Sp_j	W_S priority		S_j	G priority	
	a	b		a	b		a	b		a	b		a	b
2015	0,4	0,2	0,98	0,3	0,2	2,15	0,2	0,2	0,99	0,1	0,4	0,73	98,7	98,9
2017	0,4	0,2	0,96	0,3	0,2	3,85	0,2	0,2	0,97	0,1	0,4	0,68	98,2	98,6

Calculated indicators of the current assessment of pavement technical condition, in case of the analyzed section reached the values corresponding to very good performance condition in 2015 and 2017 - table 2.

Table 2. Determined indicators of the current technical assessment together with the requirements for very good condition (wym).

Year	Z	Z_{wym}	S	S_{wym}	R	R_{wym}	D	D_{wym}	N	N_{wym}
	%									
2015	4	< 10	0.30	< 0.57	0.09	< 1.0	0.04	< 1.0	0	< 1.0
2017	8	< 10	0.42	< 0.57	0.12	< 1.0	0.04	< 1.0	0	< 1.0

According to the conducted research concerning friction coefficient of pavement (Figure 4a), it was proved that the extended period of operation influences the reduction of the tested characteristic. However, the entire research section, independent of test date, proves satisfactory coarseness -Figure 4b.

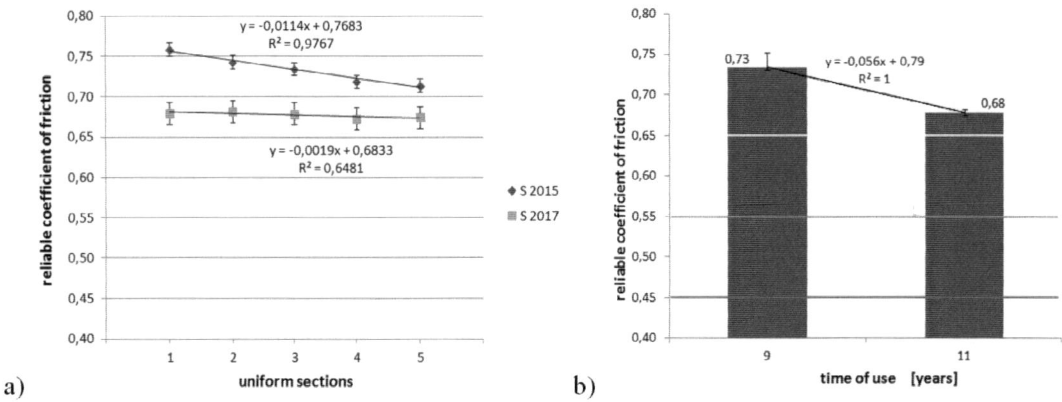

Figure 4. Average value of reliable friction coefficient for the uniform section (a) and for the entire analysed pavement section (b) (R^2 - correlation measure).

According to the determined reliable pavement rut depth it was proved that the extended period of operation influences the increase of the tested characteristic. However, the entire research section, independent of the test date, proves good condition - Figure 5.

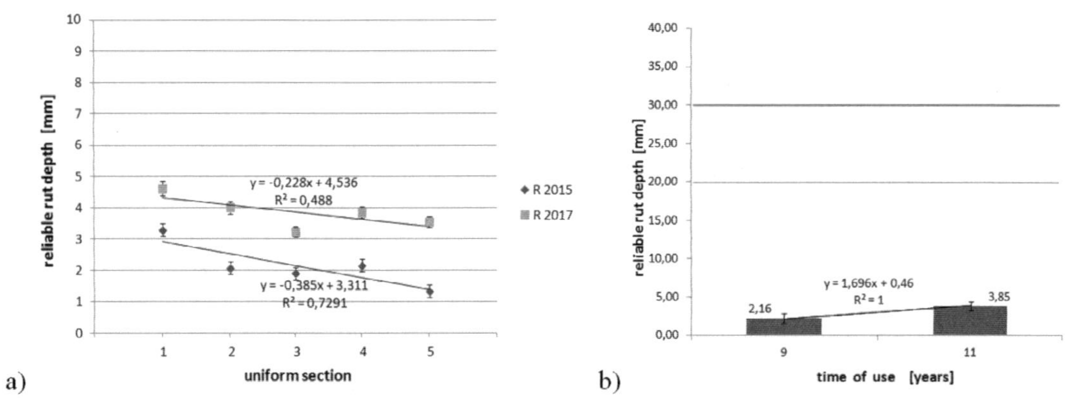

Figure 5. Average reliable rut depth for the uniform section (a) and for the entire analysed pavement section (b) (R^2 - correlation measure).

According to the conducted tests of pavement macro texture, it was proved that the extended period of operation does not significantly influence the tested characteristic. The entire test section, independent of the test date, proves average value of macro texture depth within the reliable scope – Figure 6.

IOP Conf. Series: Materials Science and Engineering **356** (2018) 012026 doi:10.1088/1757-899X/356/1/012026

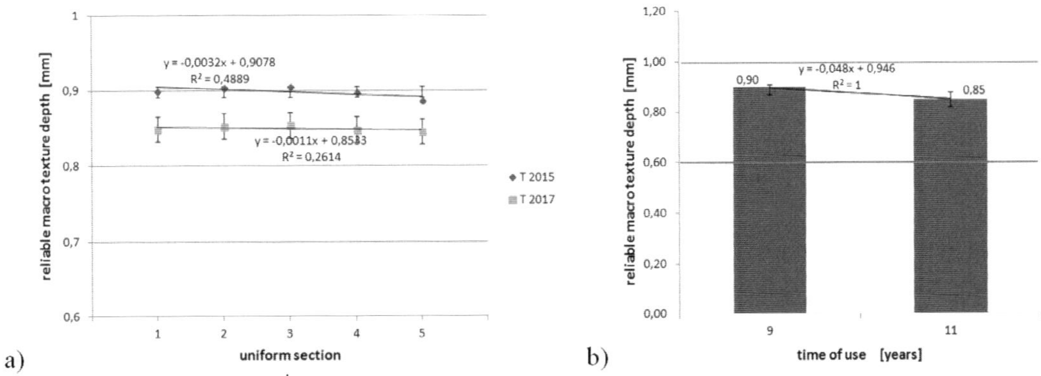

Figure 6. Average value of reliable depth of macro texture in case of uniform section (a) and entire analysed pavement section (b) (R^2 - correlation measure).

In accordance with the obtained sclerometer test results, it was proved that performance parameters of the analysed pavement have not deteriorated for two years - Figure 7.

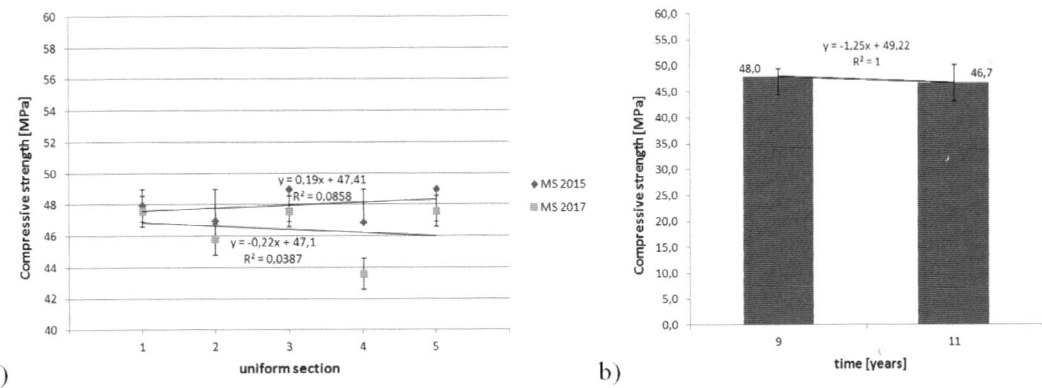

Figure 7. Average value of compressive strength - sclerometer test results - a) for uniform section (1-5), b) for the entire analysed pavement section in 2015 and 2017 (R^2 - correlation measure).

Based on the ground-penetrating radar reading, local relaxation of the structure was proved within the analysed section. The aforementioned loosening was found within sections located directly under the damaged expansion gaps (c). As a result of continuity disruption or absence of mixture filling the expansion gaps, in favourable conditions, due to migration of water and humidity deep into the pavement, degradation process of the structure progresses. Exemplary excerpts of echograms registered on the analysed road has been included in Figure 8. Within the analysed area, it was possible to determine cement concrete thickness, both in case of pavement surfacing (a) and base course (b). According to the obtained data, possible damages of structural layers of pavement were also determined. Within the entire analysed section damages within the area of layer interface (voids and delamination) (d) and damages within the area of individual layers (cracks and gaps) were identified (e).

IOP Conf. Series: Materials Science and Engineering **356** (2018) 012026 doi:10.1088/1757-899X/356/1/012026

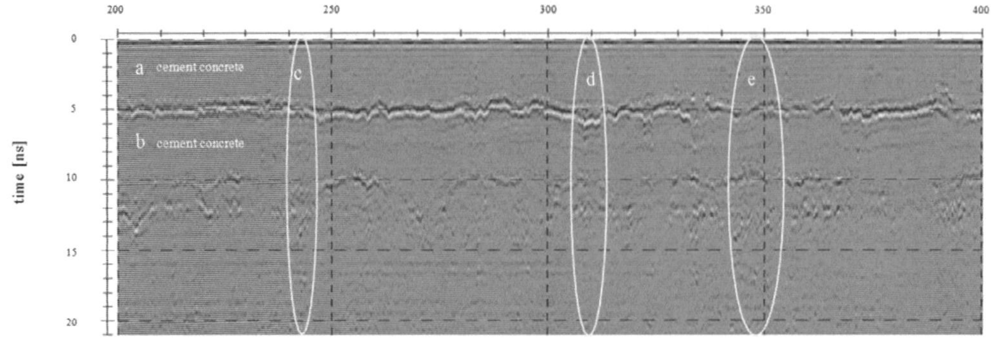

Figure 8. Exemplary echogram registered on the analysed pavement, KM 0+200 - KM 0+400.

4. Assessment of the influence of pavement operating time on the change of pavement parameters

The global pavement condition determined according to SOSN [11] and the determined indicators of the current technical assessment [4] prove very good pavement condition, which does not require any repairs. However, due to the fact that expansion gaps are the weakest element of concrete pavements and damages occurring within this area most often contribute to degradation of surface condition and the occurrence of damages within this section, the maintenance of expansion gaps filling is recommended in case of the entire analysed pavement. The occurred damages of this pavement element are the consequence of improper maintenance policy. Durability of the applied pourable sealant is maximum 5 years and it should be replaced by the new one afterwards. In the vicinity of the damaged gaps, there are rare corner damages which result from the insufficient slab support within this area. The change of slab support within corner area may be the consequence of water filtrated through untight expansion gaps filling. Damages of pourable sealant and local discontinuity thereof contribute to loosening of the structure and material washed-out and corrosion. With reference to thereof, in case of the analysed pavement, it is recommended to perform maintenance procedures concerning the replacement of the pourable sealant. Such solution will prevent the structure from damage progress.

Cracks of minor harmfulness occurring on the surface within the current period of time, do not require any repair procedures. It is recommended to apply surface treatment in the form of soaking components. However, it should be emphasised that constant monitoring of the condition of these cracks is necessary, as well as all the aspects which influence pavement technical condition. Delamination of concrete observed on the pavement surface prevailed within areas of the increased unevenness, where rainwater can cumulate during favourable weather conditions. The occurred damages did not exceed 3mm and were classified as shallow and did not require any repairing actions. However, reduction of surface destruction process is recommended. For this purpose, soaking components can be used, which will reduce surface water penetration deep into the structure through concrete sealing and reducing its absorbability.

Pavement, due to its texture parameters and roughness, does not require application of procedures which will increase this parameter in the current condition. According to the analyses of the obtained results - Figure 9 significant correlation between the examined characteristics was proved. The significance increases together with the extension of performance period. In case of the pavement after 9 years of operation, the correlation coefficient of the examined characteristics remains within the average level, while after the next 2 years of operation, it reaches the high level.

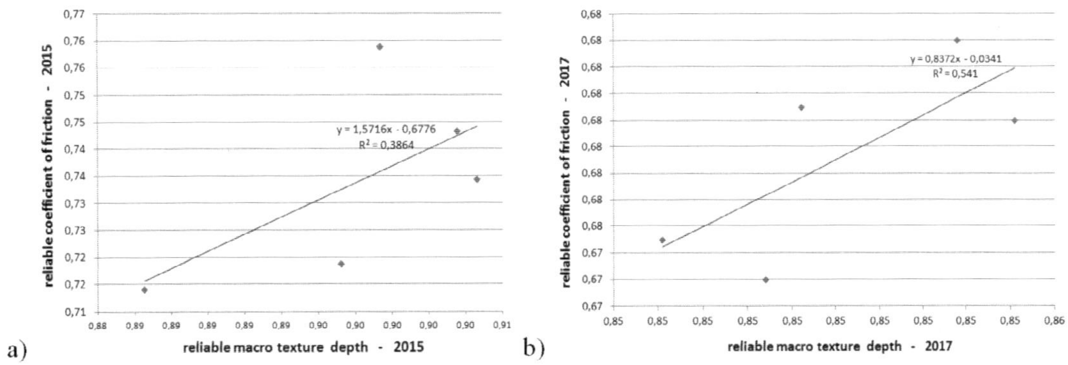

Figure 9. Correlation of pavement roughness and texture in 2015 (a) and 2017 (b).

Pavement evenness is good and it is not necessary to introduce any procedures aimed at the increase of this parameter during the current period of time.

5. Conclusions

According to the conducted research, the following conclusions were reached:

- The analysed pavement made of cement concrete is resistant to the occurrence of permanent deformations in the form of ruts, independent of operating time and weather conditions, within the analysed area and the applied load;
- The presented quantitative analysis has not proved significant damages and the global assessment of pavement condition proves good condition thereof;
- The pavement is distinguished by high performance parameters, which is referred to comfort and most of all, operation safety;
- Based on the ground-penetrating radar reading, local loosening of structure was proved within the analysed section. Which for the current situation not cause the significant threat to pavement durability.
- Advantages of concrete pavement meet users expectations, however due to the nature of expansion gaps, for proper structure operation it is recommended to clean local filler gaps and then filled in with pourable sealant.
- It is also necessary to analyse wear of the upper surface of concrete road in tribologic aspect. To justify thereof, pavement condition with respect to the change of its roughness, in the course of its operation should draw the attention, i.e. classic issue of friction occurring between vehicles moving on the pavement during slowing down.

References

[1] M. A. Glinicki, *Trwałość betonu w nawierzchniach drogowych. Wpływ mikrostruktury, projektowanie materiałowe, diagnostyka / Durability of concrete in road surfaces. Effect of microstructure, materials design, diagnostics*, IBDiM, Warsaw, 2011.

[2] J. Karczewski, *Zarys metody georadarowej / Outline of the georadar method* (in Polish), AGH, Cracow, 2007.

[3] M. Linek, *Using Scanning Electron Microscope and Computed Tomography for Concrete Diagnostics of Airfield Pavements*, World Academy of Science, Engineering and Technology, International Journal of Civil, Environmental, Structural, Construction and Architectural Engineering, 10(7), 2016, pp. 964 - 968, urn:dai:10.1999/1307-6892/10005550

[4] P. Nita, *Budowa i utrzymanie nawierzchni lotniskowych / Construction and maintenance of airport pavements* (in Polish), WKiŁ, Warsaw, 2008.

[5] P. Nita, *Betonowe nawierzchnie lotniskowe. Teoria i wymiarowanie konstrukcyjne / Concrete airfield pavements. Theories and structural dimensioning* (in Polish), Air Force Institute of Technology, Warsaw, 2005.
[6] A. Szydło, *Nawierzchnie drogowe z betonu cementowego / Road surfaces made of cement concrete* (in Polish), Polish Cement, Cracow, 2004.
[7] Polish Standard PN-EN 12504-2:2013-03 Concrete tests in structures - Part 2: Non-destructive testing - Determination of the number of reflections
[8] Polish Standard PN-EN 13036-1:2010 Surface features of road and airport pavements - Test methods - Part 1: Measurement of depth of macrotexture by volumetric method
[9] Polish Standard BN - 68 / 8931-04 Road car. Measurement of pavement equality with planograph and patch.
[10] Polish Standard PN-EN 13036-4: 2011 Roads and airports - Test methods - Part 4: Measurement method of slip resistance / slip on the surface: Pendulum test
[11] Application guidelines of „Concrete Pavement Condition Assessment System", SOSN-B.
[12] Application guidelines of „Concrete Pavement Condition Assessment System", SOSN-B. Attachment 1: Concrete Pavement Condition Assessment System.
[13] Application guidelines of „Concrete Pavement Condition Assessment System", SOSN-B. Attachment 2: Catalogue of typical concrete pavement damages for the purposes of continuous quantity survey of damage using visual method, according to Concrete Pavement Condition Assessment System.
[14] Application guidelines of „Concrete Pavement Condition Assessment System", SOSN-B. Attachment C: Principles of measurement and assessment of bituminous pavement ruts condition, according to Pavement Condition Assessment System SOSN.
[15] Application guidelines of „Concrete Pavement Condition Assessment System", SOSN-B. Attachment D: Principles of measurement of assessment of anti-skid properties of bituminous pavements, according to Pavement Condition Assessment System SOSN.

IOP Conf. Series: Materials Science and Engineering **356** (2018) 012027 doi:10.1088/1757-899X/356/1/012027

Qualitative criteria and thresholds for low noise asphalt mixture design

A Vaitkus[1], T Andriejauskas[2], J Gražulytė[3], O Šernas[4], V Vorobjovas[5], R Kleizienė[6]

[1]Director, Road research institute, Vilnius Gediminas Technical University, Lithuania
[2]Junior research fellow, Road research institute, Vilnius Gediminas Technical University, Lithuania
[3]Junior research fellow, Road research institute, Vilnius Gediminas Technical University, Lithuania
[4]Rresearch fellow, Road research institute, Vilnius Gediminas Technical University, Lithuania
[5]Deputy director, Road research institute, Vilnius Gediminas Technical University, Lithuania
[6]Research fellow, Road research institute, Vilnius Gediminas Technical University, Lithuania

E-mail: tadas.andriejauskas@vgtu.lt

Abstract. Low noise asphalt pavements are cost efficient and cost effective alternative for road traffic noise mitigation comparing with noise barriers, façade insulation and other known noise mitigation measures. However, design of low noise asphalt mixtures strongly depends on climate and traffic peculiarities of different regions. Severe climate regions face problems related with short durability of low noise asphalt mixtures in terms of considerable negative impact of harsh climate conditions (frost-thaw, large temperature fluctuations, hydrological behaviour, etc.) and traffic (traffic loads, traffic volumes, studded tyres, etc.). Thus there is a need to find balance between mechanical and acoustical durability as well as to ensure adequate pavement skid resistance for road safety purposes. Paper presents analysis of the qualitative criteria and design parameters thresholds of low noise asphalt mixtures. Different asphalt mixture composition materials (grading, aggregate, binder, additives, etc.) and relevant asphalt layer properties (air void content, texture, evenness, degree of compaction, etc.) were investigated and assessed according their suitability for durable and effective low noise pavements. Paper concluded with the overview of requirements, qualitative criteria and thresholds for low noise asphalt mixture design for severe climate regions.

1. Introduction

Traffic noise is an increasing environmental problems which is highlighted as a second major environmental problem after the air pollution. Unfortunately, traffic noise is often underestimated as it's negative impacts affect society not instantly but in a longer term. Exposure to excessive noise levels leads to various health related problems, affects work productivity, sleep quality as well as negatively affects some animal species and countries' economies.

IOP Conf. Series: Materials Science and Engineering **356** (2018) 012027 doi:10.1088/1757-899X/356/1/012027

Calculated approximate EU annual socio-economic costs because of traffic noise are higher than 40 billion EUR and expected to increase 50 % by 2050 [1]. Therefore, huge efforts are needed to decrease noise pollution in EU Member States. Implementation of Directive 2002/49/EC of the European Parliament and of the Council of 25 June 2002 relating to assessment and management of environmental noise is one of the solutions, however implementation needs to be fostered and improved in terms of completeness, comparability and adaptation of highly effective solutions.

From the road infrastructure owners'/managers' perspective, increased attention was also given to tackle the road traffic noise problem – traffic noise problem was highlighted in CEDR Strategic Plan 2014-2017 [2] as an important future road infrastructure challenge. In-line with EU Noise Directive, National road authorities are developing and implementing strategic noise management plans at national levels. However, in many cases national strategic noise management plans are focused on a narrow set of noise mitigation solutions – mainly related with the construction of noise barriers and façade insulation. Only a holistic approach and combination of different and adapted noise mitigation measures implementation can lead to the highest noise level reduction experience.

Even if noise barrier can achieve high noise level reduction rate, costs for their construction and maintenance are relatively high. Such situations are very common in new road construction and road rehabilitation projects, where it is necessary to apply noise reduction measures in order to meet environmental requirements even if there are single or few properties along the road. Additionally, their implementation is sometimes impossible due to the technical, legislative and landscape related restrictions.

Low noise asphalt pavements are cost-efficient and cost-effective alternative noise mitigation solution, which implementation has a big potential, especially in urban areas or in highways. Various research and feasibility studies has shown that low noise pavements are much more cost efficient solution comparing with noise barriers and façade insulation for every single reduced dBA or per treated km (Fig.1) [3]. Some other examples show that it is less expensive to achieve required noise reduction effect by using combination of low noise pavements and noise barriers than only barriers in a longer treated distance.

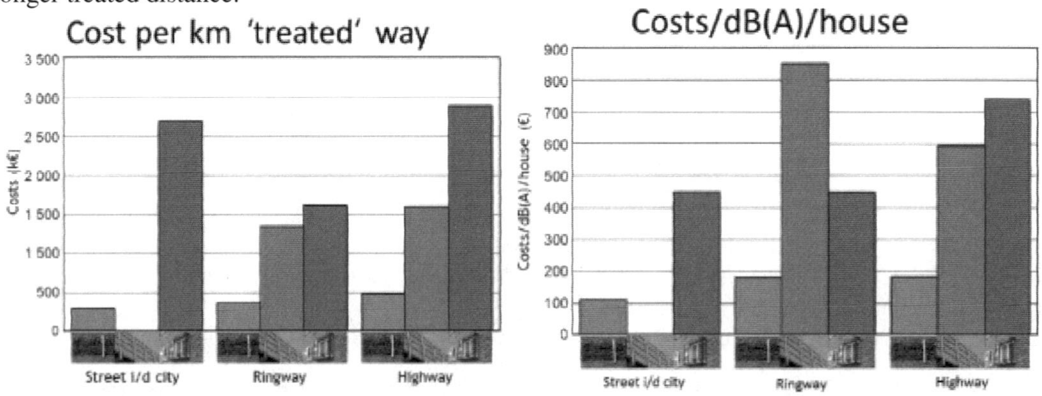

Figure 1. Costs of three noise abatement measures for three typical cases [3].

2. Low noise asphalt mixture design prerequisites

Optimisation of asphalt mixtures for tyre/road noise reduction largely depends on the potential application area of these mixtures – where it will be used, under what traffic and what climate conditions. Various conditions can be related with different set of tyre/road noise influencing factors that has to be taken into account when modelling and optimising low noise asphalt mixtures.

Design of the low noise asphalt mixtures is even more complicated for severe and clod climate regions where large temperature fluctuations and large number of frost-thaw cycles occur. In such cases, it is essential to take design low noise pavements with sufficient durability and resistance to climate conditions. To address this challenge, national research project NOPE on the design of low noise asphalt mixtures for severe climate conditions has been started in December 2017 in Lithuania.

This paper presents first stage of this project aiming to assess tyre/road noise influencing factors and their impact on the tyre/road noise generation and to combine these influencing factors with set of requirements, criteria and particular recommendations low noise asphalt mixture design. Previous research projects on this topic [4] showed that optimised thin asphalt layer mixtures of modified AC and SMA pavements have highest potential in colder climate regions, therefore this paper focuses mainly on this type of pavements.

2.1. Tyre/road noise influencing factors

Influencing factors for tyre/road noise generation can be distinguished into 4 major groups [5,6]: Road surface characteristics (road surface texture, aggregate properties, air void content, pavement stiffness, pavement colour, pavement age and deterioration), Tyre characteristics (number of tyres, tyre dimensions, tyre internal structure, tyre inequality, rubber hardness, tread pattern and structure, age and wear, tyre type, studded tyres), Environmental factors (moisture, wind, temperature, water on the surface, dust) and Factors controlled by driver (driving speed, tangential forces and acceleration, tyre load and pressure). Set of influencing factors affect noise generation mechanisms related with noise due to tyre vibrations, aerodynamic effects and amplification/reduction effects. Each of the influencing factor can be associated with specific noise frequency range.

Main criteria that needs to be evaluated before selecting appropriate low noise pavement solution in particular location are allowable speed limits, traffic volumes and composition as well as possibility to use studded tyres. With the increasing driving speed, tyre/road noise levels intend to increase too, thus it is necessary to know the speed limits for particular locations. Some low noise asphalt solutions are recommended to be used only at higher speeds (e.g. porous asphalt) due to the self-cleaning effects while at the lower speeds such solutions might not be acoustically durable due to fast clogging [7]. Since road surface texture and surface porosity are the main road pavement influencing factors that have major impact on induced tyre vibrations and air pumping noise generation mechanisms, it should be also noted that at higher speeds for smooth pavements air pumping mechanisms are more dominant than tyre vibration related mechanisms [8].

Traffic volumes (AADT) and traffic composition (percentage of heavy duty vehicles in the whole traffic flow) are also important to take into account when modelling low noise pavements. There are differences between the noise frequencies and noise generation mechanisms for passenger and heavy vehicles tyres, therefore, surface texture needs to be optimized using different maximum aggregate size. Additionally, high traffic volumes and use of studded tyres might lead to faster road surface texture deterioration and polishing what can also be linked with increasing noise levels.

2.2. Asphalt mixture optimization for tyre/road noise reduction

After the initial analysis of the traffic related influencing factors, next step is design and optimisation of asphalt mixture with its relevant characteristics for tyre/road noise reduction.

Asphalt mixture porosity and flow resistance are very important factors that impact air-flow induced noise generation mechanisms such as air pumping. High porosity also increases sound absorption properties of asphalt mixtures. It is known that air void content of at least 10% makes road surface acoustically absorbing [5]. However, the surface will be acoustically efficient only when air void content is >20%. According to the air void content and potential for tyre/road noise reduction, asphalt mixtures can be grouped to dense pavements (4-9% air void content), semi-dense (10-14% air void content), semi-porous (15-19% air void content) and porous (>19% air void content) as shown in Figure 2 [9]. The design goal is to obtain the maximum amount of sound absorption at a frequency of 1000 Hz for high speed roads and of 600 Hz for low speed roads [10] as well as to ensure the frequency spectrum bandwidth of the absorption coefficient as broad as possible. Moreover, it is desirable that the pores in the surface layer should be interconnected in order to ensure good water and air drainage [5]. Shape of the air voids may have also a significant effect on mixture fatigue, stiffness and clogging. Thus, it is recommended to design mixture with circular shape air voids rather than polygonal [11].

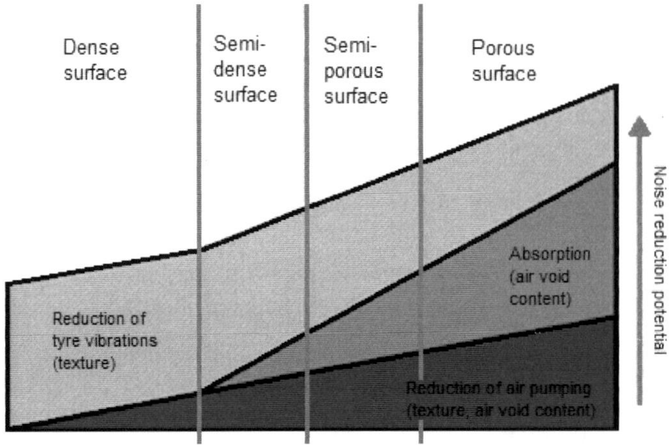

Figure 2. Noise reduction potential according to the surface type (adapted from [9]).

To minimize tyre vibration induced noise generation mechanisms road surface texture must be optimized. Texture parameters are also not only important for tyre/road noise reduction but also to minimize splash and spray, rolling resistance, reflection, glare and ensure skid resistance. Therefore, it is important to find the right balance and compatibility between these characteristics. Texture optimization is performed mostly at macrotexture scale [5] by selecting aggregate type, size and mineralogy parameters. According to [5, 12], megatexture and macrotexture is positively correlated with low frequency noise (less than 1000 Hz) while microtexture is negatively correlated with high frequency noise (more than 1000 Hz). Main principles of texture optimization are [5]:

- Macrotextures should have high amplitudes in the 1–8 mm wavelength range and low amplitudes in the 10–50 mm wavelength range for light vehicles' noise reduction.
- Macrotexture should have high amplitudes in the 0.5–12 mm wavelength range and low amplitudes in the 16–50 mm wavelength range for heavy duty vehicles' noise reduction.
- Small size maximum aggregate should be used to achieve smooth surface and reduce tyre vibrations. Max. aggregate size of 4–6 mm for light vehicle tyre noise reduction and max. aggregate size of 6–10 mm for heavy duty vehicle tyre noise reduction.
- Open and 'negative' texture that is characterized by a large amount of narrow and small spaces between the particles not like 'positive' texture that have large amount of irregularities.
- Megatexture should be minimized by avoiding large aggregate sizes and ensuring homogeneous macrotexture. Otherwise "missing aggregates" or large spaces between the aggregates will cause megatexture to increase.

Both to achieve optimised surface texture and ensure durable performance of low noise asphalt wearing layer it is essential to select proper aggregate and materials. As it was written above, selection of aggregate size is important to form "negative" macrotexture. Main recommendations for the aggregates for low noise asphalt mixtures are:

- Select aggregates with sharp edges and of a polishing resistant material to retain the sharpness as long as possible. Polishing reduce aggregate sharpness and the amplitude in general at short texture wavelengths resulting in decrease of the air drainage between the tyre tread rubber and the road surface [5, 13].
- Select cubic shape, uniform and with sharp edges aggregate to ensure even and smooth surface. It is also desirable to pack aggregates close together in surface texture [5, 14].
- Petrography of the materials is of importance as different mineralogical composition could have an impact on the durability of an asphalt mixture [13].
- Specific requirements to assess when selecting granular materials for SMA mixtures [15]: Los Angeles abrasion (<30 %), fractured faces (>90 % for particles with two faces, 100 for

particles with one face), flat and elongated particles (<5 and 20 % (rations 5:1 and 3:1 respectively)), fine aggregate angularity (FAA) (>45).

- Polishing resistance materials are recommended to avoid polished surfaces. Microtexture of the crushed stone is therefore of great influence on the skid resistance, particularly in rainy weather.
- Parameters that have significant impact on surface wear resistance [16]:
  - Aggregate quality. On a descending quality scale: Porphyry, quartzite, granite, gneiss
  - Use of mixed aggregate qualities. Wear resistance is given by proportional contributions from each aggregate source

Main recommendations for the grading of low noise asphalt mixtures [5, 17]:

- Since it is necessary to create open and "negative" type surface textures specific grading should be used – sand content has to be minimized and aggregate size should not be larger than 8 mm (for light vehicles) or 12 mm (for heavy vehicles). Texture openness between the particles will provide better air drainage, which reduces air-pumping noise generation
- In order to reduce both the vibration generated tyre-road noise at frequencies below 1500 Hz and the aerodynamic noise at higher frequencies a "steep" aggregate grading curve where filler and very small aggregates are combined with large aggregates but without aggregates of the sizes in between should be used.
- Once a mix has been designed with a given grading curve and the performance of this mix has been demonstrated, there is little margin for variations, and a need to keep the grading as constant as possible, because mixes with a stony skeleton are very sensitive to this. Changes in grading could disrupt the stony skeleton and lead to a loss of stability or a loss of cohesion.

Binder in the low noise asphalt mixtures also has a significant role for the mixture durability and performance. Specific binder recommendations for low noise asphalt mixtures [5]:

- The choice of binder can influence the ageing properties of the surface. The significant influence is on the initial porosity and clogging rate of porous surfaces. Therefore, it is recommended to use a binder that makes possible the highest initial porosity (communicating air voids) and that makes the surface as resistant to clogging as possible.
- The susceptibility to clogging is higher porous and semi-porous mixtures are made with higher binder contents. However, higher binder contents implicate mixtures with thicker binder film coating, making the mixture more resistant to weathering hence more durable [18].
- Avoid a binder that gives a very stiff surface (tentatively).
- The design binder content is obtained by adding the amount of binder drain down to the provisional binder content.
- Binders that include crumb rubber powder don't give lower noise due to reduced stiffness. However, asphalt mixtures, including crumb rubber in the binder has less problems with binder drain down (e.g. SMA mix with crumb rubber binder had 0% drain-down at 162°C and 177°C indicative of the crumb rubber's behaviour as a stabilising agent, and thus improving the resistance of SMA mix towards drain down [19])
- Major surface distress type for low noise asphalt surfaces is ravelling which is mostly a low temperature phenomenon. Adhesive failure commonly occurs at low temperatures (mostly, moisture induced damage) and cohesive failure at intermediate temperatures. The latter mechanism is caused by failure due to slow demixing (segregation) of the mastic and ageing of the binder [13]. It is recommended to assess adhesion and cohesion behaviour when selecting binder for low noise asphalt mixtures.

To improve properties and performance of low noise asphalt mixtures, various additives and modifiers can be used such as ground tyre rubber, styrene-butadiene-styrene (SBS) and styrene-butadiene-rubber for durability improvement by increasing the stiffness and ductility of the binder. If modified appropriately, the binder may prevent short-term ravelling that is caused by the shear forces between the tyre/road interface.

Additionally, cellulose fibre, hydrated lime and minerals (0.3 to 0.5 percent by mass) might be used to prevent drain down problems during transport and placing and reinforce the film thickness of the asphalt binder [20].

3. Low noise asphalt layer thickness, laboratory testing and paving techniques

Acoustical performance and noise reduction potential of semi-porous and porous low noise surfaces largely depends on the layer thickness. For the thin asphalt layers, it is recommended that layer thickness should be between 2.5 and 4 times the nominal maximum aggregate size (NMAS) [16]:

- If layer thickness is smaller than 2.5 times NMAS then there is a chance to experience problems in achieving the sufficient and desired level of compaction and thereby endangering the durability of the pavement.
- If layer thickness is larger than 4 times NMAS then there is a risk that the traffic loads will generate permanent deformation (rutting) in the layer.

In order to obtain a frequency-optimised absorption of the porous asphalt mixtures, it is recommended that [5]:

- Layer thickness should be 38 mm for high speed roads and 63 mm for low speed roads
- Layer thickness should be at least 40 mm, and preferably as high as possible.

Recommendation for the compaction of the low noise asphalt mixtures [15, 21]:

- During the compaction, aggregate particles may disintegrate due to the impact of the hammer, confining pressure, inter-particle friction, and applied boundary conditions. Among all the compaction methods used, Marshall compaction approach is more susceptible for aggregate degradation. Due to this degradation of the aggregates the actual mixture properties could be different from the designed/calculated properties such as air voids, voids in mineral aggregates, etc. Therefore, it is recommended to use gyratory compaction in the laboratory.
- For thin asphalt layers, air voids are less important than for thicker (porous) asphalt layers to border effects and its open texture. Therefore, it is more relevant to control compaction level for the thin asphalt layers.
- Inadequate compaction should be avoided as it may cause large variations in tyre/road noise levels and might foster surface layer acoustical and structural degradation.

In order to guarantee the quality and durability of low noise asphalt mixtures, the designed mixtures should be validated with respect to the performance of the low noise asphalt mixture. Asphalt mixture permeability, binder drain down, macrotexture and resistance to wear are the key performance based characteristics for low noise asphalt mixtures to be controlled [15]:

- Asphalt mixture permeability is important to evaluate both volumetric composition of the mixture and air and water drainage capabilities what has significant impact on layer acoustical performance.
- Cantabro test (for dry and moisture samples) is recommended to evaluate ageing processes and resistance to abrasion after stiffening (oxidation) of the binder.
- Resistance against wear from the tyres are recommended on high-volume roads in countries and regions with a substantial proportion of studded tyres. Resistance to wear from studded tyres could be evaluated using the Nordic abrasion test (EN 1097-9).
- Surface macrotexture are necessary to evaluate because of macrotexture's direct impact on tyre/road noise levels. Mean profile depth (MPD) and mean texture depth (MTD) are the main parameters that are recommended to be assessed during the laboratory development of low noise asphalt mixtures.
- Resistance to rutting tests are done not only evaluate the rutting resistance, but rather for the evolution of the surface texture durability after a number of wheel passes.
- As it was mentioned in previous chapter, binder drain down is an important factor that characterizes durability of the low noise asphalt mixtures, therefore it is necessary to test this parameter in the laboratory.

- Since low noise asphalt layers are design with a stony skeleton (stone-on-stone contact of the coarse aggregate), permanent deformation is not generally considered as main issue for thin low noise asphalt layers and porous asphalt mixtures.

Special attention has to be given for low noise asphalt mixture paving techniques [5, 15, 17]:

- Thin asphalt layers tend to cool faster than thicker pavements, therefore paving temperature should be controlled with higher attention than for thicker pavements. This means, that thin asphalt layers should be compacted faster. Additionally, restrictions for thin asphalt laying under low air temperatures should be made too.
- It is recommended to use vibratory screed which width is adjusted with auger's width in order to ensure the homogenous paving conditions. To eliminate risk of mixture segregation during the paving, it is necessary to use transporters/mixers.
- To ensure homogeneous, open and "negative" surface texture, paving process should be seamless, continuous and at the constant speed.
- Cold joints should be avoided.
- It is not recommended to use vibrational and pneumatic compaction rollers because it may have negative impact on surface texture and layer structure. It is recommended to use steel rollers, heavier than 6 t. Use of rollers heavier than 11 t. is not recommended because it may damage surface particles and surface texture.
- One of the most important factors is to ensure that low noise asphalt mixture is not over compacted.

4. Conclusions and recommendations

Analysis of tyre/road noise influencing factors and assessment of typical traffic and environmental conditions first step for the selection of low noise asphalt mixture solution. Such background information allows to select proper criteria for specific low noise asphalt mixture design.

Overviewed and analysed worldwide experience and best practice of low noise asphalt mixture design, testing and construction revealed main low noise asphalt mixture design criteria for severe climate regions. Taking into account structural and acoustical ageing peculiarity of various low noise asphalt pavements and relevant influencing factors, it can be stated that thin low noise asphalt layers are the most promising cost effective low noise pavement solution for colder regions. However, specific attention should be given for these mixtures design and exploitation.

In comparison to traditional asphalt mixtures additional properties must be controlled for low noise asphalt mixtures' design: binder drain down, resistance to ravelling, surface texture evolution and permeability.

This paper presents an initial stage of low noise asphalt mixture design for severe climate regions. Based on the performed assessment of various tyre/road noise influencing factors and low noise asphalt mixture design peculiarities, further step will include design and modelling of low noise asphalt mixtures.

Acknowledgment

This research is funded by a grant (No. S-MIP-17-137) from the Research Council of Lithuania.

References

[1] European Commission. Report from the Commission to the European Parliament and the Council: On the implementation of the Environmental Noise Directive in accordance with Article 11 of Directive 2002/49/EC, COM(2011) 321 final. Brussels 2011 p 13
[2] CEDR Strategic Plan 3 2013-2017 Conference of European Directors of Roads p 27
[3] Ellebjerg L L, Bendtsen H 2001 Costs and perceived noise reduction of porous asphalt pavements, Proceedings of the 2001 International Congress and Exhibition on Noise Control Engineering (INTERNOISE 2001), The Hague, The Netherlands
[4] Vaitkus A, Andriejauskas T, Vorobjovas V, Jagniatinskis A, Fiks B, Zofka E 2017 Asphalt

wearing course optimization for road traffic noise reduction. Construction and Building Materials. Oxford: Elsevier. Vol. **152** p 345-356. https://doi.org/10.1016/j.conbuildmat.2017.06.130

[5] Sandberg U, Ejsmont J 2002 Tyre/Road Noise Reference Book. Informex, Harg, SE-59040, Kisa, Sweden, p 640

[6] Kane M, Scharnnigg K, Conter M, Roe P, Schwalbe G 2009 Report on Different Parameters Influencing Skid Resistance, Rolling Resistance and Noise Emissions, TYROSAFE Project, p 95

[7] Ranieri V, Sansalone J J, Shuler S 2010 Relationships among gradation curve clogging resistance and pore-based indices of porous asphalt mixes. Road Materials and Pavement Design p 507–525, http://dx.doi.org/10.1080/14680629.2010.9690344.

[8] Winroth J, Hoever C, Kropp W, Beckenbauer T 2013 The contribution of air-pumping to tyre/road noise. Proceedings of AIA-DAGA p 1594–1597

[9] Beckenbauer T 2011 Lärmarme Fahrbahnbeläge Mehrwert bei der Straßensanierung, ALD-Herbstveranstaltung „Lärm in der Stadt“

[10] von Meier A 1992 Thin porous surface layers – design principles and results obtained. Proceedings of Eurosyposium "The mitigation of traffic noise in urban areas", LCPC, Nantes

[11] Banadaki A D, Guddati M N, Kim Y R 2016 An algorithm for virtual fabrication of air voids in asphalt concrete, International Journal of Pavement Engineering **17:3** p 225-232, https://doi.org/10.1080/10298436.2014.979822

[12] Sandberg U, Descornet G 1980 Road surface influence on tire/road noise, Proceedings of the 9th International Conference on Noise Control Engineering (Inter-noise 80), Miami, Florida, 8–10 December p 1–16.

[13] de Bondt A, Plug K, van de Water J, The P, Voskuilen J 2016. Development of a durable third generation Porous Asphalt with a high noise reduction. Proceedings of 6th Eurasphalt & Eurobitumen Congress, Prague, Czech Republic, 1-3 June 2016. dx.doi.org/10.14311/EE.2016.171

[14] Descornet G 2005 Low-noise road surfaces – European state of the art, Journal of the association of asphalt pavement technologists **74** p 1059–1084

[15] Alvarez A E, Martin A E, Estakhri C K, Button J W, Glover C J, Jung S H 2006 Synthesis of current practice on the design, construction, and maintenance of porous friction courses. Texas transportation institute, USA p 86

[16] Sandberg U, Kragh J, Goubert L, Bendtsen H, Bergiers A, Biligiri K P, Karlsson R, Nielsen E, Olesen E, Vansteenkiste S 2011 Optimization of thin asphalt layers – state of the art review. P 140

[17] Andersen B, Bendtsen H 2008 Thin Asphalt Layers for Highways - Optimised for low tyre / road noise. Danish Road Institute Report 160 p 26 ISBN 978-87-92094-42-1

[18] Hamzah M, Abdullah N, Voskuilen J, van Bochove G 2013 Laboratory simulation of the clogging behaviour of single-layer and two-layer porous asphalt, Road Materials and Pavement Design **14:1** p 107-125 https://doi.org/10.1080/14680629.2012.749803

[19] Venudharan V, Biligiri K P, Sousa J B, Way G B 2017 Asphalt-rubber gap-graded mixture design practices: a state-of-the-art research review and future perspective, Road Materials and Pavement Design **18:3** p 730-752 https://doi.org/10.1080/14680629.2016.1182060

[20] Bendtsen H, Oddershede J, Lu Q, Rezaei A 2013 Asphalt Pavement Texture and Noise: Laboratory experiment with Acoustic Optimisation Tool p 52 ISBN 9788770607506

[21] Kragh J, Nielsen E, Olesen E, Goubert L, Vansteenkiste S, De Visscher J, Sandberg U, Karlsson R 2011 OPTHINAL project - Optimization of Thin Asphalt Layers. Final Report p 82